802.11 WLANs and IP Networking

Security, QoS, and Mobility

For a listing of recent titles in the *Artech House Universal Personal Communications Series*, turn to the back of this book.

802.11 WLANs and IP Networking

Security, QoS, and Mobility

Anand R. Prasad
Neeli R. Prasad

ARTECH
HOUSE

BOSTON | LONDON
artechhouse.com

Library of Congress Cataloging-in-Publication Data
Prasad, Anand R., Neeli R. Prasad
A catalog record for this book is available from the Library of Congress.

British Library Cataloguing in Publication Data
Prasad, Anand
802.11 WLANs and IP networking: security, QoS, and mobility.—(Artech House mobile communications library)
1. Wireless LANs 2. Local area networks (Computer networks)
I. Title II. Prasad, Neeli
621.3'821

ISBN 1-58053-789-8

Cover design by Yekaterina Ratner

International Standard Book Number: 1-58053-789-8

10 9 8 7 6 5 4 3 2 1

To our parents Jyoti and Ramjee Prasad,
our brother Rajeev,
and our families Akash, Ruchika and Sneha and Jami

Contents

Preface

न वित्तेन तर्पणीयो हि मनुष्यः

(Not by wealth alone is a human satisfied)

-Rig Ved

Even after all the earthly riches are enjoyed there still remains in the heart a longing for knowledge, true knowledge. It is this longing and the desire to bring the knowledge to others that resulted in the revelation of this book.

"How do IEEE 802.11 wireless local area networks (WLANs) work together with the higher layer protocols, particularly with the IP layer? How does it really work with the mobile network? What are its issues? What is the business model of WLANs now and in the future?" were the main questions that led to the writing of this book. These questions were unanswered in our first, edited, book titled *WLAN Systems and Wireless IP for Next Generation Communications*. In this book we try to answer these questions and elaborate on them.

The first chapter introduces IEEE 802.11-based WLAN and its issues; this chapter also gives a brief overview of the complete book. In the second chapter, written by Rajeev R. Prasad, we discuss market and business for WLANs for different service providers including the mobile operator.

With this background of WLANs and market we dive deep into the WLAN standards in Chapter 3, discussing the IEEE 802.11 standard in detail. Both the medium access control (MAC) and physical layer (PHY) are covered in this chapter. The discussion of MAC enhancements for security, quality of service (QoS), and mobility are left for later chapters.

Currently the foremost issue of IEEE 802.11-based WLANs is security. The fourth chapter of the book discusses the current security solution and its issues. In this chapter various solutions being provided in the market to overcome the security issues are also discussed. Technologies discussed include Virtual Private Network (VPN), IP Security (IPSec), and Secure Session Layer (SSL). The chapter also discusses the draft IEEE 802.11i standard together with Extensible Authentication Protocol over LAN (EAPoL), which is used by IEEE 802.1x.

Having discussed the issue of security, QoS is handled in Chapter 5; several sections of the chapter are written by Mr. M. Alam. This chapter discusses the MAC layer provision for QoS including the draft IEEE 802.11e standard. The chapter also discusses QoS signaling protocols like H.323 and Session Initiation Protocol (SIP). WLAN has to interface with the Public Switched Telephone Network (PSTN), the protocol for this, including H.323, Media Gateway Control Protocol (MGCP) is presented in the chapter. Finally, transport layer solutions like the Real Time Protocol (RTP) and Real Time Control Protocol (RTCP) together with Differentiated Service (DiffServ) and Integrated Service (IntServ) are discussed in the chapter.

The issues of handover, mobility, and roaming are tackled in Chapter 6. This chapter starts with a discussion of the solution for mobility when using the original IEEE 802.11. Next the Inter Access Point Protocol (IAPP) as recommended by IEEE 802.11f is presented. Having discussed the Layer-1 and Layer-2 methods the IP layer solution, particularly Mobile IP, is discussed in detail. The Mobile IP solution is also discussed for cases where the user handovers to a different service provider. Most recent enhancements of Mobile IP and Seamless Mobility (Seamoby) are also discussed in the chapter. Mobility solution at the transport layer is briefly touched on in the chapter, while mobility when using SIP is also discussed. Currently roaming methods are being used by wireless Internet service providers (WISPs) to increase their footprint; this is also presented in Chapter 6.

Next, in Chapter 7, the major issue related to deployment of a WLAN is discussed. Deployment methods for WISPs, offices, and mobile operators are presented. This chapter also discusses the mobile and WLAN interworking/ integration methods.

A final chapter, Chapter 8, concludes the book with a vision for future. Definition for Fourth Generation (4G) mobile communications, Beyond Third Generation (B3G), and future generations are given. The need for these technologies from user, vendor, and operator perspectives is also discussed in this chapter. Technological enhancements needed from the protocol layer point of view and particularly for security, QoS, and mobility are also presented in the chapter.

In this book several draft standards are discussed which might change with time; still the information in this book should be beneficial for understanding the interaction between the IP and MAC layers. We hope that this book will be of interest to business and technical managers and also to technical novices as well as experts in this field.

Acknowledgments

The patience and support of our families was the biggest help in the completion of the book. Anand would also like to acknowledge his parents-in-law, Mr. and Mrs. Nakajima, for their support. We would like to acknowledge Rajeev R. Prasad of PCOM:I^3 for writing the second chapter on the market and business case for WLANs and Mahbubul Alam of Cisco Systems, who wrote several sections of the QoS chapter, Chapter 5. We extend our gratitude to professors R. Prasad, M. Ruggieri, and S. Hara, as we have used parts of their work in Chapter 8 and to IEEE for allowing us to use material from IEEE 802 standards.

Anand Raghawa Prasad
Neeli Rashmi Prasad
March 2005

Chapter 1

Introduction

Wireless LANs, a term that was formerly known only to a few, has in a short period of a couple of years become a layperson's term. The market penetration has been as unexpected as the growth of mobile communications and the Internet in the boom era. This growth has obviously been due to the benefits of wireless local area networks (WLANs), e.g., ease of deployment, low cost, and flexibility. However, WLANs have also brought with them several issues while opening the door to a new future of data communications. This chapter gives an introduction to WLANs and their market, requirements, and issues; the final section gives an overview of the rest [1–100].

1.1 BASIC CONCEPT OF WLANs

Two types or modes of WLANs exist, the technology that provides connectivity to the infrastructure network and the technology that provides the connectivity of one device to another or an adhoc network. This is also depicted in Figure 1.1 [1–34]. IEEE 802.11-based WLANs work in both modes. WLANs do not replace wired solutions but complement them; the same can be said about WLANs and wireless wide area networks (WWANs) and wireless personal area networks (WPANs).

WLANs provide network connectivity in difficult wiring areas; they provide flexibility to move and extend networks or make changes. WLANs allow mobile users to work with traditional wired applications. In fact WLANs are the only LAN devices that allow true mobility and connectivity. WLANs provide connectivity for slow mobility (walking speed) with high throughput for both indoor and outdoor environments. Figure 1.2 shows the place of WLANs among the different wireless communications systems.

Although WLANs came into the market almost a decade ago, standardized WLANs have been available since the late 1990s when IEEE 802.11 was born. Meanwhile several other WLAN standards came into being, for example, High

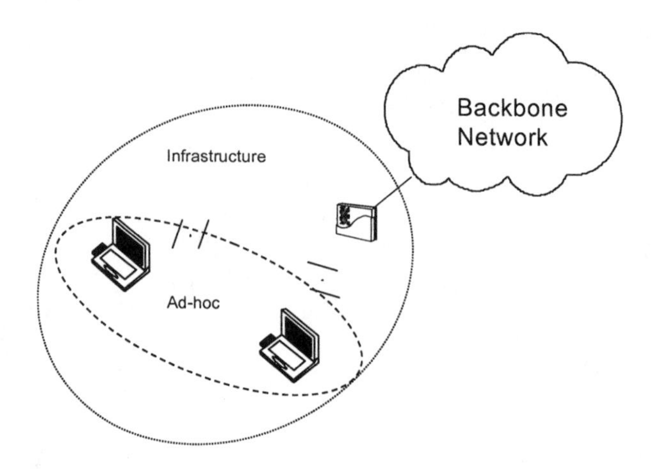

Figure 1.1 What is a WLAN?

Performance Radio Local Area Network-Type 2 (HIPERLAN/2) and HomeRF but none of them have been successful. A comparison of these technologies is given in Appendix 1A.

In 1999 the Wireless Ethernet Compatibility Alliance (WECA) was started. The purpose of WECA was to bring interoperability amongst IEEE 802.11 products of various vendors. The alliance developed a Wireless-Fidelity (Wi-Fi) interoperability test and provided logos for products that had passed the test. Today Wi-Fi has become a synonym of IEEE 802.11 and the alliance is now named as the Wi-Fi Alliance.

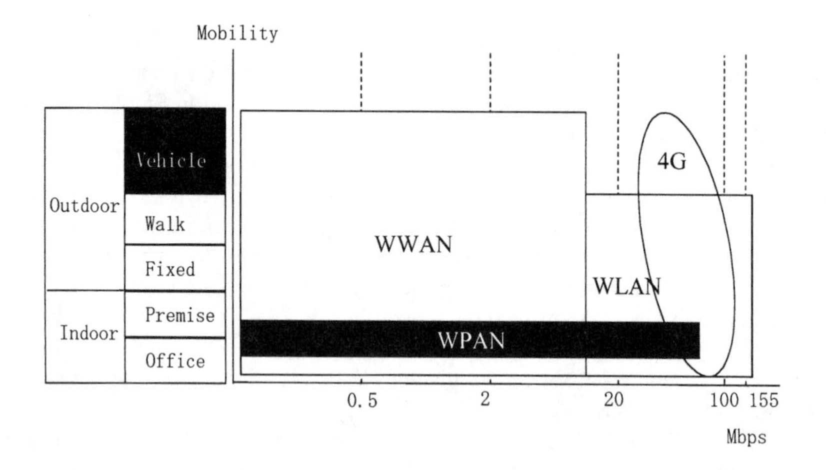

Figure 1.2 Placing Wireless LANs among wireless communications technologies.

The Wi-Fi Alliance also provides interoperability for IEEE 802.11a, 802.11b, and 802.11g [10]. The alliance has developed a wireless Internet service provider (WISP) roaming (WISPr) recommendation too. The WISPr recommendation was developed to allow WLAN users connectivity at any other WISP's hotspot while being charged at one account (i.e., a single billing solution) [11].

IEEE 802.11b works in the 2.4 GHz band, like .11g, while the IEEE 802.11a solution works in the 5 GHz frequency band. These spectrums are license free. In Table 1.1 the frequency bands in which various IEEE 802.11 standards work and the regulatory requirements on output power are given. A detailed overview of the IEEE 802.11 standards and a comparison with WPANs are given in Table 1.1 [1, 2]. A summary of IEEE 802.11 standards and current activities of IEEE 802.11 is given in the following; details are discussed in Chapter 3:

- IEEE 802.11: Carrier Sense Multiple Access with Collision Avoidance (CSMA/CA) Medium Access Control (MAC), and 1 and 2 Mbps for DSSS, FHSS in 2.4 GHz band, and Infrared, ratified in 1997;
- IEEE 802.11a: Works at 6, 9, 12, 18, 24, 36, 48 and 54 Mbps in 5GHz band, ratified in 1999;
- IEEE 802.11b: Works at 5.5 and 11 Mbps in 2.4 GHz band, ratified in 1999;
- IEEE 802.11e: MAC enhancements for Quality of Service (QoS), work ongoing;
- IEEE 802.11f: Inter Access Point Protocol (IAPP), ratified in 2003;

Table 1.1

IEEE 802.11 Standard and Worldwide Frequency Bands

Location	Regulatory Range (MHz)	Maximum Output Power (mW)	Standard
Europe	2,400–2,483.5	10 /MHz (max 100 mW)	IEEE 802.11b,g
	5,150–5,350	200	IEEE 802.11a
	5,470–5,725	1,000	
North America	2,400–2,483.5	1,000	IEEE 802.11b,g
	5,150–5,250	2.5 /MHz (max. 50 mW)	IEEE 802.11a
	5,250–5,350	12.5 /MHz (max 250 mW)	
	5,725–5,825	50 /MHz (max 1000 mW)	
Japan	2,400–2,497	10 /MHz (max 100 mW)	IEEE 802.11b,g
	5,150–5,250 4,900–5,000 (until 2007) 5,030–5,091 (from 2007)	Indoor 200	IEEE 802.11a

- IEEE 802.11g: Works at same data rates as IEEE 802.11a and other optional modes; meant for 2.4 GHz band and is backward compatible to IEEE 802.11b, ratified in 2003;
- IEEE 802.11h: Spectrum managed IEEE 802.11a; addresses European Radio Communications Committee requirements at 5GHz addition of TPC (transmit power control) and DCS (dynamic channel selection);
- IEEE 802.11i: MAC enhancements for security, work ongoing;
- IEEE 802.11j: The purpose of Task Group J is to enhance the 802.11 standard and amendments, to add channel selection for 4.9 GHz and 5 GHz in Japan and to conform to the Japanese rules on operational mode, operational rate, radiated power, spurious emissions, and channel sense;
- IEEE 802.11k: This Task Group will define Radio Resource Measurement enhancements to provide mechanisms to higher layers for radio and network measurements;
- IEEE 802.11m: The goal of the task group is to complete this review of other documents and to determine a final list of work items;
- IEEE 802.11n: Possibility to improve 802.11 to provide high throughput (100 Mbps+);
- IEEE 802.11r: The group is looking at fast roaming mainly for voice over IP service;
- IEEE 802.11s: This group is working on AP-based mesh network;
- IEEE 802.11t: The focus of this group is on wireless performance prediction.

There are also several other study groups (SGs) like the access point functionality (APF) SG, wireless internetwork and external network (WIEN) SG, wireless network management (WNG) SG, and wireless access for the vehicular environment (WAVE) SG. There are other activities on publicity, and the wireless next generation study committee is looking at globalization and harmonization. There are also thoughts on creating an advanced security SG.

1.2 BENEFITS OF WLANs

In this section some of the benefits of WLANs like mobility and speed of deployment are discussed [1, 2].

1.2.1 Mobility

The possibility to access real time information while dealing with customers is enhanced with a WLAN. In hospitals, for example, healthcare providers can improve the quality of patient care. With a WLAN, bedside inputting of data and immediate decision-making can reduce cycle times for patient care. Likewise, the reduction of errors by handling the data only once is significant. In an office situation, the ability to roam around the building while processing information is an advantage. Similarly, point-of-sale employees can circulate freely while serving customers. Insurance agents can input data directly in the customer's premises and receive realtime on-line analytical processing. If there is a business advantage in going to the customer rather than forcing the customer to come to you, the case for wireless can be compelling. Finally, WLANs permit mobile applications to be launched. Consider the WLAN-enabled student who can take his or her WLAN-connected laptop from lecture to lecture and remain connected at all times to his or her files and applications.

1.2.2 Short-Term Usage

Similar to the issue of mobility, short-term connectivity allows users to deploy capabilities on an as needed basis without concern for the cost justification for wired solutions. Financial auditors, for example, can just connect for the time necessary to conduct the audit. This allows significant operational flexibility and facilitates the formation and support of adhoc working groups. Being able to connect to the network for a short period of time in this manner can provide a competitive advantage.

1.2.3 Speed of Deployment

WLANs permit quick connectivity to the network. Forming and disbanding work groups can be done easily with WLANs. The complexity and long cycle time of moving new nodes into and out of wired LANs introduces massive ongoing operational costs compared with the flexibility of wireless attachment, where the operational costs are almost zero.

1.2.4 Difficult Wiring Environment

Many situations do not permit the easy installation of wires. Historic buildings or older buildings make the installation of LANs either impossible or very expensive. Trying to establish LANs in the outdoors is virtually impossible with legacy LANs. Consider situations in parks or athletic arenas where one wants a temporary WLAN established and removed. There are other situations where it is vital to be WLAN-enabled. Disaster recovery, for example, can make immediate and effective use of WLANs in the field to gather data and coordinate relief efforts. The use of WLANs in the battlefield is obvious. Finally, there are situations where wires cannot be laid, for example, across busy streets. Likewise, building-to-building connections can be facilitated where no existing underground cabling is present. Using wireless bridges to connect physically separated LANs or Internet connections can be very effective.

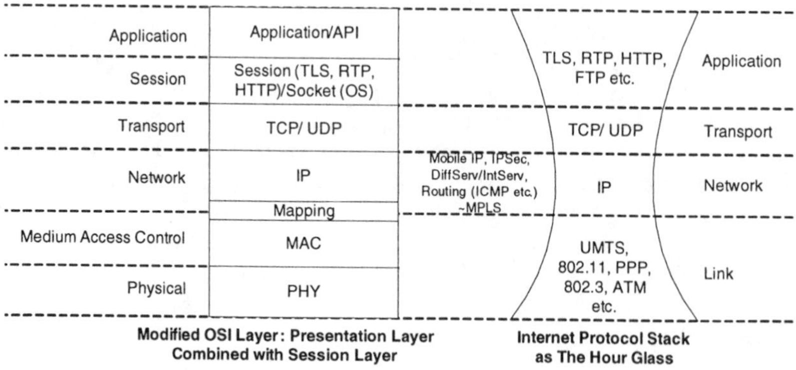

Application | Application/API
Session | Session (TLS, RTP, HTTP)/Socket (OS)
Transport | TCP/ UDP
Network | IP
Mapping
Medium Access Control | MAC
Physical | PHY

Mobile IP, IPSec, DiffServ/IntServ, Routing (ICMP etc.) ~MPLS

TLS, RTP, HTTP, FTP etc. | Application
TCP/ UDP | Transport
IP | Network
UMTS, 802.11, PPP, 802.3, ATM etc. | Link

Modified OSI Layer: Presentation Layer Combined with Session Layer

Internet Protocol Stack as The Hour Glass

Figure 1.3 IP and modified OSI protocol stack.

1.2.5 Scalability

Wireless LAN systems can be configured in a variety of topologies to meet the needs of specific applications and installations. Configurations are easily changed and range from independent networks suitable for a small number of users to full infrastructure networks of thousands of users that allow roaming over a broad area. Such opportunities are not possible with wired networks.

1.3 BASIC CONCEPT OF WIRELESS IP

Before beginning a discussion on wireless IP it is important to understand what the Internet Protocol (IP) is, although this might be well understood by the audience of the book [1, 2, 35, 36].

From the Open System Integration (OSI) layers' point of view, IP is the third layer—the network layer. The basic function of IP itself is to provide an addressing and routing solution but when one talks about IP then it is basically the IP stack. The IP stack or TCP/IP protocol suite is standardized by the Internet Engineering Task Force (IETF). In Figure 1.3 some of the protocols of the protocol suite are shown together with mapping of the protocols on the modified OSI layer, modified because the presentation layer and session layer are represented together as the session layer.

An hourglass representation of the IP protocol stack represents the functions the IP layer supports. This is a marvel in itself but can also lead to limitations.

Now with this short introduction to IP let us look at what wireless IP is. Wireless IP is nothing other than use of IP for wireless communications. The functionalities of the IP do not change but now IP must fulfill the requirements of the wireless communications, which include higher mobility, higher probability of error in the wireless medium, security issues arising from the wireless medium, and effect on QoS. Depending on the type of wireless system techniques, header compression or fragmentation of IP packets will be required.

The use of IP in the Internet, the benefit from statistical multiplexing that packet switching brings, and the (original) simplicity that IP brings have led to acceptance of IP as the common layer for mobile communications systems and future heterogeneous networks. This common layer will bring ease in management and the possibility of service access of any type, combined with various access technologies, from anywhere. Of course, as discussed above, several issues remain unsolved. Both the wireless standardization bodies and IETF are working on solutions of these problems. In this book we will discuss the use of IP level security, QoS, and mobility in WLANs.

1.4 MARKET TREND

To start to understand the market trend it is important to understand the customers of WLANs. Next it is necessary to understand why the customers need WLANs and how WLAN fulfills the need of customers. For any product to be successful it is important that there is "yes" as an answer to one of these three questions: Is there a gap in the market? Are the customers in need? Can they gain/profit from WLANs? This section tries to answer these questions briefly; details will be discussed in Chapter 2. Some of these questions also bring us to the requirements and issues of WLANs; these two topics are discussed in Sections 1.5 and 1.6, respectively.

The total WLAN market is split into four specific segments and these are divided into the consumer and business markets, as shown in Figure 1.4 [37]. This figure does not represent the penetration and growth of WLAN in the market; it only represents the usage and the fact that the growth should be expected to be

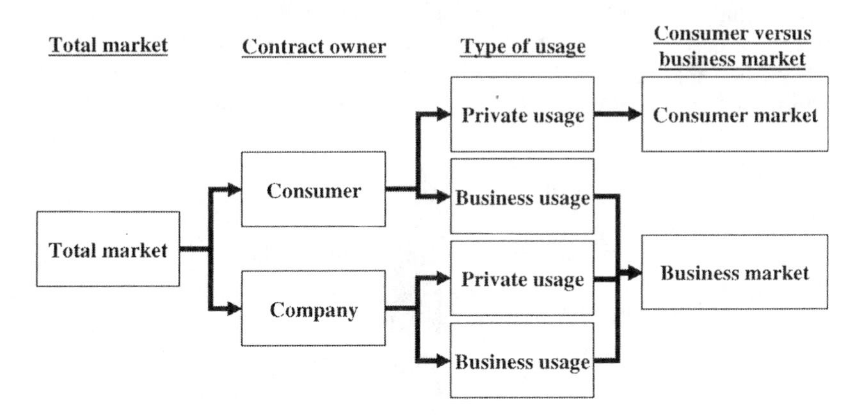

Figure 1.4 WLAN market split.

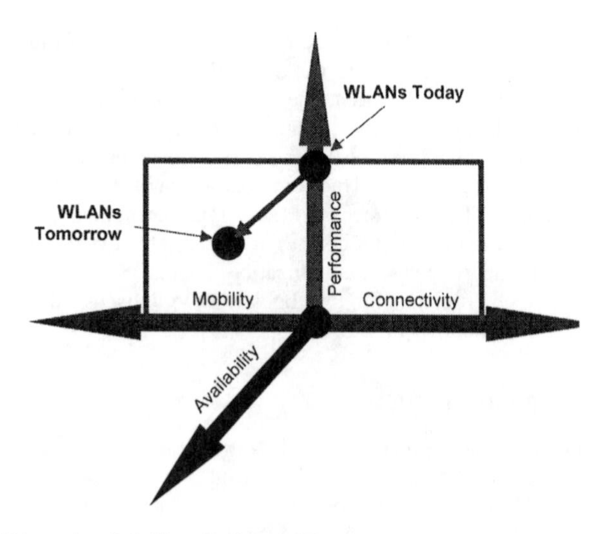

Figure 1.5 WLANs market definition: Today and Tomorrow.

high in the enterprise (company) market but that issues related to WLAN have delayed growth in the enterprise market. The biggest growth sector of WLAN is the home or residential market. Once the issues related to WLANs are solved, there will be a change in the market; see Sections 1.5 and 1.6.

The growth of WLAN has been due to the benefits discussed in Section 1.2. From the market perspective there are three main reasons for growth: People need connectivity (desktops with an Ethernet connection provides that); people need mobility (which brings us to the arena of wireless); and people need performance. WLANs can provide high throughput and thus higher performance (of course giving the required QoS is an issue but work is ongoing to solve this issue – see Section 1.6 and Chapter 3). These three factors, mobility, connectivity, and performance, give us the picture of where WLAN is today (see Figure 1.5).

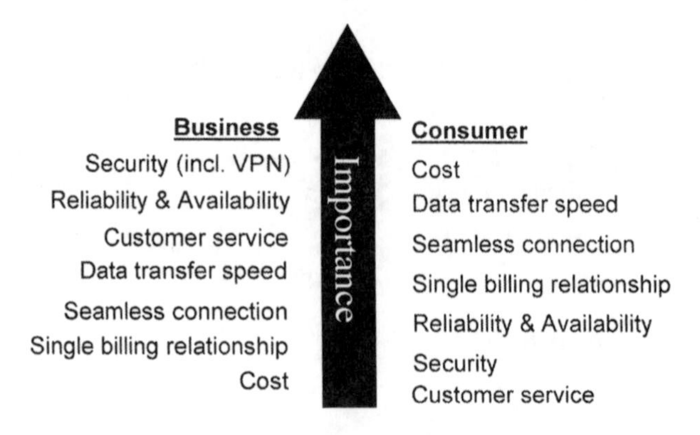

Figure 1.6 Needs of end user.

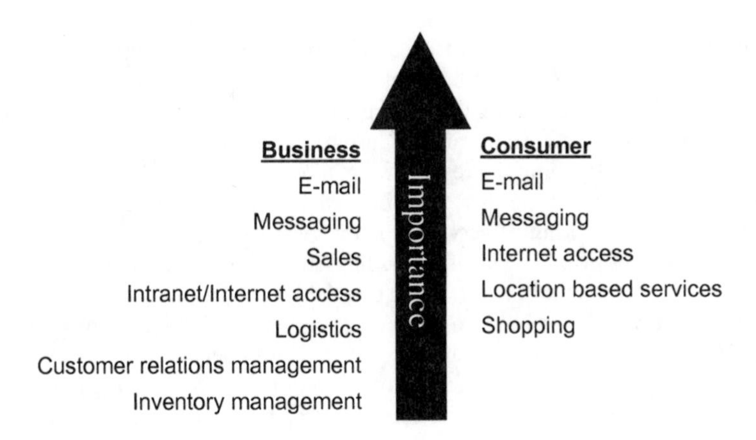

Figure 1.7 Service needs of users.

Today WLAN growth is mainly in the residential market and in the public arena, commonly known as hotspots. Incumbent Internet service providers (ISPs) have deployed WLANs, new operators like the WISPs have appeared, and Mobile Network Operators (MNOs) and fixed operators are also deploying WLANs. More and more cities are providing WLAN coverage, and several restaurants, airports, and shops/shopping zones are providing WLAN-based Internet access. This itself shows the trend towards *availability* in the market.

The future growth of WLAN is dependent on availability, so tomorrow's WLANs should not only provide the three benefits mentioned above but should also be available to users possibly anywhere and anytime. Users should in time also get the luxury of continuous connectivity all the time even if the network is owned by different stakeholders. This brings us to the keyword "seamlessness" in service at all times under any condition.

1.5 REQUIREMENTS OF WLANs

Discussion of the requirements of WLANs brings us to an important question posed in the previous section: What do the customers or users want? To some extent this question has been answered in the previous section and will be dealt with in Chapter 2. In this section some of the needs of customers from the point of view of WLANs, services, and devices are discussed.

We start with the basic needs of the users and divide the users into business and consumer types, see Section 1.4 and Figure 1.4. In the current market the business sector gives prime importance to security, after which comes reliability and availability, customer services, and data rate transfer. The consumer market on the other hand gives prime importance to cost (which is of the least importance for business) and then to data transfer speed and seamless connection. Here the term "seamless connection" should not be confused with "seamlessness" discussed at the end of Section 1.4. Seamless connection, among other things, mainly means the ease to connect. This need for the two markets is also given in Figure 1.6. This difference in need is the reason for the current difference in the penetration of the market in the enterprise and the residential markets.

Now let us look at another very important need of the users, the services. Figure 1.7 shows the kind of services the business and consumer market need. In both markets the prime need is e-mailing and then messaging. In fact it is impossible to survive today for most people without e-mail and messaging is becoming an even more common form of communication. Some companies have adopted messaging as a formal way of communication between employees.

The kind of devices users want and their requirements for these devices is the next important factor, as shown in Figure 1.8. For both the business and consumer market the main requirement lies in convenience, simplicity, and performance. The requirement for personalization is not that high at the moment but this will be a very important requirement in the future. The maturity of a product usually fulfills these requirements. Although 802.11 has passed several phases in terms of market and technology, it is still in the midst of a boom era and maturity is still to come. Technologies are being developed by different vendors to fulfill these requirements.

1.6 ISSUES

Now let us look at some of the issues WLAN makers must deal with to continue the growth momentum that the technology has achieved in the market. It is also necessary to understand the reason for the degradation of the WLAN signal. One

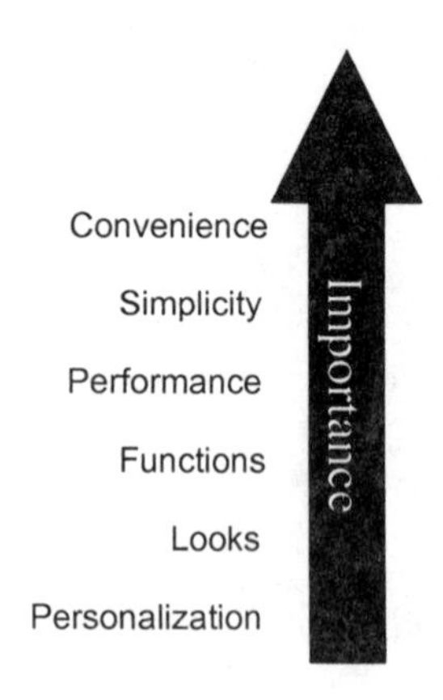

Figure 1.8 Device and service requirements of users.

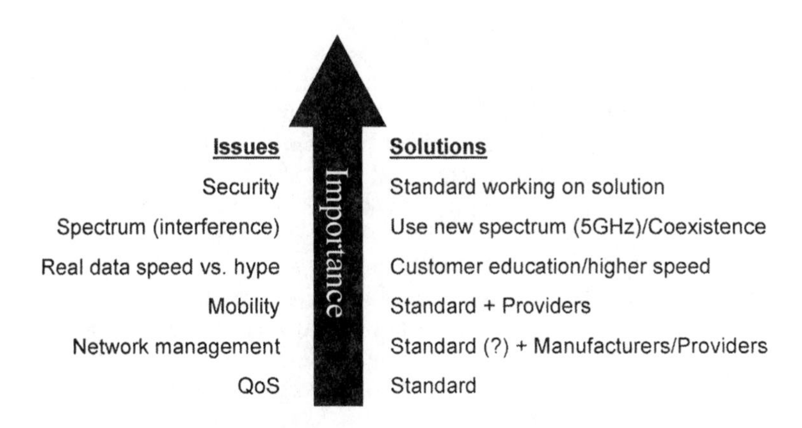

Figure 1.9 WLANs issues and solutions.

of the major issues for WLANs is the degradation of radio signals by loss and reflection. In an ideal radio channel, the received signal would consist of only a single direct path signal, which would be a perfect reconstruction of the transmitted signal [38–84]. However in a real channel, the signal is modified during transmission in the channel. The received signal consists of a combination of attenuated, reflected, refracted, and diffracted replicas of the transmitted signal. On top of all this, the channel adds noise to the signal and can shift in the carrier frequency if the transmitter, or receiver is moving (Doppler effect). Understanding these effects on the signal is important because the performance of a radio system is dependent on the radio channel characteristics.

In the following sections some general issues related to WLANs are discussed together with their solutions. Also different channel degradation conditions and health hazards for wireless systems are explained.

1.6.1 General Issues

General issues related to WLANs and solutions to these issues are shown in Figure 1.9. One of the major issues is security; the IEEE 802.11 standardization body is working on this well known issue. At the time of writing the security enhancement standard was expected to be completed by early 2005.

Another issue is spectrum. The ISM band of 2.4 GHz has become a garbage band. Several different types of equipment including microwave ovens work in this frequency band and they obviously interfere with each other which affects the performance of WLANs. IEEE 802.11a works in the 5 GHz band, which is also an unlicensed band, but efforts are ongoing towards harmonizing different wireless technologies planned to work in this frequency band.

Another issue is the actual data rate versus the hype (e.g., IEEE 802.11b promises 11 Mbps but actual data throughput is 5 Mbps at best). The reason for this is the overhead from the TCP/IP and MAC layer and collisions that occur. It is extremely important that the customers are informed or educated about it. Mobility is an issue which has been taken care of by IEEE 802.11f for within one network; mobility between different networks with service continuity remains an issue for the future. QoS is being tackled by IEEE 802.11e. Network management is also an important issue for which several vendors provide solutions.

Battery life, size of device, and integration within the device are also issues of high importance. IEEE 802.11 provides power management but vendors have also come up with good implementations and innovative ideas to save energy and thus battery life; one example is Broadcom. Intel Centrino on the other hand provides an integration of WLAN and the CPU; this kind of integration might be the direction in which we will see the future wireless/mobile communications product moving.

1.6.2 Attenuation

Attenuation is the drop in the signal power when transmitting from one point to another. It can be caused by the transmission path length, obstructions in the signal path, and multipath effects. Figure 1.10 shows some of the radio propagation effects that cause attenuation. Any objects which obstruct the line of sight signal from the transmitter to the receiver, can cause attenuation.

Shadowing of the signal can occur whenever there is an obstruction between the transmitter and receiver. It is generally caused by indoor and outdoor obstacles —in-building obstacles (e.g., furniture), buildings, and hills— and is the most important environmental attenuation factor.

Shadowing is most severe in heavily built-up areas, due to the shadowing from buildings. However, hills can cause a large problem due to the large shadow they produce. Radio signals diffract off the boundaries of obstructions, thus preventing total shadowing of the signals behind hills and buildings. However, the amount of diffraction is dependent on the radio frequency used, with low frequencies diffracting more than high frequency signals. Thus high frequency signals, especially, ultrahigh frequencies (UHFs), and microwave signals give for line-of-sight conditions the highest signal strength. To overcome the problem of shadowing, transmitters are usually elevated as high as possible to minimize the number of obstructions.

Shadowed areas tend to be large, resulting in the rate of change of the signal power being slow. It is termed *slow-fading*, or *log-normal shadowing* because the distribution of the logarithm of the amplitude is normal.

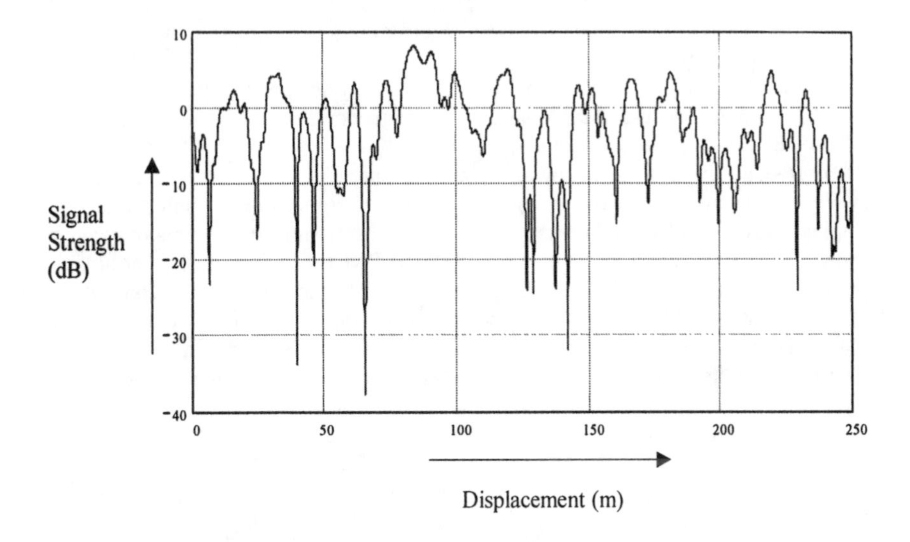

Figure 1.10 Typical Rayleigh fading while the mobile unit is moving.

1.6.3 Multipath

1.6.3.1 Rayleigh Fading

In a radio link, the RF signal from the transmitter may be reflected from objects such as hills, buildings, or vehicles. This gives rise to multiple transmission paths at the receiver.

The relative phase of multiple reflected signals can cause constructive or destructive interference at the receiver. This is experienced over very short distances (typically at half wavelength distances), and thus is given the term *fast fading*. These variations can vary from 10 to 30 dB over a short distance. Figure 1.10 shows the level of attenuation that can occur due to the fading.

The Rayleigh distribution is commonly used to describe the statistical time varying nature of the received signal power. It describes the probability of the signal level being received due to fading in case there is no LOS.

1.6.3.2 Frequency Selective Fading

In any radio transmission, the channel spectral response is not flat. It has dips or fades in the response due to reflections causing cancellation of certain frequencies at the receiver. Reflections of nearby objects (ground, buildings, trees, etc.) can lead to multipath signals of similar signal power as the direct signal. This can result in deep nulls in the received signal power spectrum due to destructive interference for some frequencies.

For narrow bandwidth transmissions, if a strong notch in the channel frequency response occurs at the transmission frequency then the entire signal can be lost. This can be partly overcome in two ways.

By transmitting a wide bandwidth signal or spread spectrum as CDMA, any dips in the spectrum only result in a small loss of signal power, rather than a complete loss. Another method is to split the transmission up into many small bandwidth carriers, as is done in a COFDM/OFDM transmission. The original signal is spread over a wide bandwidth; thus any nulls in the spectrum are unlikely to occur at all of the subcarrier frequencies. This will result in only some of the subcarriers being lost, rather than the entire signal. The information in the lost subcarriers can be recovered provided enough forward error corrections are sent.

1.6.3.3 Delay Spread

The received radio signal from a transmitter typically consists of a direct signal, plus reflections of objects such as buildings, mountains, and other structures. The reflected signals arrive at a later time than the direct signal because of the extra path length, giving rise to a slightly different arrival time of the transmitted pulse, thus spreading the received energy. Delay spread characterizes the magnitude of time spread in the received multipath signal; it is defined as the second-order moment of the channel power profile (spread-in-time of the received power).

In a digital system, multipath effects can lead to inter-symbol interference. This is due to the delayed multipath signal overlapping with the following symbols. This can cause significant errors in high bit rate systems, especially when using time division multiplexing (TDMA).

Inter-symbol interference can be minimized in several ways. One method is to reduce the symbol rate by reducing the data rate for each channel (i.e., split the bandwidth into more channels using frequency division multiplexing). Another is to use a coding scheme that is tolerant of inter-symbol interference such as CDMA.

1.6.3.4 Doppler Shift

When a wave source and a receiver are moving relative to one another the frequency of the received signal will not be the same as the source. When they are moving towards each other the frequency of the received signal is higher than the source, and when they are moving away from each other the frequency decreases. This is called the *Doppler effect*. An example of this is the change of pitch in a car's horn as it approaches then passes by. This effect becomes important when developing mobile radio systems.

The level of the frequency offset due to the Doppler effect depends on the effective speed source of the transmitter with respect to the receiver and on the speed of the propagation of the wave. Doppler shift can cause significant problems if the transmission technique is sensitive to carrier frequency offsets (for example, narrowband and OFDM) or the relative speed is higher (for example in low earth orbiting satellites). With wideband DSSS systems there is less sensitivity for this phenomenon.

1.6.4 UHF Narrowband

Narrowband is a term used to describe RF signals sent over a narrow band of spectrum, typically 12.5 KHz to 25 KHz. UHF narrowband systems transmit on both licensed and unlicensed frequencies, and systems based on this technology operate at a higher power than spread spectrum systems, typically at 1 to 2 watts. Because of the higher power, these systems have the longest transmission range of all the WLAN technologies. However, these products have been hobbled by lack of vendor interoperability, lower speeds, and the requirement for site licenses for some of the licensed frequency bands.

1.6.5 Infrared

Infrared technology is an invisible beam of light that uses signals much like those used in fiber optic links today. Infrared is reliant upon line-of-sight links between the transmitter and receiver. Physical impediments such as walls will block the transmission of signals, limiting infrared WLANs largely to in-room communications. Because of the limitations of infrared technology, it is not used in many implementations today. Infrared technology was one of the three technologies under the IEEE 802.11 specification, but under the newer 802.11b specification only Direct Sequence (one of two spread spectrum technologies) technology is used.

1.6.6 Health Consideration

Until a few years ago, the analysis of possible harmful effects of electromagnetic radiation on people was devoted mainly to power lines and radar, because of the huge power levels involved in those systems [1, 90]. Even when mobile telephone systems appeared, there was no major concern, as the antennas were installed on the roofs of cars. With the development of personal communication systems, in which users carry mobile telephones inside their coat pockets, with the antenna radiating a few centimeters from the head, safety issues gained great importance and a new perspective. Much research in the literature focuses not only on the absorption of power inside the head, but also on the influence of the head on the antenna's radiation pattern and input impedance. However, these works have addressed only the frequency bands used in today's systems—that is, up to 2-GHz (mainly on the 900- and 1,800-MHz bands)—and only very few references are made to systems working at higher frequencies, as it is in the case of wireless broadband communications like WLANs.

The problems associated with infrared technology are different from those posed by microwaves and millimeter waves. Eye safety, rather than power absorption inside the head, is the issue here, because the eye acts as a filter to the electromagnetic radiation, allowing only light and near-frequency radiation to enter into it, and the amount of power absorption inside the human body is negligible. Exposure of the eye to high levels of infrared radiation may cause cataract-like diseases, and the maximum allowed transmitter power seems to limit the range to a few meters. If this is the case, safety restrictions will pose severe limitations on the use of infrared in wireless broadband systems, as far as general applications are concerned. The question in this case is not that there are always problems during system operation (e.g., mobile telephones), but the damage that may be caused if someone looks at the transmitter during operation.

Microwaves and millimeter waves have no special effect on eyes, other than power absorption. In WLANs, antennas do not radiate very near (1 or 2 cm) to the user as in the mobile telephone case, thus enabling power limitations to be less restrictive (also the case if mobile multimedia terminals are used as they are in PDAs). However, if terminals are used in the same form as mobile telephones, then maximum transmitter powers have to be established, similar to those for the current personal communication systems. The standards for safety levels have already been set in the United States and Europe, as the ones used for UHF extend up to 300 MHz (IEEE/ANSI and CENELEC recommendations are the references). Thus, it is left to researchers in this area to extend their work to higher frequencies, by evaluating SAR (the amount of power dissipated per unit of mass) levels inside the head (or other parts of the human body very near the radiating system), from which maximum transmitter powers will be established. This may not be as straightforward as it seems, however, because the calculation of SAR is usually done by solving integral or differential equations using numerical methods (method of moments or finite difference), which require models of the head made of small elements (e.g., cubes) with dimensions on the order of a tenth of the wavelength. This already requires powerful computer resources (in memory and CPU time) for frequencies in the high UHF band, and may limit the possibility of analyzing frequencies much higher than UHF. On the other hand, the higher the frequency, the smaller the penetration of radio waves into the human body, hence making it possible to have models of only some centimeters deep. This is an area for further research.

1.7 FUTURE DIRECTIONS

Today basically three wireless technologies, besides satellite communications, have made an impact: WLANs, WPANs, and WWANs. WLANs complement LANs while WPANs are used for short distance communications and WWANs cover wide areas and are most commonly known as mobile or cellular communications. Recently the WLANs are being seen as a threat to the WWANs but in fact these two are complementary technologies. Another set of technology is the Fixed Wireless Access (FWA) or Broadband Wireless Access (BWA). The current standardization trend shows that the FWA technologies will get mobility functionalities; if this happens then FWA can be a threat to the WWANs. Development of 802.20 a Mobile BWA (MBWA) could surely be a threat for the WWANs in the future. In the following the future direction of WLANs, WWANs, and WPANs are presented; an overview of wireless technology standards is given in Table 1.2 [1–4, 13, 35–38, 91–99].

1.7.1 WLANs

LANs mostly make use of the Internet IP. The growth in wireless and the benefits it provides has brought forward changes in the world of LANs in recent years. WLANs provide much higher data rates as compared to WWANs for slow mobile or static systems. The IEEE 802.11b-based WLANs are already widely being used while the IEEE 802.11g and IEEE 802.11a are also available.

WLAN technologies are mainly used for wireless transmission of IP packets. Until now, in contrast to the WWANs, the WLANs provided network access as a complement to the wireline LANs. In the near future QoS-based WLANs are expected to come onto the market.

Table 1.2

Wireless Technologies

<table>
<tr><th>WWAN</th><th>WLAN</th><th>WPAN</th><th>Cordless</th><th>FWA/BWA</th></tr>
<tr><td>GSM -HCSD, GPRS, EDGE-(WAP)</td><td>IEEE 802.11</td><td>IEEE 802.15</td><td>PHS</td><td>IEEE 802.16, IEEE 802.20 (MBWA)</td></tr>
<tr><td>IS-95</td><td>HIPERLAN/2</td><td>Bluetooth</td><td>DECT</td><td>HIPERACCESS</td></tr>
<tr><td>IS-54/IS-136</td><td>MMAC Ethernet WG & ATM WG (HiSWAN)</td><td>HIPERPAN</td><td>CT2/CT2+</td><td>High speed Wireless access</td></tr>
<tr><td>PDC (I-mode)</td><td>MBS</td><td></td><td></td><td>BWIF</td></tr>
<tr><td>3G</td><td colspan="2">MMAC Wireless Homelink
HomeRF 1.0 & 2.0</td><td></td><td>LMDS
MMDS</td></tr>
</table>

IEEE 802.11e is working towards MAC enhancements. The purpose of the MAC enhancement is to enable the present MAC, CSMA/CA, to provide QoS. The current draft has accepted two variations for QoS enhancements: central control- and distributed control-based. For security in IEEE 802.11i, the main direction is towards applying the IEEE 802.1X-based solution with stronger and more choice of encryption algorithms. The IEEE 802.11 Working Group (WG) has also accepted a mobility solution known as Inter Access Point Protocol (IAPP), IEEE 802.11f. Another group in IEEE 802.11 is working on radio resource management (IEEE 802.11j); The IEEE 802.11 committee has approved IEEE 802.11h, dynamic frequency assignment, and transmit power control. Due to the success of the standard there are several other study groups looking at higher data rate solutions (IEEE 802.11n 110 Mbps+) and next generation technologies including standardization work with 3G standardization committees.

WiFi Alliance, an industry alliance, is providing interoperability specifications and tests of the IEEE 802.11 products for better acceptance in the market. This alliance also provides recommendations for roaming between different WISPs so that a user, or customer of one WISP, can access WLAN services when in another WISP's hotspot and still receive one bill.

Other known WLAN technologies are HIPERLAN Type 2 and HomeRF. HIPERLAN Type 2 is already standardized; it provides hooks for QoS and security for different environments. HomeRF developed several solutions but since the beginning of 2003 HomeRF has been thought to be dead.

The direction for WLANs at present would be to move towards a common international standard. Harmonization in 5-GHz band technologies is a must so as to avoid making the 5-GHz band a garbage band. Although harmonization is a solution it is possible that the market will be a deciding factor and choose one technology. For the time being the success of a standard will depend on pricing, performance, availability, and marketing of the standards.

Besides the work being done by the standardization committees there should be a study on providing top-to-bottom mapping. The correct mapping of higher layer protocols to lower layer protocols is a must to provide optimum service. Especially in the case of IEEE 802.11 where the standard only defines the bottom two layers, relations must be created with IETF, the committee developing layer three, and some higher layer protocols.

Basically most of the current development will lead to providing users different services within WLANs; in other words it is the integration of services within one system. Another step currently becoming visible is towards integration with WWAN technologies like 3G.

1.7.2 WWANs

Growth in the field of wireless wide area networks (WWANs), more commonly known as mobile communications has been tremendous in the past decade. Second generation (2G), 2.5G, and Third Generation (3G) standards of mobile systems are being used while efforts are going on towards development and standardization of Beyond 3G (B3G) systems. The existing (2G) systems are mainly for voice purposes. Due to the tremendous growth in the Internet some support for data services like Wireless Application Protocol (WAP) and i-mode has been developed. 2G supplement systems, 2.5G, like GPRS (General Packet Radio Systems) and now 3G systems provide further possibilities for data services with varying QoS requirements.

At present the main application for data services over mobile communications systems is Internet access. The future is towards a full multimedia type application providing various levels of QoS using an IP (Internet Protocol)-based backbone. Thus WWAN is also moving towards integration of services.

Further work is being done by the standardization committees to integrate WLANs with 3G. Another development in the standardization of WWAN is towards an IP network. All this shows us that the WWANs are moving towards packet-switched solutions and integration of technologies now that integration of services has almost been achieved.

1.7.3 WPANs

Besides the WLANs, the wireless personal area networks (WPANs) like BLUETOOTH, HIPERPAN, and IEEE 802.15 are standardized. These technologies will be used for short distance (~10m) communications with low data rates for different QoS service. It is envisaged that the WPANs will exist in all the mobile terminals in the near future. The WPAN standards, IEEE 802.15.3 and .3a, have developed and work is ongoing on higher data rates of about 55 Mbps thus

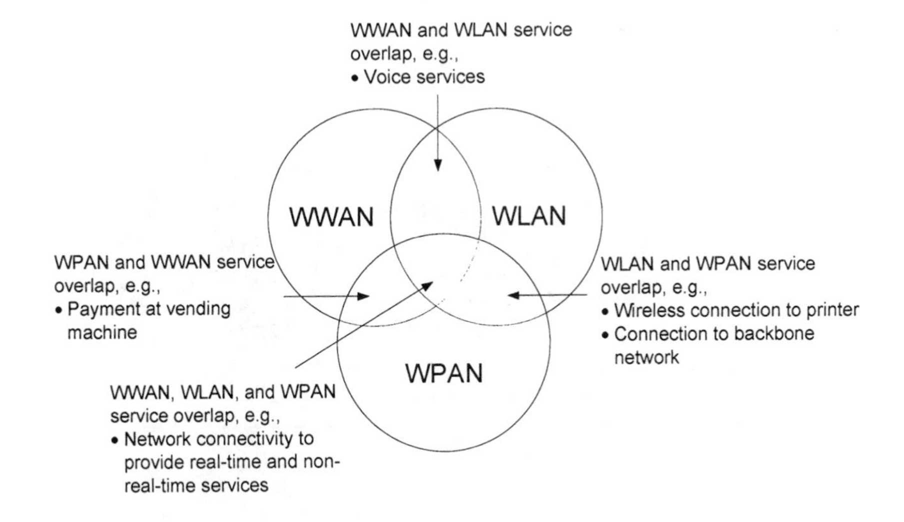

Figure 1.11 WWAN, WPAN, and WLAN overlap.

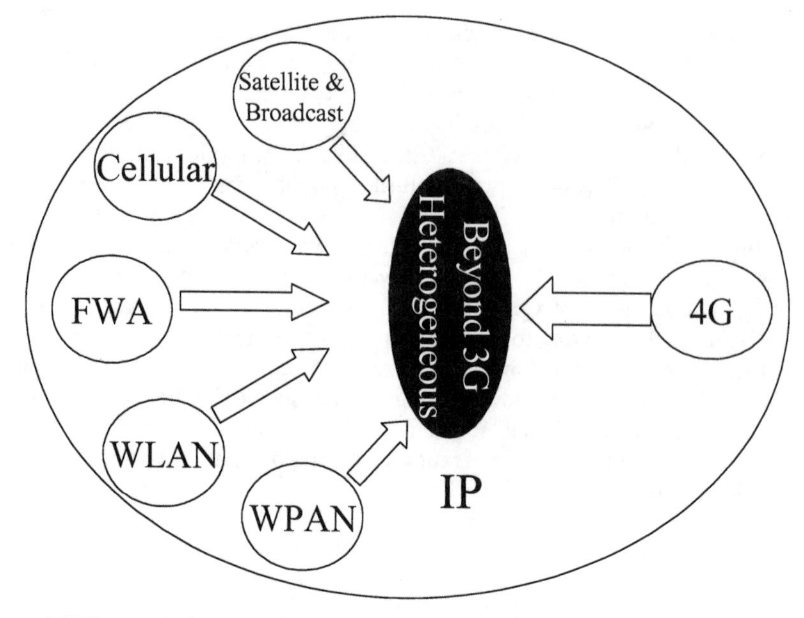

Figure 1.12 Future of telecommunications.

paving the path towards Broadband WPANs. IEEE 802.15.4 is focusing on very low data rate solutions, which will work at a few or a few hundred Kbps which is a first step towards the development of body area networks. Several companies have reached consensus ultra-wideband (UWB) as a low data rate solution for IEEE 802.15.

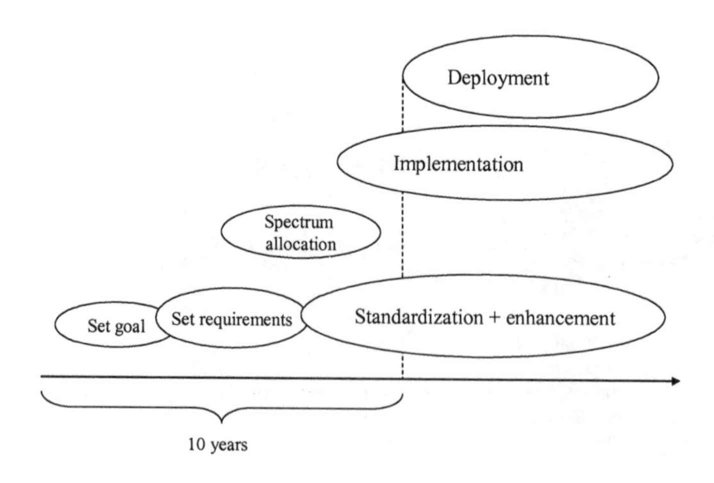

Figure 1.13 Time required for new technology development and deployment.

1.8 THE NEXT GENERATION

Each wireless technology is moving towards future standardization. This standardization work is mainly focusing on wireless IP-based QoS provision for any type of data. Here data is everything, be it audio, video, games, or any other application. Basically this means an integration of services. All these technologies overlap each other's areas of services to some extent. This is illustrated in Figure 1.11. Thus a move towards integration of technology is a logical next step to provide service continuity and higher user experience (Quality of Experience) (see Figure 1.12).

The ITU-R vision for 4G also calls for integration of technologies, which is commonly known as heterogeneous systems or B3G systems (to some people B3G could mean any standard or technology developed after 3G). Integration of technology will provide adequate services to a user depending on mobility and availability. Of course this brings along several new challenges (for example, handover/handoff or mobility, security, and QoS). These issues should be resolved without changing the existing standards. Seamless handover should be provided while a user moves from the network of one access technology to the other and domain of one stakeholder to the other. Seamless handover means provision of seamless service while the user is mobile, i.e., the user does not perceive any disruption in service or quality even during handover. IEEE 802 Handoff Executive Committee Study Group (ECSG) is working on the issue of handover for 802-based technologies.

The ITU-R vision also talks about a new air interface, also known as 4G. As any new system takes about 10 years to develop and deploy (see Figure 1.13), work on B3G and 4G has already started; a possible solution is given in [96]. The current market shows that 3G is going through trouble (recent news show that 3G services are expanding); lessons will be learnt from it while developing 4G.

There are several issues that should be solved for future wireless communications; one of them is work on seamless handover, which brings in the study of several issues like security and QoS, which, in turn, should be done at each protocol layer and network element. This topic itself will require further study on development methods and technologies including hardware, software, and firmware and technologies like Application Specific Integrated Circuits (ASICs). Another important research topic is Software Defined Radio (SDR), which includes reconfigurability at every protocol layer [2].

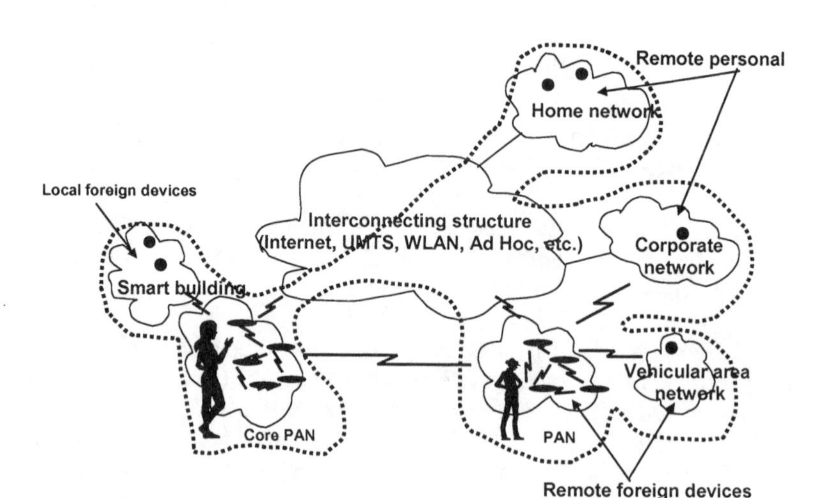

Figure 1.14 Personal network [98].

WLANs provide roaming within LANs, and work is going on towards further enhancement in this field. While WWANs provide roaming too, the challenge now is to provide seamless roaming from one system to another, from one location to another, and from one network provider to another. In terms of security, again both WLANs and WWANs have their own approach. The challenge is to provide the level of security required by the user while roaming from one system to another, user must get end-to-end security independent of any system, service provider, or location. Security also incorporates user authentication that can be related to another important issue: billing. Both security and roaming must be based on the kind of service a user is accessing [3]. The required QoS must be maintained when a user roams from one system to another. Besides maintaining the QoS it should be possible to know the kind of service that can be provided by a particular system, service provider, and location. Work on integration of the WLANs and the WPANs must also be done. The biggest technical challenge here will be the coexistence of the two devices as both of them work in the same frequency band. FWA is a technology that should be watched as it develops; depending on its market penetration and development of standards it should also be integrated together with other technologies.

Another area of research for the next generation communications will be in the field of personal networks (PNs) [98]. PN provides a virtual space to the users that spans over a variety of infrastructure technologies and adhoc networks. In other words PNs provide a personal distributed environment where people interact with various companions, embedded or invisible computers not only in their vicinity but potentially anywhere. Figure 1.14 portrays the concept of PNs. Several technical challenges arise with PNs, besides interworking between different technologies, some of which are security, self-organization, service discovery, and resource discovery [6].

1.9 OVERVIEW OF THE BOOK

This book has multiple goals, and each chapter tries to fulfill one of the goals. The first goal is to explain the WLAN market and bring in some business models for WLANs. Other goals are to explain the current situation of IEEE 802.11 security, QoS, mobility, and integration with other technologies like 3G and broadband wireless access (BWA) and to examine the future of wireless communications. The use of IP is considered in all the chapters.

Chapter 2 discusses market and business cases for WLANs. The chapter gives a brief description of the current market situation and builds a business case for WLANs. The important issue of billing is also discussed in the chapter as are business cases together with future growth directions and factors.

The IEEE 802.11 standard will be presented in Chapter 3. The different physical layers (PHY) and the Medium Access Control (MAC) layer will be discussed. The IEEE 802.11 standardization committee is working towards MAC enhancement for QoS, security, and several other issues. The PHY committee is also working towards a higher data rate (above 100 Mbps). These standardization efforts are discussed in this chapter.

Security is the most widely discussed and well known issue of the IEEE 802.11-based WLANs. In Chapter 4 of the book this issue is dealt with in detail. The past and current security solutions of IEEE 802.11 are presented in the chapter. Security issues related to these solutions are also discussed. IP-based solutions that can be used by WLANs and future solutions (IEEE 802.11i) are presented in the chapter.

The original IEEE 802.11 standard did not have any good solution for QoS. Seeing the future need of WLANs to fulfill different service requirements, the IEEE 802.11 standardization committee started working on MAC enhancements for QoS. This standard cannot by itself provide the required QoS; IP is needed. Chapter 5 of the book presents IEEE 802.11e and the IP layer solutions that can fulfill the QoS needs of WLANs.

The issue of mobility for WLANs is discussed in Chapter 6 of the book. Mobility using the IAPP, IEEE 802.11f, and IP-based solutions is described in this chapter.

Good deployment fulfilling the requirement of the client is extremely important for the success of any wireless network including WLANs. These deployment issues are discused in Chapter 7. This chapter also discusses WLAN deployment together with Universal Mobile Telecommunications System (UMTS) and GPRS.

The final chapter looks into the future of WLANs and wireless/mobile communications in general. Chapter 8 presents the direction the future mobile communications will take, which is normally known as 4G or B3G. The chapter also discusses issues to be solved for future wireless/mobile communications.

References

[1] Prasad, N. R., and A. R. Prasad, eds., *WLAN Systems and Wireless IP for Next Generation Communication*, Norwood, MA: Artech House, 2002.

[2] Prasad, A. R., *Wireless LANs: Protocols, Security and Deployment*, Ph.D. Thesis, Delft University of Technology, Delft University Press, 2003.

[3] Prasad, N. R., *Adaptive Security for Heterogeneous Networks*, Ph.D. Thesis, University of Rome "Tor Vergata," Rome, Italy, 2004.

[4] Prasad, N. R., and H. P. Schwefel, "A state-of-the-art of WLAN and WPAN," ECWT 2003, 6–10 October, Munich, Germany.

[5] IEEE, "802.11, Wireless LAN Medium Access Control (MAC) and Physical Layer (PHY) specifications," November 1997.

[6] IEEE, "Supplement to Standard for Telecommunications and Information Exchange Between Systems—LAN/MAN Specific Requirements—Part 11: Wireless MAC and PHY Specifications: Higher Speed Physical Layer Extension in the 2.4-GHz Band," P802.11b/D7.0, July 1999.

[7] IEEE, "Supplement to Standard for Telecommunications and Information Exchange Between Systems—LAN/MAN Specific Requirements—Part 11: Wireless MAC and PHY Specifications: High Speed Physical Layer in the 5-GHz Band," P802.11a/D7.0, July 1999.

[8] IEEE 802.11i, *IEEE 802.11: Specification for Robust Security*, D5.0, August 2003.

[9] IEEE 802.11g, IEEE 802.11: Further High Data Rates Extension in 2.4 GHz band, D8.2, April 2003.

[10] IEEE 802.11f, IEEE802.11: Recommended Practice for Multi-Vendor Access Point Interoperability via an Inter-Access Point Protocol Across Distribution Systems Supporting IEEE 802.11 Operation, D5, January 2003.

[11] Wi-Fi Alliance, Web URL: http://www.wi-fi.org/, 8 October 2003.

[12] WISPr, *Best Current Practices for Wireless Internet Service Provider (WISP) Roaming*, February 2003, v1.0, Wi-Fi Alliance.

[13] IEEE 802.11, Web URL: http://grouper.ieee.org/groups/802/11/, 8 October 2003.

[14] Munoz, L., and R. Prasad, *WLANs and WPANs towards 4G Wireless*, Norwood, MA: Artech House, March 2003.

[15] Tuch, B., "Development of WaveLAN, an ISM Wireless LAN," *AT&T Technical Journal*, Vol. 72, No. 4, July/August 1993, pp. 27–37.

[16] Kamerman, A., and L. Monteban, "WaveLAN-II: A High-Performance Wireless LAN for the Unlicensed Band," *Bell Labs Technical Journal*, Vol. 2, No. 3, 1997, pp. 118–133.

[17] van Nee, R., G. Awater, M. Morikura, H. Takanashi, M. Webster, and K. Halford, "New High Rate Wireless LAN Standards," *IEEE Communications Magazine*, December 1999.

[18] Kamerman, A., "Spread Spectrum Schemes for Microwave-Frequency WLANs," *Microwave Journal*, Vol. 40, No. 2, February 1997, pp. 80–90.

[19] *IEEE Personal Communications*, Vol. 7, No. 1, February 2000.

[20] Prasad, N.R., et al, "A state-of-the-art of HIPERLAN/2," *VTC 1999 Fall*, 19–22 September 1999, Amsterdam, The Netherlands, pp. 2661–2666.

[21] ETSI BRAN, "HIPERLAN Type 2 Functional Specification Part 1–Physical Layer," DTS/BRAN030003-1, June 1999.

[22] Kamerman, A., and A. R. Prasad, "IEEE 802.11 and HIPERLAN/2 Performance and Applications," *ECWT 2000*, 2–6 October 2000, Paris, France.

[23] Prasad, A. R., H. Moelard, and J. Kruys, "Security Architecture for Wireless LANs: Corporate & Public Environment," *VTC 2000 Spring*, 15–18 May 2000, Tokyo, Japan, pp. 283–287.

[24] Visser, M. A., and M. El Zarki, "Voice and Data Transmission over an 802.11 Network," *Proc. PIMRC'95*, Sept. 1995, Toronto, Canada, pp. 648–652.

[25] Prasad, A. R., "Performance Comparison of Voice of IEEE 802.11 Schemes," *VTC 1999 Fall*, 19–22 September 1999, Amsterdam, The Netherlands, pp. 2636–2640.

[26] Prasad, A. R., N. R. Prasad, A. Kamerman, H. Moelard, and A. Eikelenboom, "Indoor Wireless LANs Deployment," *VTC 2000 Spring*, Tokyo, Japan, 15–18 May 2000.

[27] Prasad, A. R., A. Eikelenboom, H. Moelard, A. Kamerman, and N. R. Prasad, "Wireless LANs Deployment in Practice," In *Wireless Network Deployments*, R. Ganesh and K. Pahlavan (eds.), Norwell, MA: Kluwer Academic Publishers, 2000.

[28] Prasad, R., and L. M. Correia, "Wireless Broadband Multimedia Communications," *International Wireless and Telecommunications Symposium*, Shah Alam, Malaysia, Vol. 2, pp. 55–70, May 14–16, 1997.

[29] Prasad, R., and L. M. Correia, "An Overview of Wireless Broadband Multimedia Communications," *Proc. MoMuC'97*, Seoul, Korea, September–October 1997, pp. 17–31.

[30] Prasad, R., "Wireless Broadband Communication Systems," *IEEE Comm. Mag.*, Vol. 35, January 1997, p. 18.

[31] Correia, L. M., and R. Prasad, "An Overview of Wireless Broadband Communications," *IEEE Comm. Mag.,* Vol. 35, January 1997, pp. 28–33.

[32] Morinaga, M., M. Nakagawa, and R. Kohno, "New Concepts and Technologies for Achieving Highly Reliable and High Capacity Multimedia Wireless Communications System," *IEEE Comm. Mag.* Vol. 35, January 1997, pp. 34–40.

[33] Fernandes, J. J., P. A. Watson, and J. C. Neves, "Wireless LANs: Physical Properties of Infrared Systems Versus mmWave Systems," *IEEE Comm. Mag.,* Vol. 32, No. 8, August 1994, pp. 68–73.

[34] ETSI, Web URL: http://www.etsi.org/, 8 October 2003.

[35] MMAC, Web URL: http://www.arib.or.jp/mmac/e/index.htm, 8 October 2003.

[36] IETF, Web URL: http://www.ietf.org/, 8 October 2003.

[37] Dixit, S., and R. Prasad, *Wireless Internet*, Norwood, MA: Artech House, 2002.

[38] R. R. Prasad, *Strategic Analysis of Ben & Ben's Strategic Option in WLAN*, M.Sc. Thesis, Finance and International Business, Aarhus School of Business, August 2002.

[39] Prasad, R., *CDMA for Wireless Personal Communications,* Norwood, MA: Artech House, 1996.

[40] Prasad, R., *Universal Wireless Personal Communications,* Norwood, MA: Artech House, 1998.

[41] van Nee, R., and R. Prasad, *OFDM for Wireless Multimedia Communications*, Norwood, MA: Artech House, 1999.

[42] Prasad, R., W. Mohr, and W. Konhäuser (eds.), *Third Generation Mobile Communication Systems*, Norwood, MA: Artech House, 2000.

[43] Prasad, R. (ed.), *Towards a Global 3G System: Advanced Mobile Communications in Europe*, Volume 1, Norwood, MA: Artech House, 2001.

[44] Prasad, R. (ed.), *Towards a Global 3G System: Advanced Mobile Communications in Europe*, Volume 2, Norwood, MA: Artech House, 2001.

[45] Farserotu, J., and R. Prasad, *IP/ATM Mobile Satellite Networks*, Norwood, MA: Artech House, 2001.

[46] Harada, H., and R. Prasad, *Simulation and Software Radio for Mobile Communications*, Norwood, MA: Artech House, 2002.

[47] Prasad, R., and M. Ruggieri, *Technology Trends in Wireless Communication*, Norwood, MA: Artech House, 2003.

[48] Hara, S., and R. Prasad, *Multicarrier Techniques for 4G Mobile Communications*, Norwood, MA: Artech House, 2003.

[49] Ojanpera, T., and R. Prasad (eds.), *Wideband CDMA for Third Generation Mobile Communications*, Norwood, MA: Artech House, 1998.

[50] Steel, R., (ed.), *Mobile Radio Communications*, New York: IEEE Press, 1994.

[51] Prasad, R., "Overview of Wireless Personal Communications: Microwave Perspectives," *IEEE Comm. Mag.*, Vol. 35, April 1997, pp. 104–108.

[52] Hafesi, P., D. Wedge, M. Beach, and M. Lawton, "Propagation Measurements at 5.2 GHz in Commercial and Domestic Environments," *PIMRC '97*, Helsinki, September 1–4, 1997, pp. 509-513.

[53] Johnson, I. T., and E. Gurdenelli, "Measurements of Wideband Channel Characteristics in Cells Within Man Made Structures of Area Less Than 0.2 km^2," COST 231 TD(90)083.

[54] Lahteenmaki, J., "Indoor Measurements and Simulation of Propagation at 1.7 GHz," COST 231 TD(90)084.

[55] Anderson, P. C., O. J. M. Houen, K. Kladakis, K. T. Peterson, H. Fredskild, and I. Zarnoczay, "Delay Spread Measurements at 2 GHz," COST 231 TD(91)029.

[56] Hashemi, H., "The Indoor Radio Propagation Channel," *Proceedings of the IEEE*, Vol. 81, No. 7, July 1993, pp. 943-968.

[57] Saleh, A. A. M., and R. A. Valenzuela, "A Statistical Model for Indoor Multipath Propagation," *IEEE Journal on Selected Areas in Communications*, Vol. SAC-5, No. 2, February 1987, pp. 128-137.

[58] Bultitude, R. J. C., et al., "The Dependence of Indoor Radio Channel Multipath Characteristics on Transmit/Receive Ranges," *IEEE Journal on Selected Areas in Communications*, Vol. 11, No. 7, September 1993, pp. 979-990.

[59] Bultitude, R. J. C., S. A. Mahmoud, and W. A. Sullivan, "A Comparison of Indoor Radio Propagation Characteristics at 910 MHz and 1.75 GHz," *IEEE Journal on Selected Areas in Communications*, Vol. 7, No. 1, January 1989, pp. 20-30.

[60] Pahlavan, K., and S. J. Howard, "Frequency Domain Measurements of Indoor Radio Channels," *Electronics Letters*, Vol. 25, No. 24, November 23, 1989, pp. 1645-1647.

[61] Davies, R., and J. P. McGeehan, "Propagation Measurements at 1.7 GHz for Microcellular Urban Communications," *Electronics Letters*, Vol. 26, No. 14, July 5, 1990, pp. 1053-1055.

[62] Zollinger, E., and A. Radovic, "Measured Time Variant Characteristics of Radio Channels in the Indoor Environment," COST 231 TD(91)089.

[63] Rappaport, T. S., and C. D. McGillem, "Characterization of UHF Multipath Radio Channels in Factory Buildings," *IEEE Transactions on Antennas and Propagation*, Vol. 37, No. 8, August 1989, pp. 1058-1069.

[64] Nobles, P., and F. Halsall, "Delay Spread and Received Power Measurements Within a Building at 2 GHz, 5 GHz, and 17 GHz," *IEE Tenth International Conference on Antennas and Propagation*, Edinburgh, UK, April 14-17, 1997.

[65] Hall, M.P.M., and L.W. Barclay, Radiowave Propagation, IEE Electromagnetic Waves Series 30, London, 1991.

[66] Correia, L. M., and P. O. Frances, "A Propagation Model for the Estimation of the Average Received Power in an Outdoor Environment at the Millimeter Wave Band," *Proc. VTC'94 — IEEE VTS 44th Vehicular Technology Conference*, Stockholm, Sweden, pp. 1785–1788, July 1994.

[67] Lovnes, G., J. J. Reis, and R. H. Raekken, "Channel Sounding Measurements at 59 GHz in City Streets," *PIMRC'94*, The Hague, The Netherlands, September 1994, pp. 496–500.

[68] Devasirvatham, D. M. J., "Multi-Frequency Propagation Measurements and Models in a Large Metropolitan Commercial Building for Personal Communications," *PIMRC '91*, London, UK, September 23-25, 1991, pp. 98-103.

[69] Chang, R. W., "Synthesis of Band Limited Orthogonal Signals for Multichannel Data Transmission," *Bell Syst. Tech. J.,* Vol. 45, pp. 1775–1796, December 1996.

[70] Salzberg, B. R., "Performance of an Efficient Parallel Data Transmission System," *IEEE Trans. Comm.,* Vol. COM-15, pp. 805–813, December 1967.

[71] Mosier, R. R., and R.G. Clabaugh, "Kineplex, a Bandwidth Efficient Binary Transmission System," *AIEE Trans.,* Vol. 76, pp. 723–728, January 1958.

[72] "Orthogonal Frequency Division Multiplexing," U.S. Patent No. 3, 488,455, filed November 14, 1966, issued January 6, 1970.

[73] Weinstein, S. B., and P .M. Ebert, "Data Transmission by Frequency Division Multiplexing Using the Discrete Fourier Transform," *IEEE Trans. Comm.,* Vol. COM-19, pp. 628–634, October 1971.

[74] Zou, W. Y., and Y. Wu, "COFDM: an overview," *IEEE Trans. Broadc.,* Vol. 41, No. 1, pp. 1–8, March 1995.

[75] Hirosaki, B., "An Orthogonally Multiplexed QAM System Using the Discrete Fourier Transform," *IEEE Trans. Comm.,* Vol., COM-29, pp. 982–989, July 1981.

[76] Halls, G., "HIPERLAN Radio Channel Models and Simulation Results," RES10TTG 93/58.

[77] Fernandes, L., "R2067 MBS - Mobile Broadband System," *ICUPC'93*, Ottawa, Canada, October 1993.

[78] European Commission, "Communications for Society Visionary Research," *DG XIII/B*, Brussels, Belgium, February 1997.

[79] Rokitansky, C. –H., and M. Scheibenbogen (eds.), "Updated Version of System Description Document," *RACE Deliverable R2067/UA/WP215/DS/P/068.b1*, RACE Central Office, European Commission, Brussels, Belgium, December 1995.

[80] Prasad, A. R., "Asynchronous Transfer Mode Based Mobile Broadband System," *Proc. IEEE 3rd Symposium on Comm. & Vech. Technology in Benelux*, Eindhoven, The Netherlands, October 1995, pp. 143–148.

[81] Prasad, R., and L. Vandendorpe, "An Overview of Millimeter Wave Indoor Wireless Communication System," *Proc. 2nd Int. Conf. on Universal Personal Communications*, Ottawa, Canada, 1993, pp. 885–889.

[82] Prasad, R., and L. Vandendorpe, "Cost 231 Project: Performance Evaluation of a Millimetric-Wave Indoor Wireless Communications System," *RACE Mobile Telecommunications Workshop*, Metz, France, June 16–18, 1993, pp. 137–144.

[83] Leslie, I. M., D. R. McAuley, and D. L. Tennenhouse, "ATM Everywhere?" *IEEE Network*, March 1993, pp. 40–46.

[84] Lawrey, E., *The Suitability of OFDM as a Modulation Technique for Wireless Telecommunications, with a CDMA Comparison,* Report Bachelor

of Engineering with Honours, Computer Systems Engineering, James Cook University, October 1997.

[85] Rappaport, T. S., *Wireless Communication Principles & Practice*, Upper Saddle River, NJ: Prentice-Hall PTR, 1996.

[86] IEEE 802.1X, "IEEE Standard for Local and Metropolitan Area Networks–Port-Based Network Access Control," July 2001.

[87] IEEE 802.15, Web URL: http://grouper.ieee.org/groups/802/15/, 8 October 2003.

[88] IEEE 802.16, Web URL: http://grouper.ieee.org/groups/802/16/, 8 October 2003.

[89] IEEE 802.20, Web URL: http://grouper.ieee.org/groups/802/20/, 8 October 2003.

[90] Cahner In-Stat Group, Web URL: http://www.instat.com/, 8 October 2003.

[91] FCC, Office of Engineering and Technology, Radio Frequency Safety, Web URL: http://www.fcc.gov/oet/rfsafety/, 8 October 2003.

[92] IEEE Handoff ECSG, http://www.ieee802.org/handoff/.

[93] IETF Seamoby WG, http://www.ietf.org/html.charters/seamoby-charter.html.

[94] Broadcom Singlechip solution: http://www.broadcom.com, 8 October 2003.

[95] Intel Centrino: http://www.intel.com, 8 October 2003.

[96] i-Mode: http://www.nttdocomo.com/i/, 8 October 2003.

[97] Farserotu, J., G. Kotrotsios, I. Kjelberg, and A. R. Prasad, "Scalable, Hybrid Optical-RF Wireless Communication System for Broadband and Multimedia Service to Fixed and Mobile Users," Invited Paper, *International Journal on Wireless Personal Communications*, Kluwer Academic Publishers, January 2003, Vol. 24, No. 2, pp. 327-339.

[98] Lauridsen, O.M., and A. R. Prasad, "User Needs for Services in UMTS," Invited Paper, *International Journal on Wireless Personal Communications*, Kluwer Academic Publishers, August 2002, Vol. 22, No. 2, pp. 187-197.

[99] Niemegeers, I.G., and S. M. Heemstra de Groot, "Research Issues in Ad-Hoc Distributed Personal Networks," *International Journal on Wireless Personal Communications*, Kluwer Academic Publishers, September 2003.

[100] IEEE 802.15, "Part 15.1: Wireless Personal Area Network Medium Access Control (MAC) and Physical Layer (PHY) Specifications," May 2000.

APPENDIX 1A: COMPARISON OF WLAN AND WPAN TECHNOLOGIES

Standard	IEEE 802.11/b/n	IEEE 802.11a/g	HIPERLAN/2
Frequency Range (MHz)	2,400–2,483 (North America/ Europe) 2,470–2,499 (Japan) n: also 5GHz as for IEEE 802.11a	a: 5,150–5,250 (Europe, North America, Japan) 5,250–5,350 (Europe, North America) 5,470–5,725 (Europe) 5,725–5,825 (North America) 4,900–5,000 (Japan) g: same as IEEE 802.11/b	Same as 802.11a
Multiple Access Method	CSMA/CA	CSMA/CA	TDMA
Duplex Method	TDD	TDD	TDD
Number of Independent Channels	FHSS: 79 DSSS: 3 to 5	a: 12 g: 3 to 5	12
Modulation	FHSS GFSK (0.5 Gaussian Filter) DSSS DBPSK (1MB/s), DQSK (2MB/s) b DSSS: CCK n 110 Mbps+	a/g: OFDM 48 carriers 6 Mbps BPSK 1/2 9 Mbps BPSK 3/4 12 Mbps QPSK 1/2 18 Mbps QPSK 3/4 24 Mbps 16QAM 1/2 36 Mbps 16QAM 3/4 48 Mbps 64QAM 2/3 54 Mbps 64QAM 3/4 g PBCC and DSSS OFDM optional	OFDM 48 carriers 6 Mbps BPSK 1/2 9 Mbps BPSK 3/4 12 Mbps QPSK 1/2 18 Mbps QPSK 3/4 24 Mbps 16QAM 1/2 36 Mbps 16QAM 9/16 48 Mbps 64QAM 3/4 54 Mbps 64QAM 3/4
Channel Bit Rate (Mbps)	1 or 2 b:5.5 or 11 n:100	a/g: 6, 9, 12, 18, 24, 36, 48, and 54	6, 9, 12, 18, 24, 36, 48, and 54

Standard	**IEEE 802.15 1.0 and Bluetooth**	**IEEE 802.15 3.0 and .3a**	**IEEE 802.15 4.0 and SG 4a**
Frequency Range (MHz)	Same as 802.11b	Same as 802.11b	Same as 802.11b 868 915
Multiple Access Method	TDMA	CSMA/CA and TDMA	CSMA/CA
Duplex Method	FDD	TDD	TDD
Number of Independent Channels	FHSS: 79	.3: High density 4 and .11b coexistence 3	.4: 11–20 in 2.4GHz 1 in 868MHz and 10 in 915 MHz
Modulation	FHSS GFSK (0.5 Gaussian Filter)	.3: 11 Mbps QPSK-TCM 22 Mbps DQPSK 33 Mbps 16 QAM-TCM 44 Mbps 32 QAM-TCM 55 Mbps 64 QAM-TCM .3a: 110 Mbps+	0.4: 250 Kbps 16-ary orthogonal 20 Kbps BPSK .4a: -
Channel Bit Rate (Mbps)	1	.3: 11, 22, 33, 44, 55 .3a: 110 +	.4: .250 and .020 .4a: -

Chapter 2

Market and Business Cases

Wireless local area network (WLAN) based on IEEE 802.11 has seen an explosive growth in recent years. Still the market and different factors affecting the market are not well understood. This chapter starts with a study of WLAN market and the factors that impact market. The chapter also proposes business cases for WLAN.

2.1 INTRODUCTION

At this point it is safe to assume that most of our readers are fairly well aware of the WLAN market. Therefore, we will not dwell too long on the statistics and emergence of the WLAN market but rather on the ensuing business model. We noticed the technology being aggressively pushed since the whole 3G fiasco. WLAN very clearly has become the tool that is delivering broadband mobile data services as was earlier promised by 3G or even the technologies bridging 3G. The intense worldwide competition among the WLAN vendors has brought the price of the product rapidly down; hence the margins for WLAN vendors are wafer thin [1–15].

This cascading effect (bandwagon effect) has led to an early maturity of the first WLAN standard, 802.11b. Compared to the market penetration 802.11b has arguably reached early maturity; we are at a point where the advent of newer and better versions of WLAN are being introduced into the market (e.g., IEEE 802.11a/g, etc.). However, market demand is falling behind the heralding technical enhancements being made.

This form of cyclical movement has been elaborated by Carlota Perez in her book *Technological Revolutions and Financial Capital: The Dynamics of Bubbles and Golden Ages* (Edward Elgar 2002) [2]. According to Perez's model it can be interpreted that WLAN is at the installation period and only once that WLAN reaches the deployment period will the technology become mature as illustrated in Figure 2.1. This implies that the market still has to evolve and that the dominant player is yet to be determined.

The evolution of the WLAN market has to be assessed also from the dynamism of the other actors in the value chain. Currently, most market studies' emphasis has been on vendors; however there are other actors without whom success of this technology in the marketplace could be compromised. Value chain, nonetheless, is still not explicitly defined by the market, and there are various possible shapes the market model can take. In this chapter we will begin with the market development of WLAN, followed by assessing the factors that are impacting the evolution of a marketing model for WLAN. This will be followed by the proposal of a business model and concluding remarks.

2.2 MARKET DEVELOPMENT

When the Wireless LAN market development is put under the microscope it can be argued that it is following the same road as has been followed by the automobile and the railway industries. Christopher Freeman, in his book, *Economics of Industrial Innovation*, has extensively discussed the dynamics of various industries, such as the automobile and the railway markets; an analogy can be drawn from the development of these mature industries to the path being followed by Wireless LAN.

A much more recent analogy would be the Internet. The advent of the Internet saw a major cascading effect; many new dot coms spawned during the installation period of the Internet market. This is attributable to the rush to be an early starter and hence be the first at procuring the expected revenue from a promising technology. The sudden surge of Internet start-ups led to the obvious fall in the prices of services provided; companies were working on wafer-thin margins.

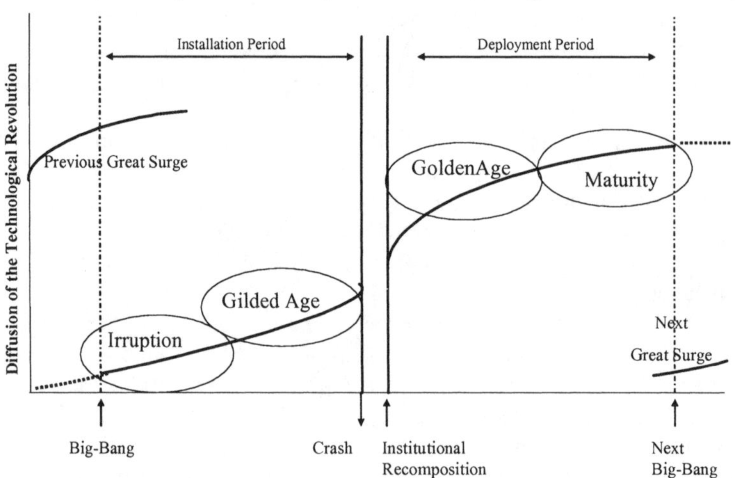

Figure 2.1 Product life cycle.

Owing to the fall in cash flow many start-ups perished whereas larger companies acquired some of the others. WLAN is going through a similar stage; with the technology hitting the market many start-ups sprouted all over the world, and they have been offering IEEE 802.11b chips for the last three to four years. Now, many of these WLAN start-ups have disappeared and quite a few of them have been acquired by larger competitors, such as Cisco. Nonetheless, the frenzy is still not over; research and development work still continues in the WLAN. Companies are still vying for market share by introducing newer and faster technologies in WLAN. Until recently we have been talking about 802.11b, and now for sometime there has been continuous news about a newer and better version of 802.11a OFDM–based technology hitting the market.

The continuous development in WLAN can be identified as the early stage of its product life cycle. However, the case of WLAN development defies the conventional trend; it has been observed in other industries, such as the semiconductor industry, that the selling price fell after the growth. However, in the case of WLAN the selling price has dropped during the products' introductory stage while the average cost of production of 802.11b chips has dropped tremendously in the last couple of years; the price of 802.11b chips is expected to drop by 75% by the end of 2003, according to a report by DigiTimes and other reports with similar projections.

How is the market reacting to the news of new technological entrants? If forecasts of the growth of WLAN are anything to go by, the growth of the hotspot worldwide is illustrated in Figure 2.2.

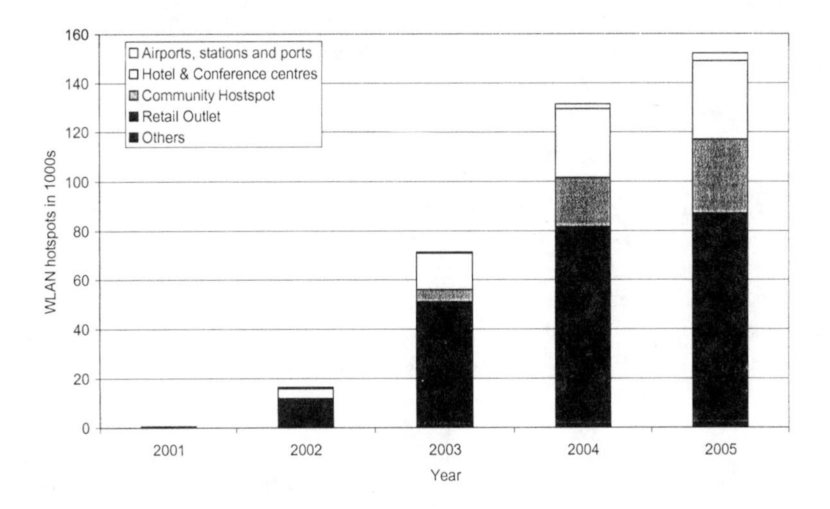

Figure 2.2 Market growth of WLAN [1].

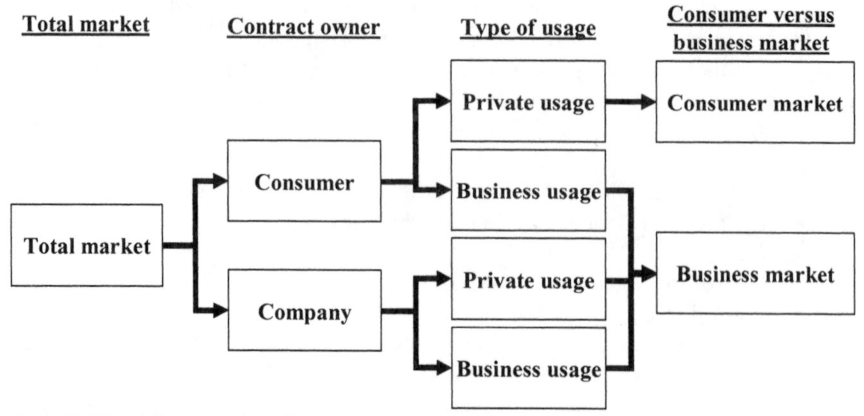

Source: R.R. Prasad, *Strategic Analysis of Ben & Ben's Strategic Option in WLAN*, M.Sc. Thesis, Finance and International Business, Århus School of Business, August 2002.

Figure 2.3 Market segmentation.

Retail outlets will have the most hotspots followed by hotels; in the category of community hotspots, airports are expected to have less than the other public hotspots. Lacking in the comparison are the private hotspots, those such as in enterprises and homes. A report in *The Economist* has shown that there are more private hotspots than public. However, the public hotspots are growing and are expected to outgrow private hotspots. This study, however, fails to give any insight into the market response to new WLAN technologies such as the current growth in IEEE 802.11a and IEEE 802.11g. The WLAN providers along with the vendors and software companies have become very dynamic at the prospect of capturing a large market globally with this new technology.

2.2.1 WLAN Target Market

The target market for WLAN can be divided into two categories: private and public sectors. Each of the sectors is the composite of various forms of hotspots; public hotspots refers to hotspots where the general public has access to WLAN (e.g., cafes, libraries, etc.). Private hotspots are a composite of enterprises, hospitals, and the educational institutions. Figure 2.3 depicts the market segmentation, illustrating the two markets and the types of usage, private and business usage.

The business and private market segments require different business models. Historically, the marketing of hi-techs in the early stages has been largely geared towards the business market and eventually addressed to the consumer (private) market. The business model adapted by the provider of the hi-tech products has been greatly influenced by the profit margin in either scenario.

In general, the business market is an early adapter of new technologies as compared to the consumer market. The business market offers in general a larger profit margin and lower churn rate than the consumer market. The requirements or needs of the two markets vary, due to the applications needed from WLAN. The recent requirements of WLAN users is illustrated in Figures 1.6–1.9. These requirements at large address all the WLAN applications we can imagine. The more the requirements are met the more applications can be envisioned for WLAN and hence increase the availability of the hotspots, in turn increasing the market size.

2.2.1.1 Applications

The most commonly talked about public hotspots are airports, hotels, and cafés/restaurants. Clearly, by examining these hotspots we can tell that the target market is a business user. In some countries such as The Netherlands and Germany WISPs are deploying in gas stations on the highways, making more accessible the consumer markets. Nonetheless, the market still needs to fully develop. WLAN has brought new possibilities to realize efficient productivity. The usage in the enterprise environment is tremendous; for example inventory management can be improved in hotels and factories. Lately, wireless deployment in hospitals is widely talked about, since the cost savings can be tremendous.

Wireless enabling of healthcare is a widely discussed topic; the United States has taken the lead in this application, with the total revenue generated in hospitals in 2001 $193 million and projected to be near $295 million by 2005. However security issues are still a paramount concern, in order to maintain patient confidentiality.

Another recently addressed application is Intelligent Transportation Systems (ITS), which essentially enables transportation with wireless. An example of connecting transportation with Wi-Fi is that in India the government has equipped unlicensed devices on public buses and allowed all Wi-Fi devices to operate at high power. Thus, the bus while driving through neighborhoods updates all the Internet accounts in the neighborhood. Similarly when driving past backbone node it wirelessly updates and is updated by the rest of the network. This is applicable only for asynchronous services such as e-mail and downloading applications. Nonetheless, it is very inexpensive to create ubiquitous Internet and broadband access. Among other growing concerns the means of handover is also being strongly addressed: in a mobile environment the user can roam from one network to the other; for WLAN to be similarly ubiquitous it is necessary to create roaming possibilities. This aspect will be dealt with in Section 2.4.

2.2.2 WLAN Providers

Since 2002, mobile operators around the world have entered the WLAN market and tried to replicate their existing business model on the wireless data only to realize the differences between wireless voice and wireless data services is significant. *Brain Heart* magazine in its article "New Clouds in the Mobile Operator's Way" has argued that mobile operators around Europe in order to hedge their investment in 3G are investing in Public WLAN. The mobile operators Teleno, Sonera, Telia, TDC, mmO2, Vodafone D2, Orange, T-Mobile, and Swisscom have made significant WLAN rollouts. However, it is now being realized that it is the WISPs, such as Megabeam, that are in a better position to address this market.

Mobile operators such as Swisscom and Telecom Italia have resorted to using WISPs to rollout the network and they are focusing now on the bundling of services. Hence, the idea is for WISPs to have "neutral host providers" as their core rolling out the infrastructure. Swisscom has acquired WLAN GMBH, Megabeam, Hubbhop.com, and Aervik B.V. With these acquisitions and joint ventures with WISPs there are still crucial issues to solve, such as roaming and handover. We will address these issues in the section "Market Model".

Among WLAN vendors, WISPs, and mobile operators, there are yet other players that are important to ensure the success of this product; venues and integrators are also very important actors in the value chain. Venues are the hotspot locations and integrators are responsible for aggregating and rolling out the network.

Public hotspot locations include airports, hotels, cafes, etc., as is illustrated in Figure 2.4. Copenhagen, Denmark, planned for the airport and a conference center to share the WLAN network so that users have a roaming possibility. Similarly, parts of several cities in Europe such as Paris, Rome, Amsterdam, and Århus have been wireless enabled.

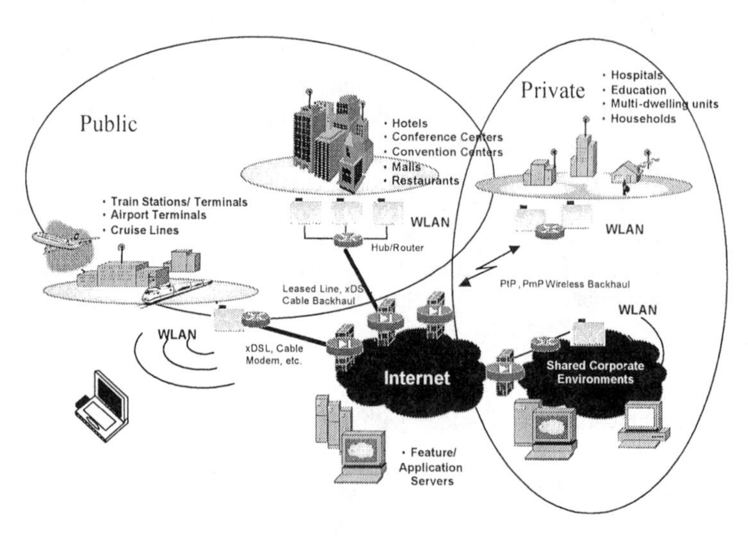

Figure 2.4 Public hotspots.

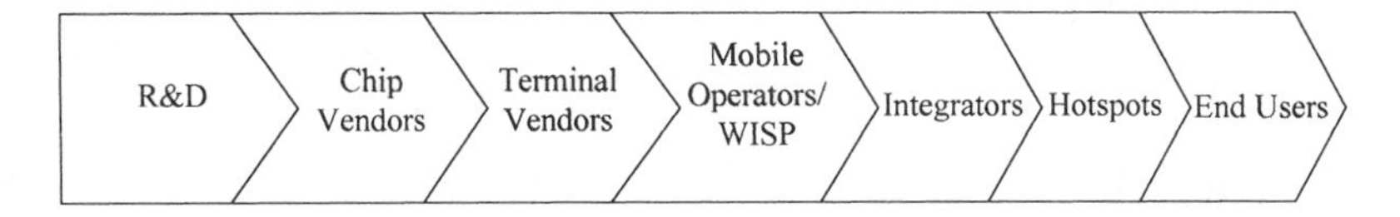

Figure 2.5 Value chain.

To provide new services and improve the quality of life, all major cities in Europe are becoming wireless (Source: Wireless Integrated Billing and Security.).

Paris – 10 Access Points for every metro station,
38 Bus routes are now wireless enabled
Rome and Milan – Public Demonstration,
all major museum are wireless, more than
500 Access Points covering Airport, Library,
Hotels, Shopping Centers, Tourist Places
London – BT is covering motorway service stations,
airports, rail stations, conference centres,
hotels, and cafes
Oslo – Coverage of the Airport and City Centre in
the near future

These citywide deployments have in major part occurred due to the efforts of mobile operators. Such citywide efforts can enable roaming and be successfully delivered when provided by mobile operators. This is due to the mass market that they own. Roaming here is referred to as the possibility of having roaming for billing. Although WISPs have as their core business WLAN expertise, they are still too few to actually provide this service; according to the *Brain Heart* magazine there are 200 WISPs too few to achieve this feat.

By coupling with mobile operators WLAN can be available to the masses. Currently, the value chain delivering WLAN can be expressed as in Figure 2.5.

This value chain reflects the order in which the WLAN is being serviced to end users in most cases. For instance TDC, the Danish Mobile Operator, is the biggest WLAN player in Denmark and its model for providing the WLAN is as follows: it deploys the infrastructure in various hotspots and keeps the ownership of the hardware. Further, it has opted for voucher-based billing, the revenue of which they share with the hotspot owners. The billing issue for WLAN has been regularly addressed; it is an issue that is a consequence of the business model taken up by the WLAN providers.

2.2.3 Billing

Here we will mention the possible forms of billing that can be "provided" to end users. First of all, it should be noted that the authors treat billing as a value added service. An extensive study has been conducted on billing for broadband IP networks; several pricing techniques have been suggested, some of which are alluded to in [3]. So far, the proposals are the following:

- Flat pricing;
- Priority pricing;
- Paris-Metro pricing;
- Smart-Market pricing;
- Edge pricing;
- Expected Capacity pricing;
- Responsive pricing;
- Effective Bandwidth pricing;
- Proportional Fairness pricing.

In [3], the authors assess these pricing schemes in light of economic, social, and technical issues. Essentially, [3] investigated pricing from two perspectives: recovery of expansion charges and a means of congestion control. Ensuing from this paper another work by R. R. Prasad [4] comments, "*However, the price elasticity of demand of various forms of users has not been addressed in [3]; this information is vital to accurately facilitate any of the aforementioned pricing strategies.*"

The price elasticity is a crucial issue to consider for WLAN; as it is still a new product for end users the pricing of this should be developed based on the environment. There are various factors that influence the competitive strategy from which pricing policy ensue other resultants. These will be discussed in section 2.3. Porter's [5] five competitive forces have influenced the concept of assessing factors in motion for WLAN.

Currently, the pricing scheme practice, in general, is by the means of voucher payments, where the user pays based for a certain amount of time; usually the payment is based on blocks of hours (e.g., 1, 3, 6, or 24 hours). The pricing scheme for data in the mobile industry such as for Mobile Multimedia Services (MMS) and Short Messaging System (SMS) based on flat fees as opposed to Wireless Access Protocol (WAP) where the pricing is time dependent and packet size dependent. This is to illustrate the various forms of pricing techniques for data communications that exist. Since WLAN is still establishing itself in the consumer market the providers still have to determine the best form of billing for the user.

In the paper by R. R. Prasad they propose billing that is independent of the access technique; one of the manifestations of such a proposal could be foreseen as a universal invoice for all forms of communications. Implicating that when the user gets his or her phone bill in the same invoice all other fees for other forms of communications are attached. WLAN billing could be based on this model, where the invoice for WLAN will be attached in the phone bill of the user.

This concept is leading towards SMS-based billing, which requires cooperation between mobile operators and WISPs; however most of the WISPs are usually mobile operators. The mobile operators are best suited to offer this because they already have preexisting roaming contracts around the world; by providing a platform such as this inherently will assist in boosting the WLAN into becoming ubiquitous. Obviously, the form of billing is not the only factor that might determine the global appeal of WLAN; pricing is a result of the environment.

So far we have discussed the possible forms of pricing that exist and that could be addressed under various strategic scenarios. Now our attempt will be to assess the pricing to be levied upon the end users; we all are aware of the general law of price equilibrium, the point where the quantity demanded meets the quantity supplied. This would be the simplistic way of assessing the concept. There are other actors in the environment who create the forces that influence pricing; the most important is the maturity of the product. The immediate factor one focuses upon is the competitors; in this market it's the WISPs that are setting the price for the end users. In the European market most of the hotspots are provided by mobile operators, which determine the fee for the end users; finally the revenue is divided between the hotspot owners and the mobile operator, such as the case with Tele Denmark.

The industrial structure [6–8] of mobile operators can be defined as an oligopoly, meaning that the barrier to entry is high and that there are few firms that are sharing the economic gains from WLAN. The pricing strategy, which we have witnessed in the mobile phone industry, can be assumed to be an accurate replication of the road WLAN will be following. What sets WLAN apart from the conventional mobile phone industry, however, is that the barrier to entry among WISPs is lower; there are smaller players still vying to get a share in the economic gains. Largely the target market for the mobile operators is large hotspots such as airports, hotels, and conference centers since this is economically more justifiable for their size than the smaller hotspots. In Denmark the market distribution is informally visible between the two, small WISPs and mobile operators. WISPs such as Redspot in Denmark has largely small hotspots such as cafes and restaurants; however they are also responsible for the deployment of WLAN in the city of Århus, Denmark, where it is being offered to the end users free of cost. There is a market for smaller players to enter with the expectation of high investment returns but currently the product is still new to the consumer market; the market still has to be developed in order for the smaller players to start getting their return on investment. Therefore, the pricing to the end users should be focused such that they are interested in discovering and adopting the new product; therefore a lower barrier of entry is favorable for the market as a whole.

The historical trend of new products usually started as high while the industry was in the early phase and then with increasing competition the prices fall; this is when the market was maturing as is discussed about the semiconductor industry and the case with the mobile phone industry. The prices of the products started to decline as the competition increased; competition led to further research and development, which made the products inexpensive for the end users in the semiconductor industry. In the mobile industry the prices for voice and data declined as competition increased and as the market became nearly saturated. The WLAN market is in a sense a hybrid of the two examples; as was mentioned earlier the technology for the product has matured more and the price of the hardware has dropped faster than the maturity of WLAN among the end users.

2.3 FORCES IN MOTION

Forces in motion establish the factors in the environment that are responsible for the strategies firms pursue. Identification of these forces prepares the company in recognizing the dynamism in the industry and outside the industry. Furthermore, it is essential for firms to streamline the internal environment of the firm with that of the industry, as the two reinforce one another; a breach could result in failure of the firm to react to needs of the market.

Porter (1998) proposed five forces driving industry competition, the suppliers, the buyers, threats from new entrants, threats from substitute products, and existing rivalry with incumbents, illustrated in Figure 2.6. These are generic forces that apply to firms in all industries. For WLAN we have identified the regulators as another very essential force; regulators could be the national telecommunications regulators and international regulators such as the International Telecommunications Union (ITU) and standardization bodies such as IEEE.

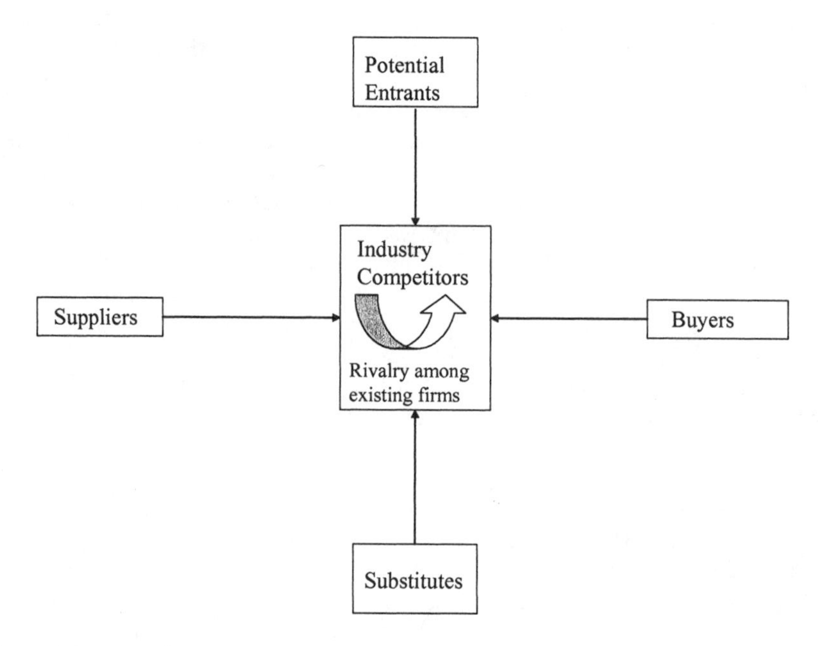

Figure 2.6 Porter's five competitive forces. (Source: [5].)

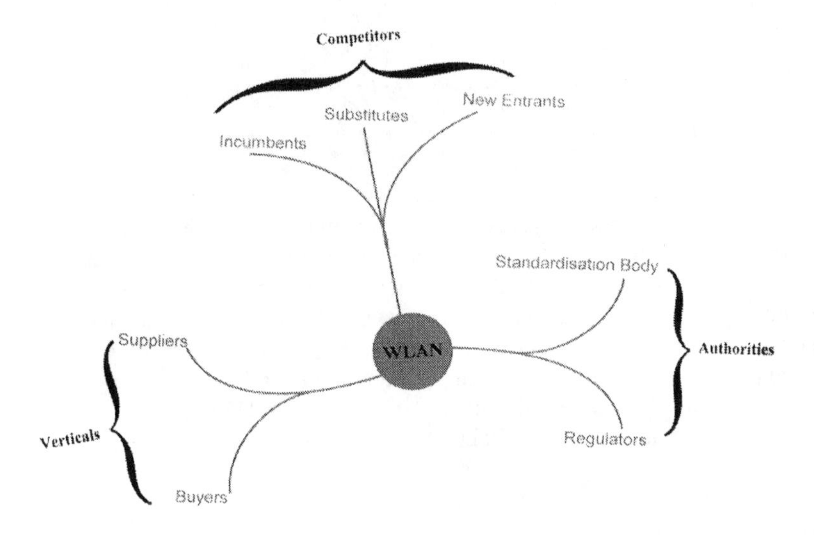

Figure 2.7 WLAN forces in motion.

In Figure 2.7 we illustrated the forces impacting the WLAN and the ways that they are reinforcing each other. This model of forces is a composite of Buyers, Suppliers, Substitute Technologies, Regulators, Standardization Bodies, Incumbents, and New Entrants.

These forces are applicable for all companies at every level of the WLAN value chain. If we look at the history of WLAN it is a product push, much like any other telecommunications technology. It was initially standardized by the regulatory body IEEE; hence the IEEE 802.11b,a,g, etc. came about and the technology enabling this standard is Direct Sequence Spread Spectrum (DSSS) and Orthogonal Frequency Division Multiplexing (OFDM), respectively. When the WLAN products needed to be interoperable, the WiFi was instigated. Once having laid the groundwork it was up to the national regulators to decide whether the product would be allowed to operate in ISM bandwidth, 2.4GHz. Countries around the world made this frequency available for WLAN deployment all at different times. Once the national regulators allowed the public use of WLAN it was up to the market to determine the "fate" of the product. Therefore, the power of buyers and suppliers is very important factor to comprehend as well; these are described as the verticals or recognition of the power. To maintain harmonious symbiosis is crucial for the survival and the success of the WLAN.

In order to develop the market for WLAN, know-how needs to trickle down from the companies conducting the research and development to the chip vendors, and terminal vendors all the way through to the end user as illustrated in Figure 2.6. These verticals influence the competitive forces of a company in the value chain.

Figure 2.8 Value chain.

The competitive forces are substitutes, incumbents, and the entrants; each of these factors is also influenced by the verticals and the authorities and vice versa. Substitutes are the alternate technologies, which might threaten to cannibalize the WLAN market (for example, EDGE, UMTS, etc.). New entrants are always to be looked out for as usually in WLAN we notice that new startups are coming up with more enhanced WLANs. Furthermore, the threat could be indirectly coming from a vertical (e.g., Intel has invested on the order of $150 million on WLAN companies). Finally, there is the constant threat from the peers.

2.4 BUSINESS CASE

In this section we will propose a business model wherein we suggest two methods for making WLAN readily available with ease to the end users. These methods are the following:

- Roaming
- Handover

Furthermore, an issue to consider is the evolution of the value chain, which will be largely determined by the product life cycle of WLAN. Currently, most of the market is served as shown in Figure 2.8.

As alluded to earlier in the chapter mobile operators such as TDC in Denmark are directly catering to the end user. This business model could take any of the following forms as shown in Figure 2.9 and Figure 2.10. Figure 2.9 illustrates the model, which has been discussed earlier where the Mobile operator services the end users by also deploying a network. Another option could be for mobile

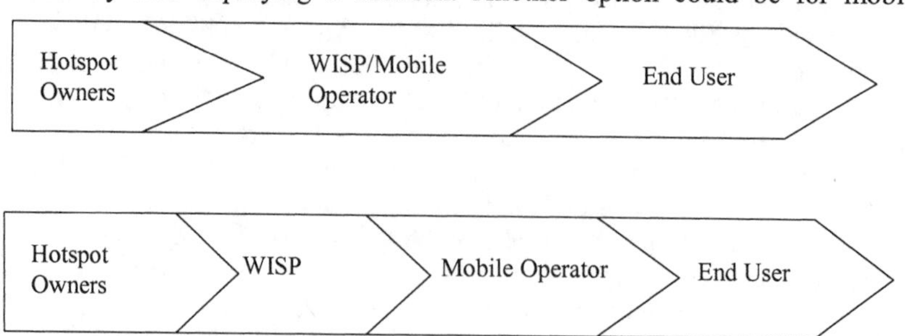

Figure 2.9 Emerging business model.

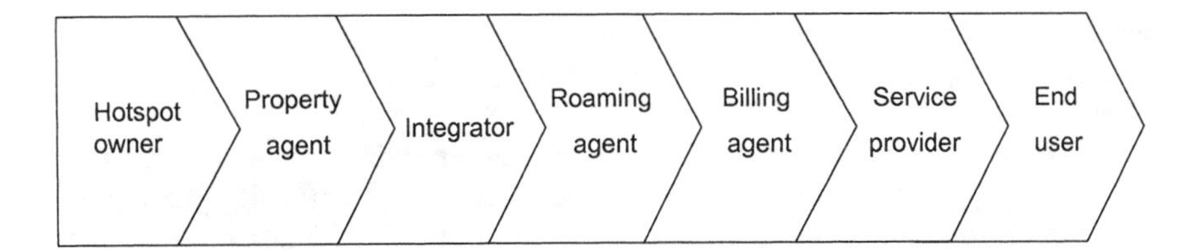

Figure 2.10 Highly fragmented model.

operators to package the product and bring it to the end users while the WISP focuses on the network deployment. Essentially, there is a major difference in the ways in which the Europeans have modeled the business and U.S. firms. As mentioned earlier, in Europe the mobile operators are playing the main role of providing Wireless ISP; however in the United States WISPs are mainly new startups that are poorly funded.

Figure 2.10 illustrates a highly fragmented solution, where the complexity is higher. Nonetheless, the proposal here is what leads to our next point: In roaming, where a user can move from one network to another the billing does not change and the handover remains seamless.

Several companies are working on developing technology to enable this. Start-up Wireless Integrated Billing & Security (WIBS) offers a software solution, which is independent of the access technique; Nokia and Ericsson are developing chips that can enable handover from WLAN to GPRS. The work continues in this field. However, the time to enter the market is essential.

In order to enable roaming and handover, other than the technology, an interoperability agreement is mandated among the WLAN operators and mobile operators. To get a business perspective let us examine the following scenarios.

2.4.1 Business Assessment of Various Hotspot Scenarios

The scenario pertains to a chain of commercial centers, M Plazas; the chain is owned by an investors group, M Makers & Shakers. There are 12 of these centers around the country. The corporation decides to modernize the centers so it hires Wireless LAN company WIBS to advise it on the deployment of WLAN and the management of this network. WIBS describes two possibilities:

1. Decentralized scenario
2. Centralized scenario

The *decentralized* solution refers to the WLAN management system, handover, and roaming, where each of the hotspots (commercial centers) is isolated from one another and any other networks. Hence, in our case these commercial centers will each be a hotspot and will act as islands.

The *centralized* solution refers to the WLAN system where a group of hotspots is managed from a single administration. In the case of M Plazas this would be realized by administering all the 12 centers from one location. Thus, the users can enjoy the benefit of a single invoice regardless of whichever center they might be in.

Currently, most of the deployments are decentralized solutions in Europe; most common public hotspots are "islands" and do not offer the possibility of mobility thorough different networks nor do they offer a common billing scheme. Most of the mobile operators are offering the services, even though the façade consists of the hotspot owners. T-Mobile is one of the largest providers of WLAN in Europe and the United States[1]. The pricing scheme that they have introduced is voucher-based, in Starbucks, gas stations, and all around their hotspots. TDC provides a similar billing feature. Sweden is well ahead of most of the European countries in embracing WLAN, where its users are treated to various forms of billing systems, from flat-fee to volume-based payment.

The business models currently for hotspots still in the making show the real growth in the market is coming in from growth in the sales in the enterprise market. In order to make the WLAN ubiquitous larger companies such as AT&T, Intel, HP, etc. are entering the market. However, the mobile operators will have to play a major role in pushing the market for WLAN in the hotspots. Interoperability and an agreement among WISPs will allow users to move from one WLAN network to the other. However, to make the WLAN much more attractive for the users, we need a solution that seamlessly provides handover from one access technique to the other—for instance, mobility from a WLAN network to a GRPS network.

In order to attain such a feat a combination of wireless ISPs and mobile operators is inevitable. Now that the network solution will be provided it is equally important that terminals should be able to handle such solutions; this brings vendors into the picture. A very important role has to be played by vendors too in providing terminals such as laptops, PDAs, etc. with a user-friendly interface, size, power consumption, etc. Finally, potential hotspot owners will have to be more willing to provide the venue. Essentially, in order for success of this technology in the market a congruent effort is required.

WAP was such an example, requiring multitudes of players to work together. I-Mode, an extremely successful story in Japan, required a similar business model. Basically as we move more towards data centric telecommunications potential applications grow. This implies opportunity for more players to enter the market and hence create a flourishing economic ecosystem.

2.5 FUTURE GROWTH AREAS AND FACTORS

What we have established in this chapter is the current state of the WLAN market and its ecosystem and how it fits into the historical frameworks of other

---

[1] However there was no information about its hotspots anywhere on its home page, until this chapter was written 19 September 2003.

technologies. What could be foreseeable in the medium run of this industry is that the number of hotspots around the world will grow and during this growth phase newer players will emerge. Meanwhile, changing of hands more IEEE 802.11a/g will be pervasive. The applications requiring WLAN empowerment will be more and more common to witness.

The unlicensed status of WLAN gives tremendous business opportunities to smaller players. The unlicensed nature of WLAN and the fact that it operates in low power has made it ideal to scale it for the purpose of a last-mile broadband solution; this is illustrated in the example of ITS. The social and economic benefit of WLAN is far more than single hotspots due to the possibility to scale wireless broadband from hotspot to last-mile. This notion can bring rural areas online; deprived parts of the world can have access to high-speed Internet. Some argue the pervasiveness of WLAN in the future will be such that it will become a necessity. It will not only be for the purpose of keeping a human user online all the time but also to present the possibility for a machine to communicate with another machine; the famous example comes of your fridge informing your car about the grocery list, which is then transmitted to you on the way from work to home. However, the argument persists over whether WLAN can actually substitute the huge capacity of fiber. Some see it as a supplement that might be true at present, but the rate at which technological development is occurring means that it could very well one day be a substitute for fiber. Companies such as PCOM:I^3 have developed technology that as an add on to other technologies improves their capacity at least two times without even increasing the power or the bandwidth. The application of WLAN in the future is massive; it gives access to individuals to enter the market as actors in the value chain, usually as a WISP because it is inexpensive. Nonetheless, in order for this exponentially growing mosaic to create a seamless WLAN all the smaller WISPs must build a consensus and at the moment it is the mobile operators that provide the most favorable alternative to this solution. What might happen is that at the point of market saturation there will be room for fewer players, and those that are not competitive will have to find other markets.

References

[1] The Economist, *Bubble Trouble*, June 26th 2003.

[2] Perez, C., *Technological Revolutions and Financial Capital: The Dynamics of Bubble and Golden Ages,* EE 2002.

[3] Falkner, M., M. Devitsikiotis, and I. Lambadaris, "An Overview of Pricing Concept for Broadband IP Network," *IEE Communications Survey*, Second Quarter 2000.

[4] Sherlekar, S., and R. R. Prasad, "A Global Access Independent Billing Model: Users' Perspective," *Wireless Personal Communications*, Kluwer Academic Publisher, Vol. 26, Issue 2, Jan 2003, Pages 169-178.

[5] Porter, M. E., "Competitive Strategy Techniques for Analyzing Industries and Competitors," *Free Press*, 1998.

[6] Dasgupta, P., and J. Stiglitz, "Industrial Structure and the Nature of Innovative Activity," *Economic Journal*, 1980a, Vol. 90, pp. 266–93.

[7] Dasgupta, P., and J. Stiglitz, "Uncertainty, Industrial Structure and the Speed of R&D," *Bell Journal of Economics*, 1980b, pp.1–28.

[8] Demsetz, H., "Industry Structure, Market Rivalry and Public Policy," *Journal of Law and Economics*, 1973, Vol.16, No. 1, pp. 1–10.

[9] Stead, R., P. Curwen, and K. Lawler, *Industrial Economics Theory, Applications and Policy*, London: McGraw-Hill, 1996.

[10] Barras, R., "Towards Theory of Innovation in Services," *Research Policy*, 1986, Vol. 15, pp. 161–173.

[11] Barras, R., "Interactive Innovation in Financial and Business Services: The Vanguard of Service Revolution," *Research Policy*, 1990, Vol. 19, No. 3, pp.215–239.

[12] Blair, M. J., *Economic Concentration: Structure, Behaviour and Policy*, New York: Harcourt Brace Jovanovic 1972.

[13] Acs, J. Z., and B.D. Audretsch, "Innovation in Large and Small Firms: Empirical Analysis," *American Economic Review*, September 1988.

[14] Baldwin, L. W., and T. J, Scott, "Market Structure and Technological Change," *Harwood Academic*, 1987, Switzerland.

[15] Connolly, A. R., and M. Hirschey, "R&D, Market Structure and Profits: A Value Based Approach," *Review of Economics and Statistics*, 1984, 66, pp. 678–681.

Chapter 3

IEEE 802.11

The IEEE 802.11 for the wireless local area network (WLAN) standard focuses on the medium access control (MAC) and physical layer (PHY) for access point (AP)-based networks and ad-hoc networks. The original standard supported three PHY infrared (IR), frequency hopping spread spectrum (FHSS), and direct sequence spread spectrum (DSSS) [1–3]. The 802.11b extension of the standard supports DSSS in the 2.4-GHz band with data rates of 1, 2, 5.5, and 11 Mbps. The last two bit rates are achieved through complementary code keying (CCK) [2–4]. Extension 802.11a and g are for a high bit rate orthogonal frequency division multiplexing modulation (OFDM) PHY standard providing bit rates in the range of 6 to 54 Mbps in the 5-GHz and 2.4-GHz band, respectively (see [2, 3, 5, 6]). All the PHY have the same MAC layer, carrier sense multiple access/collision avoidance (CSMA/CA), of which enhancements are being finalized particularly for security and quality of service (QoS).

In this chapter an overview of various IEEE 802 wireless related activities is given and the original IEEE 802.11 standard is explained in detail. Details of IEEE 802.11 standards on security, QoS, and mobility are explained in later chapters.

3.1 IEEE 802 STANDARDIZATION PROCESS

As this chapter discusses various IEEE 802 standards and ongoing activities it is good to have an understanding of the standardization process [7].

The Local Area Network (LAN) Metropolitan Area Network (MAN) Standardization Committee (LMSC) or IEEE Project 802 develops LAN and MAN standards, mainly for the lowest two layers (MAC and PHY) of the reference model for open systems interconnection (OSI). There are several different groups developing MAC and PHY for different purposes. The unifying part for all standards is the common upper interface to the logical link control (LLC) sublayer, common data framing elements, and some commonality in media interface.

Once a project is approved within a group it is assigned a letter (e.g., 802.11a). A study group (SG) is formed when a new area is first investigated for standardization. The SG can be within an existing working group (WG) or technical advisory group (TAG), or it can be independent of the WGs. A new project in an existing group is developed by a task force, while a new independent project creates a new WG.

For each project a project authorization request (PAR) is normally submitted for approval within six months of the start of work. A new project should have supporting material in the form of "5 criteria" to show that it meets the charter of LMSC. The draft PAR is voted on by the sponsor executive committee (SEC), and then goes to the IEEE standards board new standards committee (NesCom) which recommends it for approval as an official IEEE standards project. The PAR also identifies which outside standards groups there will be liaisons with. The liaisons help avoid conflicts or duplication of effort in one area.

Proposals are evaluated by the WG, and a draft standard is written and voted on by the WG. The work progresses from technical to editorial/procedural as the draft matures. When the WG reaches enough consensus on the draft standard, a WG letter ballot is done to release it from the WG. It is next approved by the SEC and then goes for sponsor letter ballot.

After the sponsor letter ballot has passed and "No" votes are answered, the draft standard is sent to the IEEE standards board standards review committee (RevCom). Once recommended by RevCom and approved by the Standards Board, it can be published as an IEEE standard. Most draft standards in LMSC are also sent to ISO at or before the time they go to sponsor letter ballot. A parallel approval path is followed in ISO JTC1/SC6 (joint technical committee 1, subcommittee 6—responsible for LANs) that leads to publication as an ISO standard. The process from start to finish can take several years for new standards, and less for revisions or addenda.

3.2 OVERVIEW OF IEEE 802 ACTIVITIES

Several wireless standards are being developed within IEEE 802 [8]. In this section an overview is given of these standards before discussing their activities in the next section.

The groups working on wireless technology are listed as follows; these are also mapped in Figure 3.1, a standard figure representing various technologies now and in the near future.

- IEEE 802.11: Develops standards for wireless local area network;
- IEEE 802.15: Develops standards for wireless personal area network (WPAN);

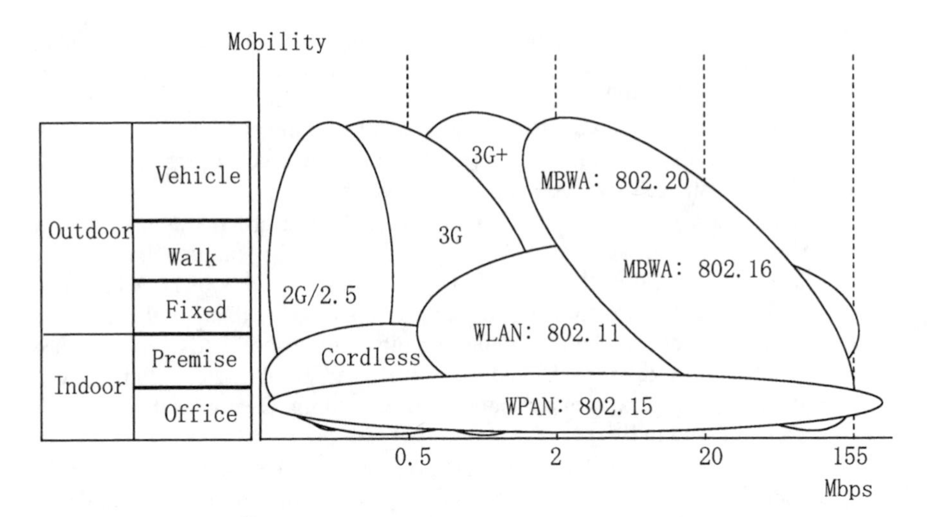

Figure 3.1 Standards, data rate, and mobility.

- IEEE 802.16: Develops standards for broadband wireless access (BWA) or wireless metropolitan area network (WMAN). One of its groups is also looking into mobility;
- IEEE 802.20: Develops standards for mobile broadband wireless access (MBWA);
- IEEE 802.21: Develops standards for handover between different technologies and administrative domains.

3.3 IEEE 802 CURRENT ACTIVITIES

Currently IEEE 802 is active in many fields. Naming and discussing all of them is simply far too much for one chapter if not a book. To give an overview of the broad range of activities that IEEE 802 is carrying out, this section provides a brief description of the ongoing activities related to the IEEE 802 wireless air interface.

3.3.1 802.15

The IEEE 802.15 working group is divided into four main task groups, with a fifth one added recently. A short description of each of the TGs is given in the following:

- TG1: This TG was the starting point of IEEE 802.15. The preliminary task of the group was to standardize Bluetooth.
- Another group, TG1a, is working on standardization of Bluetooth 1.2.

- TG2: Task Group 2 is developing recommended practices to facilitate coexistence of wireless personal area network (802.15) and wireless local area networks (802.11). The task group is developing a coexistence model to quantify the mutual interference of a WLAN and a WPAN™. The Task Group is also developing a set of Coexistence Mechanisms to facilitate the coexistence of WLAN and WPA devices.
- TG3: Task Group 3 or the 3 high rate (HR) task group was tasked with developing an ad-hoc MAC layer suitable for multimedia WPAN applications and a PHY capable of data rates in excess of 20 Mbps. The 802.15.3 standard specifies data rates up to 55 Mbps in the 2.4GHz unlicensed band. The technology employs an ad-hoc PAN topology not entirely dissimilar to Bluetooth, with roles for "master" and "slave" devices. Complexity is compared with Bluetooth. Details of this standard can be found in [9].
- TG3a is looking at alternative PHY to provide higher data rates. Currently there are two proposals under consideration: one of them uses direct sequence–ultra wideband (DS-UWB) and other uses multiband orthogonal frequency division multiplexing (OFDM). The achievable data rates with both the proposals are fewer than 100 Mbps. The PHYs will work in unlicensed band of 3.1–10.6 GHz in the United States. Significant parts of this band are and will remain licensed even though UWB is allowed to operate here. The key term is coexistence and this is achieved by allowing UWB to operate on a non interference basis with the existing systems in this band. The MAC layer will be the same as that of TG3.
- TG4: The IEEE 802.15 TG4 standardized a low data rate solution with the goal of multi month to multi year battery life and very low complexity. It is operates in an unlicensed frequency band of 2.4 GHz and 900 MHz using DSSS. Potential applications are sensors, interactive toys, smart badges, remote controls, and home automation. The data rates achievable are 250 Kbps, 40 Kbps and 20 Kbps. The MAC used is CSMA/CA (the industry name for this standards is Zigbee).
    - A new task group, TGb, was formed to write an amendment PAR for specific enhancements and clarifications to the IEEE 802.15.4-2003 standard, such as resolving ambiguities, reducing unnecessary complexity, increasing flexibility in security key usage, and considerations for newly available frequency allocations.
    - There is also a low rate alternative PHY task group (TG4a). The principal task of this group is to provide communications and high precision ranging/location capability (1 meter accuracy or better), high aggregate throughput, and ultra low power, as well as adding scalability to data rates, longer range, and lower power consumption and cost. The group plans to use UWB as PHY.

- TG5: The IEEE 802.15 Task group 5 is chartered to determine the necessary mechanisms that must be present in the PHY and MAC layers of WPANs to enable mesh networking. A mesh network is a PAN that employs one of two connection arrangements, full mesh topology, or partial mesh topology. In the full mesh topology, each node is connected directly to each of the others. In the partial mesh topology, some nodes are connected to all the others, but some of the nodes are connected only to those other nodes with which they exchange the most data. Mesh networks have the capability to provide (1) extension of network coverage without increasing transmit power or receive sensitivity, (2) enhanced reliability via route redundancy, (3) easier network configuration, and (4) better device battery life due to fewer retransmissions.

3.3.2 802.16

The IEEE 802.16 working group on broadband wireless access standards develops standards and recommended practices to support the development and deployment of broadband wireless metropolitan area networks. The IEEE 802.16 standard has drawn much notice since the existence of the WiMax forum. The purpose of this forum is to enable broadband wireless access network deployment by certifying interoperability of products and technologies.

IEEE 802.16 has developed (OFDM-based) solutions for 2–11GHz licensed and license-exempt bands or 802.16a and for 10–60 GHz or 802.16/802.16c. Another group, IEEE 802.16e, is looking into enhancements to provide mobility. IEEE 802.16e is an enhancement of IEEE 802.16a in the 2–6GHz band and thus will provide backward compatibility. It will provide local/regional mobility. IEEE 802.16 provides mesh networking and QoS.

3.3.3 802.18

As IEEE 802 is involved in various wireless standards development it is important that a group monitors and participates in regulatory activities in the international arena [10]. This is exactly the task of the IEEE 802.18 radio regulatory technical advisory group (RR-TAG).

Thus this is the group that can provide information concerning the hot item related to China's new security solution for IEEE 802.11. Basically China has developed a WLAN authentication and privacy infrastructure (WAPI) protocol standard [11]. Any vendor selling a solution in China must have the WAPI solution. Further the exact WAPI standard will be known by Chinese companies only; thus any foreign company has to share its technology with one of the Chinese companies to have the WAPI standard incorporated in its product, although recent news suggests that China has agreed to shelve the WAPI solution due to U.S. protests [12].

3.3.4 802.19

The IEEE 802.19 coexistence technical advisory group (TAG) will develop and maintain policies defining the responsibilities of 802 standards developers to address issues of coexistence with existing standards and other standards under development. It will also, when required, offer assessments to the sponsor executive committee (SEC) regarding the degree to which standards developers have conformed to those conventions. The TAG may also develop coexistence documentation of interest to the technical community outside 802.

3.3.5 802.20

IEEE 802.20 or mobile broadband wireless access (MBWA) is working on a fully mobile broadband access solution. The mission of IEEE 802.20 is to develop the specification for an efficient packet-based air interface that is optimized for the transport of IP-based services. The goal is to enable worldwide deployment of affordable, ubiquitous, always-on, and interoperable multivendor mobile broadband wireless access networks that meet the needs of the business and residential end user markets.

This group will specify physical and medium access control layers of an air interface for interoperable mobile broadband wireless access systems, operating in licensed bands below 3.5 GHz, optimized for IP-data transport, with peak data rates per user in excess of 1 Mbps. It will support various vehicular mobility classes up to 250 Km/h in a MAN environment and targets spectral efficiencies, sustained user data rates, and numbers of active users that are all significantly higher than achieved by existing mobile systems.

3.3.6 802.21

The IEEE 802.21 or the handoff executive committee is looking at methods of handover between different technologies and administrative domains. The group wants to provide a solution that will also work within a technology. This solution will provide seamless handover and thus maintain the required quality of service (QoS). This also means that handover should not happen to a technology/network where QoS cannot be maintained.

3.4 BASIC IEEE 802.11

The initial 802.11 PAR (project authorization request) states, "...the scope of the proposed [wireless LAN] standard is to develop a specification for wireless connectivity for fixed, portable, and moving stations within a local area." The PAR further says, "...the purpose of the standard is to provide wireless connectivity to automatic machinery and equipment or stations that require rapid deployment, which may be portable, handheld, or which may be mounted on moving vehicles within a local area."

The resulting standard, which is officially called "IEEE Standard for Wireless LAN Medium Access (MAC) and Physical Layer (PHY) Specifications," defines over-the-air protocols necessary to support networking in local areas. As with other IEEE 802-based standards (e.g. 802.3, and 802.5), the primary service of the 802.11 standard is to deliver MAC service data units (MSDUs) between peer logical link controls (LLCs). Typically, a radio card and AP provide functions of the 802.11 standard [1].

3.4.1 IEEE 802.11 Features

The 802.11 standard provides MAC and PHY functionality for wireless connectivity of fixed, portable and moving stations moving at pedestrian and vehicular speeds within a local area [1–6, 13–15]. The 802.11 standard takes into account the following significant differences between wired and WLANs:

- Power management: Most WLAN network interface cards (NICs) are available in PCMCIA format; thus one can outfit portable and mobile handheld computing equipment with WLAN connectivity. The problem, though, is that these devices must often rely on batteries to power the electronics within them. The addition of a WLAN NIC to a portable computer can quickly drain batteries. The 802.11 working group (WG) struggled with finding solutions to conserve battery power; however, they found techniques enabling wireless NICs to switch to lower-power standby modes periodically when not transmitting, reducing the drain on the battery. The MAC layer implements power management functions by putting the radio to sleep (i.e., lowering the power drain) when no transmission activity occurs for some specific or user-definable time period. The problem, though, is that a sleeping station can miss critical data transmissions. 802.11 solves this problem by incorporating buffers to queue messages. The standard calls for sleeping stations to awaken periodically and retrieve any applicable messages.
- Bandwidth: The industry, scientific, and medical (ISM) spread spectrum (SS) bands do not offer a great deal of bandwidth, keeping data rates lower than desired for some applications. The 802.11 WG, however, dealt with methods to compress data, making the best use of available bandwidth.
- Security: WLAN signals can be received within a certain area and do not need a galvanic connection with wired media, such as twisted-pair, coaxial cable, and optical fiber. In terms of privacy, therefore, wireless LANs have a much larger area to protect. To employ security, the 802.11 group coordinated its work with the IEEE 802.10 standards committee

responsible for developing security mechanisms for all 802 series LANs. This IEEE 802.11 security solution is enhanced by IEEE 802.11.

- Addressing: The topology of a wireless network is dynamic; therefore, the destination address does not always correspond to the destination's location. This raises a problem when routing packets through the network to the intended destination. The 802.11f recommendation on inter access point protocol (IAPP) solves this issue.

3.4.2 IEEE 802.11 Topology

The IEEE 802.11 topology consists of components, interacting to provide a WLAN that enables station mobility transparent to higher protocol layers, such as the LLC. A station (STA) is any device that contains functionality of the 802.11 protocol (i.e., MAC layer, PHY layer, and interface to a wireless medium). The functions of the 802.11 standard reside physically in a radio NIC, the software interface that drives the NIC, and AP. The 802.11 standard supports the following two topologies [1]:

1. Independent basic service set (IBSS) networks
2. Extended service set (ESS) networks

These networks utilize a basic building block the 802.11 standard refers to as a BSS, providing a coverage area in which stations of the BSS remain fully connected. A station is free to move within the BSS, but it can no longer communicate directly with other stations if it leaves the BSS.

The 802.11 standard defines the distribution system (DS) as an element that interconnects BSSs within the ESS via APs. The DS supports the 802.11 mobility types by providing logical services necessary to handle address-to-destination mapping and seamless integration of multiple BSSs. An AP is an addressable station, providing an interface to the DS for stations located within various BSSs. The independent BSS and ESS networks are transparent to the LLC layer.

The 802.11 standard does not constrain the composition of the DS; therefore, it may be 802-compliant or some nonstandard network. If data frames need transmission to and from a non-IEEE 802.11 LAN, then these frames, as defined by the 802.11 standard, enter and exit through a logical point called a portal. The portal provides a logical integration between existing wired LANs and 802.11 LANs. When the DS is constructed with 802-type components, such as 802.3 (Ethernet) or 802.5 (token ring), then the portal and the AP become one and the same. In the following subsection all these are explained in detail.

3.4.2.1 Basic Service Set

The BSS is the fundamental building block of the IEEE 802.11 architecture. A BSS is defined as a group of stations that are under the direct control of a single

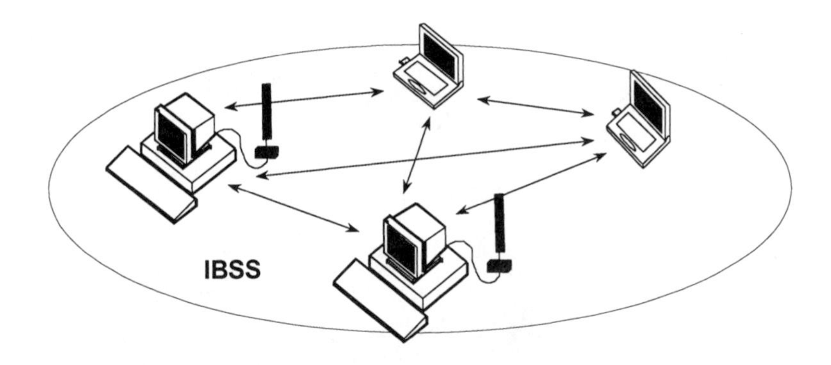

Figure 3.2 Independent basic service set (IBSS).

coordination function, i.e., a distributed coordination function (DCF) or point coordination function (PCF) which is defined below. The geographical area covered by the BSS is known as the basic service area (BSA), which is analogous to a cell in a cellular communications network. Conceptually, all stations in a BSS can communicate directly with all other stations in a BSS. However, transmission medium degradations due to multipath fading, or interference from nearby BSSs reusing the same physical-layer characteristics (e.g., frequency and spreading code, or hopping pattern), can cause some stations to appear "hidden" from other stations [1–3, 13–15].

An ad-hoc network is a deliberate grouping of stations into a single BSS for the purposes of internetworked communications without the aid of an infrastructure network. Figure 3.2 is an illustration of an independent BSS (IBSS), which is the formal name of an ad-hoc network in the IEEE 802.11 standard. Any station can establish a direct communications session with any other station in the BSS, without channeling all traffic through a centralized AP.

The following procedure is followed while starting an IBSS:

1. One station is configured to be "initiating station," and is given a service set ID (SSID);
2. Starter sends beacons;
3. Other stations in the IBSS will search the medium for a service set with SSID that matches their desired SSID and act on the beacons and obtain the information needed to communicate;
4. There can be more stations configured as "starter."

3.4.2.2 Extended Service Set

In contrast to the ad-hoc network, infrastructure networks are established to provide wireless users with specific services and range extension. Infrastructure

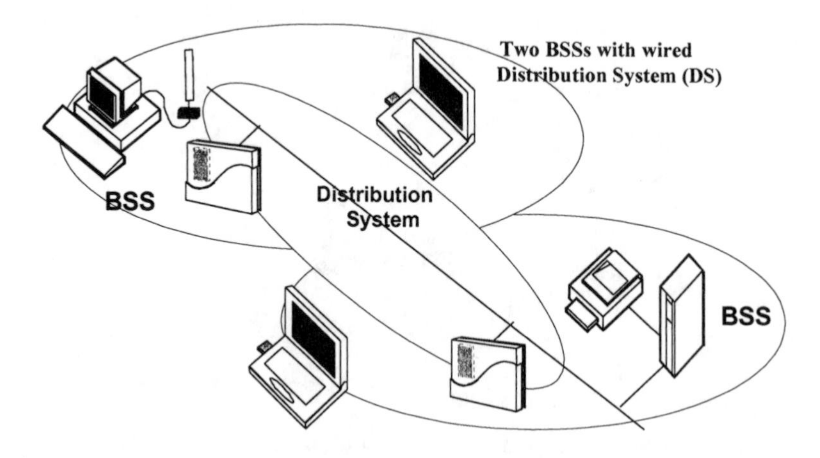

Figure 3.3 ESS with wired distribution system.

networks in the context of IEEE 802.11 are established using APs. The AP is analogous to the base station in a cellular communications network. AP supports range extension by providing the integration points necessary for network connectivity between multiple BSSs, thus forming an extended service set (ESS). The ESS has the appearance of one large BSS to the LLC sublayer of each station. The ESS consists of multiple BSSs that are integrated together using a common DS. The DS can be thought of as a backbone network that is responsible for MAC-level transport of MAC service data units (MSDUs). The DS, as specified by IEEE 802.11, is implementation-independent. Therefore, the DS could be a wired IEEE 802.3 Ethernet LAN, IEEE 802.4 token bus LAN, IEEE 802.5 token ring LAN, fiber distributed data interface (FDDI) metropolitan area network (MAN), or another IEEE 802.11 wireless medium. Note that while the DS could physically be the same transmission medium as the BSS, they are logically different, because the DS is solely used as a transport backbone to transfer packets between different BSSs in the ESS [1–3, 13–15].

An ESS can also provide gateway access for wireless users into a wired network such as the Internet. This is accomplished via a device known as a *portal*. The portal is a logical entity that specifies the integration point on the DS where the IEEE 802.11 network integrates with a non-IEEE 802.11 network. If the network is an IEEE 802.X, the portal incorporates functions that are analogous to a bridge; that is, it provides range extension and the translation between different frame formats. Figures 3.3 and 3.4 illustrate ESS developed with two BSSs, and wired and wireless DS respectively.

The following procedure is followed to start an ESS:

1. The infrastructure network is identified by its extended service set ID (ESSID);
2. All APs will have been set according to this ESSID;

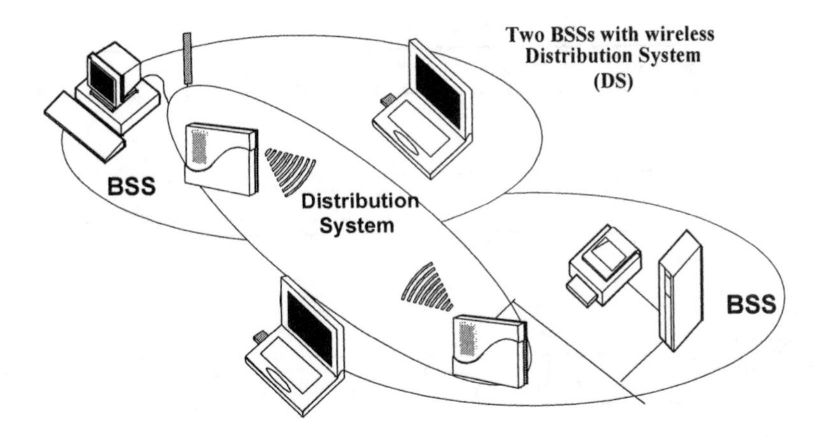

Figure 3.4 ESS with wireless distribution system.

3. On power up, stations will issue probe requests and will locate the AP that they will associate with.

3.4.3 IEEE 802.11 Logical Architecture

A topology provides a means of explaining necessary physical components of a network, but the logical architecture defines the network's operation. The logical architecture of the 802.11 standard that applies to each station consists of a single MAC and one of multiple PHYs (Figure 3.5: as shown in DSSS, OFDM, IR, and FHSS [1–5, 13–15].)

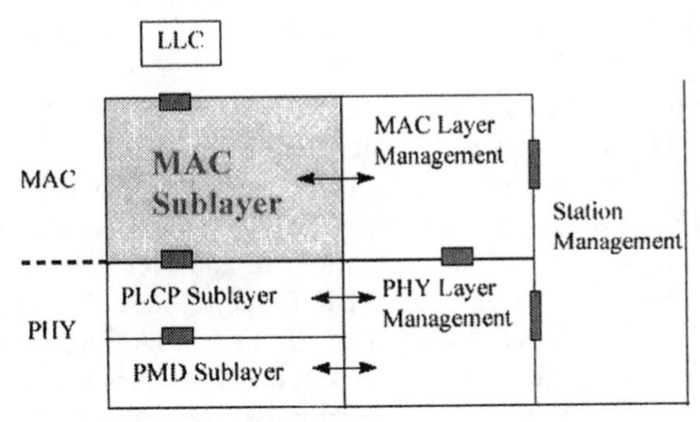

Figure 3.5 IEEE 802.11 entities [1, 4] (reprinted with permission from IEEE 802.11-1999).

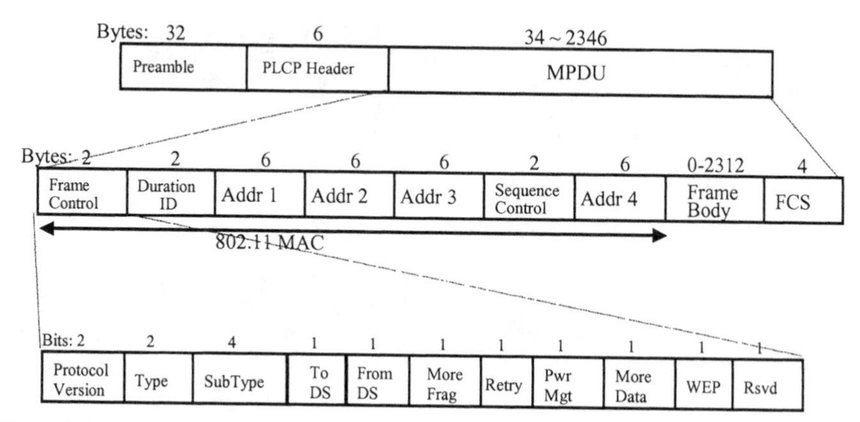

Figure 3.6 MAC frame format [1, 4].

The goal of the MAC layer is to provide access control functions (such as addressing, access coordination, frame check sequence generation and checking, and LLC PDU delimiting) for shared-medium PHYs in support of the LLC layer. The MAC layer performs the addressing and recognition of frames in support of the LLC. The 802.11 standard uses CSMA/CA, whereas standard Ethernet uses carrier sense multiple access with collision detection (CSMA/CD). It is very impractical to transmit and receive on the same radio channel at the same time due to the extreme ratio between the transmitted and received power. Therefore, an 802.11 wireless LAN only takes measures to avoid collisions, not detect them. In Figure 3.5 IEEE 802.11 entities are given.

3.5 MEDIUM ACCESS CONTROL LAYER

In this section the 802.11 basic MAC protocol is explained; MAC enhancements for security and QoS are explained in later chapters.

The 802.11 basic medium access behavior allows interoperability between compatible PHYs through the use of the CSMA/CA protocol and a random back-off time following a busy medium condition. In addition, all directed traffic uses immediate positive acknowledgment (ACK frame), where the sender schedules a retransmission if no ACK is received. The 802.11 CSMA/CA protocol is designed to reduce the collision probability between multiple stations accessing the medium at the point in time where collisions would most likely occur. Collisions are most likely to happen just after the medium becomes free, just after busy medium conditions. This is because multiple stations would have been waiting for the medium to become available again. Therefore, a random back-off arrangement is used to resolve medium contention conflicts. The 802.11 MAC also describes the way beacon frames are sent by the AP at regular intervals (like 100 ms) to enable stations to monitor the presence of the AP. The MAC also gives a set of management frames that allow a station to actively scan for other APs on any available channel. Based on this information the station may decide on the best suited AP. In addition, the 802.11 MAC defines special functional behavior for fragmentation of packets, medium reservation via request-to-send/clear-to-send (RTS/CTS) polling interaction, and point coordination (for time-bounded services) [1,2, 15–17].

The MAC sublayer is responsible for the channel allocation procedures, protocol data unit (PDU) addressing, frame formatting, error checking, and fragmentation and reassembly. The transmission medium can operate in the contention mode exclusively, requiring all stations to contend for access to the channel for each packet transmitted. The medium can also alternate between the contention mode, known as the contention period (CP), and a contention-free period (CFP). During the CFP, medium usage is controlled (or mediated) by the AP, thereby eliminating the need for stations to contend for channel access. IEEE 802.11 supports three different types of frames: management, control, and data. The management frames are used for station association and disassociation with the AP, timing and synchronization, and authentication and deauthentication. Control frames are used for handshaking during the CP, for positive acknowledgments during the CP, and to end the CFP. Data frames are used for the transmission of data during the CP and CFP, and can be combined with polling and acknowledgments during the CFP. The standard IEEE 802.11 frame format is illustrated in Figure 3.6. Note that the frame body (MSDU) is a variable-length field consisting of the data payload and seven octets for encryption/decryption if the optional wired equivalent privacy (WEP) protocol is implemented. The IEEE standard 48-bit MAC addresses are used to identify source and destination stations. The two octets for the duration field indicate the time (in microseconds) the channel will be allocated for successful transmission of a MAC protocol data unit (MPDU). The type bits identify the frame as either control, management, or data. The subtype bits further identify the type of frame (e.g., Clear to Send control frame). A 32-bit cyclic redundancy check (CRC) is used for error detection.

In this section we will discuss the different functions of the MAC layer.

3.5.1 Inter Frame Spacing

In Figure 3.7 inter frame spacing (IFS) required for the original 802.11 MAC protocol are given while values for different IFS are given in Table 3.1. Enhanced MAC with QoS has other IFS also as dealt with in Chapter 5. Their lengths are [1, 2]:

- SIFS = Short inter frame space = as in Table 3.1 dependent on PHY
- PIFS = point coordination function (PCF) inter frame space = SIFS + slot time

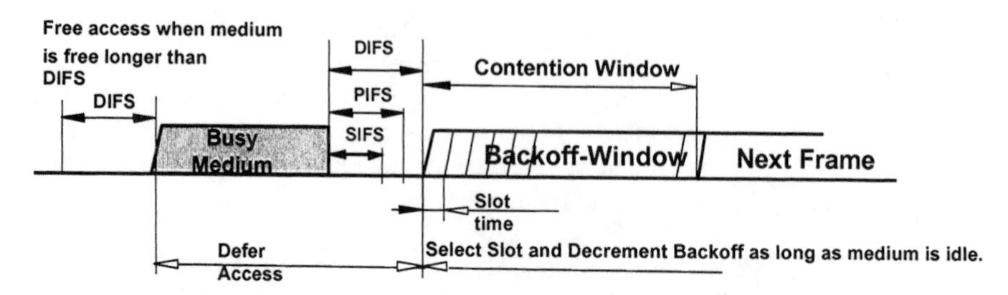

Figure 3.7 IEEE 802.11 interframe space [1, 4] (reprinted with permission from IEEE 802.11-1999).

- DIFS = distributed coordination function (DCF) inter frame space = PIFS + slot time

The back-off timer is expressed in terms of number of time slots.

3.5.2 Distributed Coordination Function

The DCF is the fundamental access method used to support asynchronous data transfer on a best effort basis. As identified in the specification, all stations must support the DCF. The DCF operates solely in the ad-hoc network, and either operates solely or coexists with the PCF in an infrastructure network. The MAC architecture is depicted in Figure 3.8, where it is shown that the DCF sits directly on top of the physical layer and supports contention services. Contention services imply that each station with a MSDU queued for transmission must contend for access to the channel and, once the MSDU is transmitted, must recontend for access to the channel for all subsequent frames. Contention services promote fair access to the channel for all stations [1, 2, 15–17].

The DCF is based on CSMA/CA. In IEEE 802.11, carrier sensing is performed at both the air interface, referred to as physical carrier sensing, and at the MAC sublayer, referred to as virtual carrier sensing. Physical carrier sensing detects the presence of other IEEE 802.11 WLAN users by analyzing all detected packets, and detects activity in the channel via relative signal strength from other sources.

Table 3.1

MAC Values in Microseconds for Different PHYs

PHY	SIFS	DIFS	Slot Time	CWmin
802.11a	16	34	9	15
802.11b	10	50	20	31
802.11g	10	50	20	15

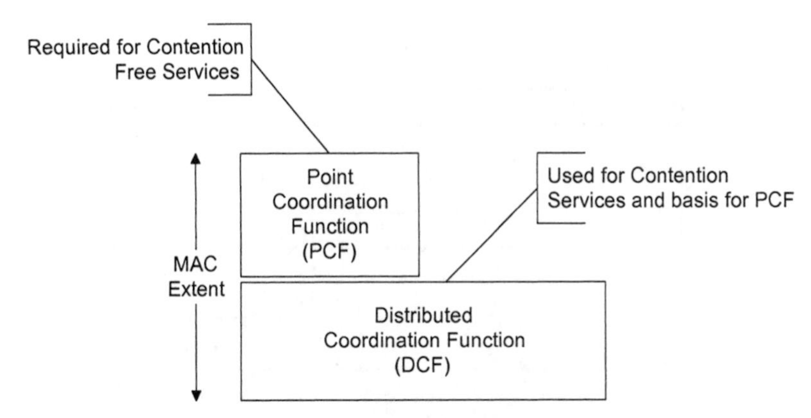

Figure 3.8 MAC architecture [1, 2] (reprinted with permission from IEEE 802.11-1999).

A source station performs virtual carrier sensing by sending MPDU duration information in the header of request to send (RTS), clear to send (CTS), and data frames. A MPDU is a complete data unit that is passed from the MAC sublayer to the physical layer. The MPDU contains header information, payload, and a 32-bit CRC. The duration field indicates the amount of time (in microseconds) after the end of the present frame the channel will be utilized to complete the successful transmission of the data or management frame. Stations in the BSS use the information in the duration field to adjust their network allocation vector (NAV), which indicates the amount of time that must elapse until the current transmission session is complete and the channel can be sampled again for idle status. The channel is marked busy if either the physical or virtual carrier sensing mechanisms indicates the channel is busy.

Priority access to the wireless medium is controlled through the use of interframe space (IFS) time intervals between the transmission of frames. The IFS intervals are mandatory periods of idle time on the transmission medium. Three IFS intervals, as explained in Section 3.5.1, are specified in the standard: short IFS (SIFS), point coordination function IFS (PIFS), and DCF-IFS (DIFS). The SIFS interval is the smallest IFS, followed by PIFS and DIFS, respectively. Stations only required to wait a SIFS have priority access over those stations required to wait a PIFS or DIFS before transmitting; therefore, SIFS has the highest priority access to the communications medium. For the basic access method, when a station senses the channel is idle, the station waits for a DIFS period and samples the channel again. If the channel is still idle, the station transmits an MPDU. The receiving station calculates the checksum and determines whether the packet was received correctly. Upon receipt of a correct packet, the receiving station waits a SIFS interval and transmits a positive acknowledgment frame (ACK) back to the source station, indicating that the transmission was successful. Figure 3.9 is a timing diagram illustrating the successful transmission of a data frame. When the data frame is transmitted, the duration field of the frame is used to let all stations in the

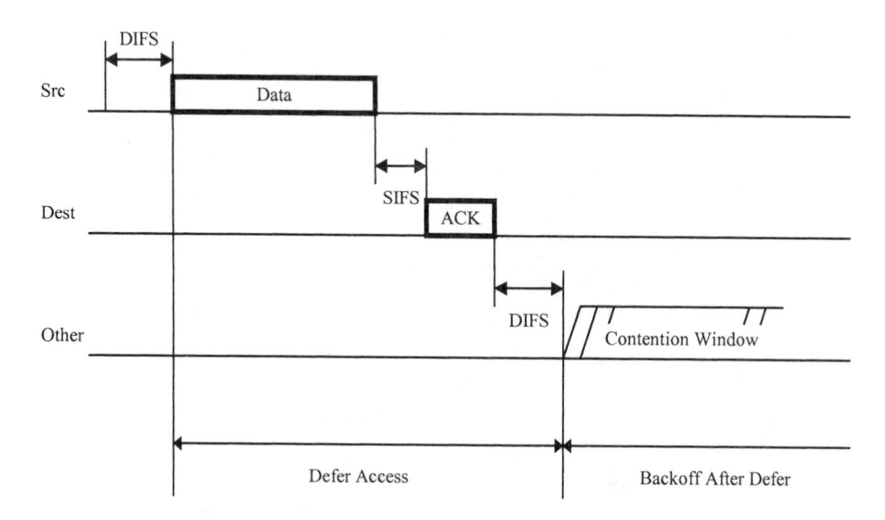

Figure 3.9 Transmission of a MPDU without RTS/CTS [1, 4] (reprinted with permission from IEEE 802.11-1999).

BSS know how long the medium will be busy. All stations hearing the data frame adjust their NAV based on the duration field value, which includes the SIFS interval and the ACK following the data frame.

The collision avoidance portion of CSMA/CA is performed through a random backoff procedure [1, 2, 4, 5, 15]. If a station with a frame to transmit initially senses the channel to be busy, then the station waits until the channel becomes idle for a DIFS period, and then computes a random backoff time. For IEEE 802.11, time is slotted in time periods that correspond to a Slot_Time. The Slot_Time used in IEEE 802.11 is much smaller than an MPDU and is used to define the IFS intervals and determine the backoff time for stations in the CP. The Slot_Time is different for each physical layer implementation. The random backoff time is an integer value that corresponds to a number of time slots. Initially, the station computes a backoff time in the range 0–7. After the medium becomes idle after a DIFS period, stations decrement their backoff timer until the medium becomes busy again or the timer reaches zero. If the timer has not reached zero and the medium becomes busy, the station freezes its timer. When the timer is finally decremented to zero, the station transmits its frame. If two or more stations decrement to zero at the same time, a collision will occur, which leads to missing ACKs, and each station will have to generate a new backoff time in the range 0–63 (for 802.11b, and 0–31 for 802.11a) times the slot time period. The generated backoff time correpsonds to an uniform distributed integer multiple of slot time periods. For the next retransmission attempt, the backoff time grows to 0–127 (for 802.11b, and 0–63 for 802.11a) Slot Time periods and so on with at maximum the range 0–1023. The idle period after a DIFS period is referred to as the contention window (CW). The advantage of this channel access method is that it promotes fairness among stations, but its weakness is that it probably could not support time-bound services. Fairness is maintained because each station must recontend for the channel after every transmission of an MSDU. All stations have equal probability of gaining access to the channel after each DIFS interval. Time-bounded services typically support applications such as packetized voice or video that must be maintained with a specified minimum delay. With DCF, there is no mechanism to guarantee minimum delay to stations supporting time-bounded services.

3.5.3 RTS/CTS

Since a source station in a BSS cannot hear its own transmissions, when a collision occurs, the source continues transmitting the complete MPDU. If the MPDU is large (e.g., 2,300 octets), a lot of channel bandwidth is wasted due to a corrupt MPDU [1, 2, 4, 5, 15]. RTS and CTS control frames can be used by a station to reserve channel bandwidth prior to the transmission of an MPDU and to minimize the amount of bandwidth wasted when collisions occur. RTS and CTS control frames are relatively small (RTS is 20 octets and CTS is 14 octets) when compared to the maximum data frame size (2,346 octets). The RTS control frame is first transmitted by the source station (after successfully contending for the channel) with a data or management frame queued for transmission to a specified destination station. All stations in the BSS, hearing the RTS packet, read the duration field (Figure 3.10) and set their NAVs accordingly. The destination station responds to the RTS packet with a CTS packet after an SIFS idle period has elapsed. Stations hearing the CTS packet look at the duration field and again update their NAV. Upon successful reception of the CTS, the source station is virtually assured that the medium is stable and reserved for successful transmission

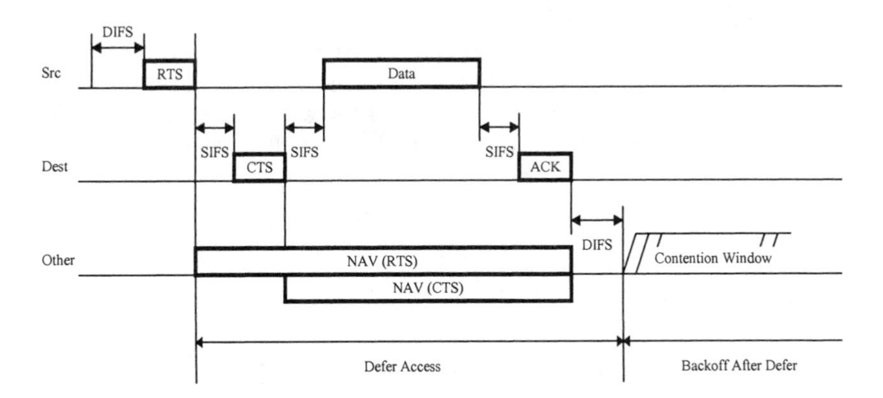

Figure 3.10 Transmission of a MPDU using RTS/CTS [1, 4] (reprinted with permission from IEEE 802.11-1999).

of the MPDU. Note that stations are capable of updating their NAVs based on the RTS from the source station and CTS from the destination station, which helps to combat the "hidden terminal" problem. Figure 3.10 illustrates the transmission of an MPDU using the RTS/CTS mechanism. Stations can choose to never use RTS/CTS, use RTS/CTS whenever the MSDU exceeds the value of RTS_Threshold (manageable parameter), or always use RTS/CTS. If a collision occurs with an RTS or CTS MPDU, far less bandwidth is wasted when compared to a large data PDU. However, for a lightly loaded medium, additional delay is imposed by the overhead of the RTS/CTS frames.

3.5.4 Fragmentation

IEEE 802.11 also defines fragmentation and reassembly. Implementation of fragmentation is optional but reassembly is mandatory. Fragmentation can be very helpful in a noisy environment but will create overhead under good channel conditions [1, 2].

Large MSDUs handed down from the LLC to the MAC may require fragmentation to increase transmission reliability. To determine whether to perform fragmentation, MPDUs are compared to the manageable parameter Fragmentation_Threshold. If the MPDU size exceeds the value of Fragmentation_Threshold, the MSDU is broken into multiple fragments. The resulting MPDUs are of size Fragmentation_Threshold, with the exception of the last MPDU, which is of variable size not to exceed Fragmentation_Threshold. When an MSDU is fragmented, all fragments are transmitted sequentially (Figure 3.11). The channel is not released until the complete MSDU has been transmitted successfully, or the source station fails to receive an acknowledgment for a transmitted fragment. The destination station positively ACKs each successfully received fragment by sending a DCF ACK back to the source station. The source station maintains control of the channel throughout the transmission of the MSDU by waiting only a SIFS period after receiving an ACK and transmitting the next fragment. When an ACK is not received for a previously transmitted frame, the source station halts transmission and recontends for the channel. Upon gaining access to the channel, the source starts transmitting with the last unacknowledged fragment.

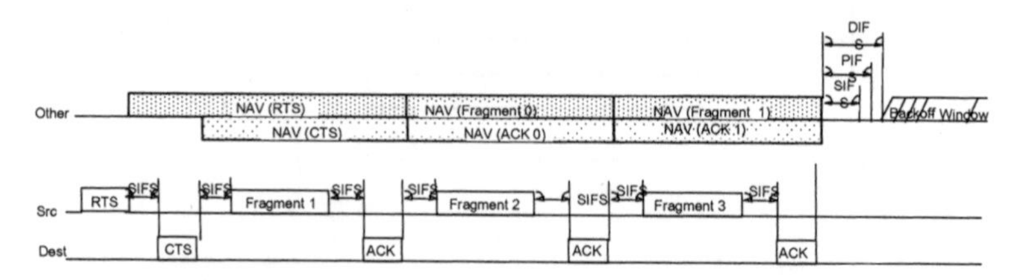

Figure 3.11 Transmission of fragmented MPDU [1, 4] (reprinted with permission from IEEE 802.11-1999).

If RTS and CTS are used, only the first fragment is sent using the handshaking mechanism. The duration value of RTS and CTS only accounts for the transmission of the first fragment through the receipt of its ACK. Stations in the BSS thereafter maintain their NAV by extracting the duration information from all subsequent fragments.

3.5.5 Point Coordination Function

The PCF is an optional capability that is connection-oriented and provides contention-free (CF) frame transfer. The PCF relies on the point coordinator (PC) to perform polling, enabling polled stations to transmit without contending for the channel. The function of the PC is performed by the AP within each BSS. Stations within the BSS that are capable of operating in the CF period (CFP) are known as *CF-aware* stations. The method by which polling tables are maintained and the polling sequence is determined is left to the implementor [1, 2, 15–16].

The PCF is required to coexist with the DCF and logically sits on top of the DCF (Figure 3.8). The CFP repetition interval (CFP_Rate) is used to determine the frequency with which the PCF occurs. Within a repetition interval, a portion of the time is allotted to contention-free traffic, and the remainder is provided for contention-based traffic. The CFP repetition interval is initiated by a beacon frame, where the beacon frame is transmitted by the AP. One of its primary functions is synchronization and timing. The duration of the CFP repetition interval is a manageable parameter that is always an integral number of beacon frames. Once the CFP_Rate is established, the duration of the CFP is determined. The maximum size of the CFP is determined by the manageable parameter CFP_Max_Duration.

The minimum value of CFP_Max_Duration is the time required to transmit two maximum-size MPDUs, including overhead, the initial beacon frame, and a CF-End frame. The maximum value of CFP_Max_Duration is the CFP repetition interval minus the time required to successfully transmit a maximum-size MPDU during the CP (which includes the time for RTS/CTS handshaking and the ACK).

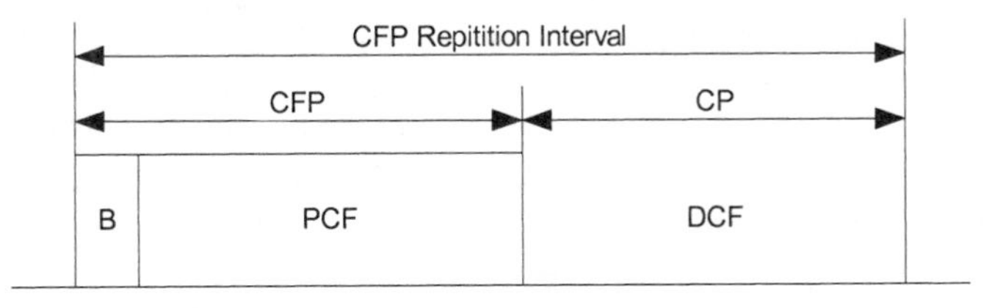

B: Beacon
CFP: Contention Free Period
CP: Contention Period
PCF: Point Coordination Function
DCF: Distributed Coordination Function

Figure 3.12 Coexistence of PCF and DCF [1, 4].

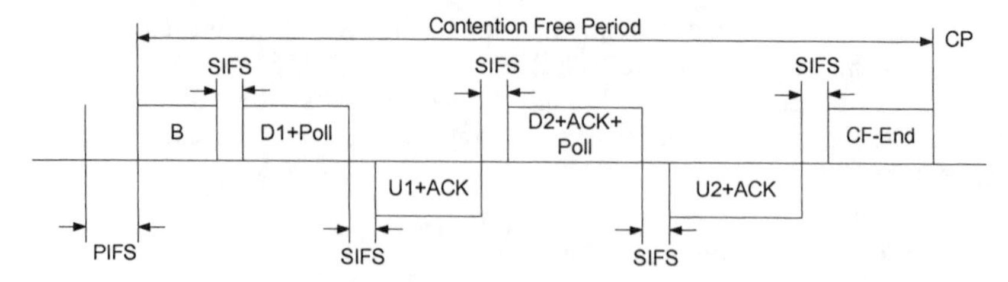

Figure 3.13 PC to station transmission [1, 4] (reprinted with permission from IEEE 802.11-1999).

Therefore, time must be allotted for at least one MPDU to be transmitted during the CP. It is up to the AP to determine how long to operate the CFP during any given repetition interval. If traffic is very light, the AP may shorten the CFP and provide the remainder of the repetition interval for the DCF. The CFP may also be shortened if DCF traffic from the previous repetition interval carries over into the current interval. The maximum amount of delay that can be incurred is the time it takes to transmit an RTS/CTS handshake, maximum MPDU, and ACK. Figure 3.12 is a sketch of the CFP repetition interval, illustrating the coexistence of the PCF and DCF.

At the nominal beginning of each CFP repetition interval, all stations in the BSS update their NAV to the maximum length of the CFP (i.e., CFP_Max_Duration). During the CFP, the only time stations are permitted to transmit is in response to a poll from the PC or for transmission of an ACK a SIFS interval after receipt of an MPDU. At the nominal start of the CFP, the PC senses the medium. If the medium remains idle for a PIFS interval, the PC transmits a beacon frame to initiate the CFP. The PC starts CF transmission a SIFS interval after the beacon frame is transmitted by sending a CF-Poll (no data), Data, or Data+CF-Poll frame. The PC can immediately terminate the CFP by transmitting a CF-End frame, which is common if the network is lightly loaded and the PC has no traffic buffered. If a CF-aware station receives a CF-Poll (no data) frame from the PC, the station can respond to the PC after a SIFS idle period, with a CF-ACK (no data) or a Data + CF-ACK frame. If the PC receives a Data + CF-ACK frame from a station, the PC can send a Data + CF-ACK + CF-Poll frame to a different station, where the CF-ACK portion of the frame is used to acknowledge receipt of the previous data frame. The ability to combine polling and acknowledgment frames with data frames, transmitted between stations and the PC, was designed to improve efficiency. If the PC transmits a CF-Poll (no data) frame and the destination station does not have a data frame to transmit, the station sends a Null Function (no data) frame back to the PC. Figure 3.13 illustrates the transmission of frames between the PC and a station, and vice versa. If the PC fails to receive an ACK for a transmitted data frame, the PC waits a PIFS interval and continues transmitting to the next station in the polling list.

After receiving the poll from the PC, as described above, the station may choose to transmit a frame to another station in the BSS. When the destination station

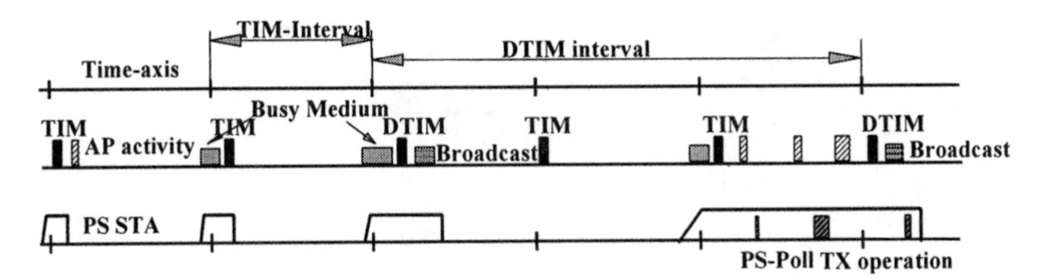

Figure 3.14 IEEE 802.11 power management [1, 4].

receives the frame, a DCF ACK is returned to the source station, and the PC waits a PIFS interval following the ACK frame before transmitting any additional frames. The PC may also choose to transmit a frame to a non-CF-aware station. Upon successful receipt of the frame, the station would wait a SIFS interval and reply to the PC with a standard ACK frame. Fragmentation and reassembly are also accommodated with the Fragmentation_Threshold value used to determine whether MSDUs are fragmented prior to transmission. It is the responsibility of the destination station to reassemble the fragments to form the original MSDU.

3.5.6 Scanning

Scanning is required for many functions like the following [1, 2]:

- Joining a network;
- Initializing an ad-hoc network;

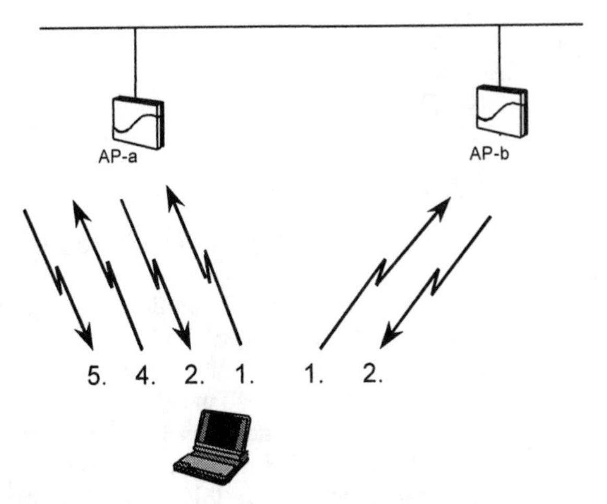

Figure 3.15 Association process.

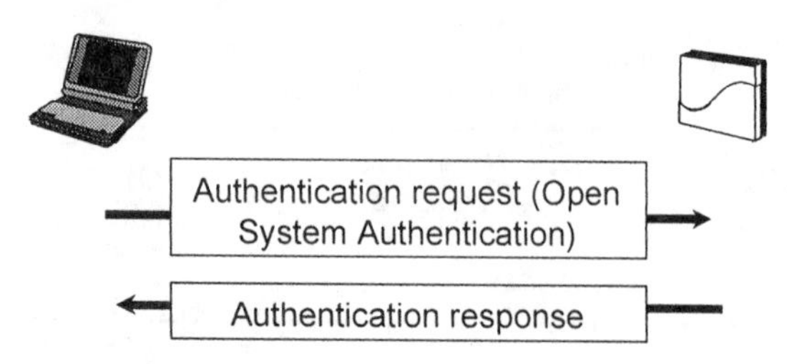

Figure 3.16 Open system authentication.

- Maintaining a single ad-hoc network;
- Finding a new AP while roaming.There are two types of scanning: passive and active scanning. In passive scanning a station listens to beacons in each channel; when a beacon is received the ESS-ID, BSS-ID, and Timestamp are saved. This is good for a network with few channels and short beacon intervals. Active scanning is a faster process; in this scheme a station sends a probe in each channel and waits for a response. When a response is received similar information as in passive scanning is saved. After scanning the best AP is chosen, and the station joins the network.

3.5.7 Association

After scanning, a station must associate with an AP. This procedure together with scanning is explained in Figure 3.15; the communication between the AP and station is given in the following [1, 2]:

1. Station sends probe (active scanning);
2. APs send probe response;
3. Station selects best AP;
4. Station sends association request;
5. AP sends Association Response.

3.5.8 Authentication

IEEE 802.11 defines two subtypes of authentication service: open system and shared key [1, 2].

Open system authentication is the simplest of the available authentication algorithms. Essentially it is a null authentication algorithm. Any station that requests authentication with this algorithm may become authenticated if the recipient station is set to open system authentication (Figure 3.16).

Shared key authentication supports authentication of stations as either a member of those who know a shared secret key or a member of those who do not. IEEE 802.11 shared key authentication accomplishes this without the need to transmit the secret key in the clear, requiring the use of the wired equivalent privacy (WEP) mechanism. Therefore, this authentication scheme is only available if the WEP option is implemented. The required secret, shared key is presumed to have been delivered to participating stations via a secure channel that is independent of IEEE 802.11. During the shared key authentication exchange, both the challenge and the encrypted challenge are transmitted. This facilitates unauthorized discovery of the pseudorandom number (PRN) sequence for the key/IV (initialization vector) pair used for the exchange. Therefore the same key/IV pair for subsequent frames should not be used. The shared key authentication process is given in Figure 3.17.

3.5.9 Encryption

The WEP algorithm is a form of electronic code book in which a block of plaintext is bitwise XORed with a pseudorandom key sequence of equal length. The key sequence is generated by the WEP algorithm.

Referring to Figure 3.18 and viewing from left to right, encipherment begins with a secret key that has been distributed to cooperating stations by an external key management service. WEP is a symmetric algorithm in which the same key is used for encipherment and decipherment.

The secret key is concatenated with an IV and the resulting seed is input to a pseudorandom number generator (PRNG). The PRNG outputs a key sequence k of pseudorandom octets equal in length to the number of data octets that are to be transmitted in the expanded MPDU plus 4 [since the key sequence is used to protect the integrity check value (ICV) as well as the data]. Two processes are applied to the plaintext MPDU. To protect against unauthorized data modification, an integrity algorithm operates on P to produce an ICV. Encipherment is then accomplished by mathematically combining the key sequence with the plaintext concatenated with the ICV. The output of the process is a message containing the IV and ciphertext.

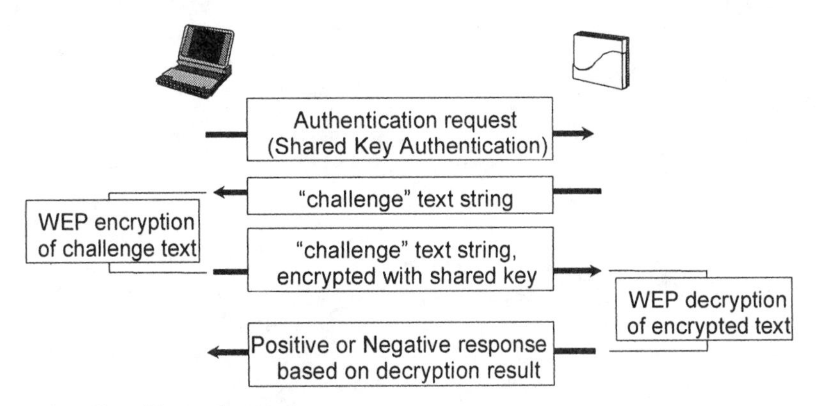

Figure 3.17 Shared key authentication.

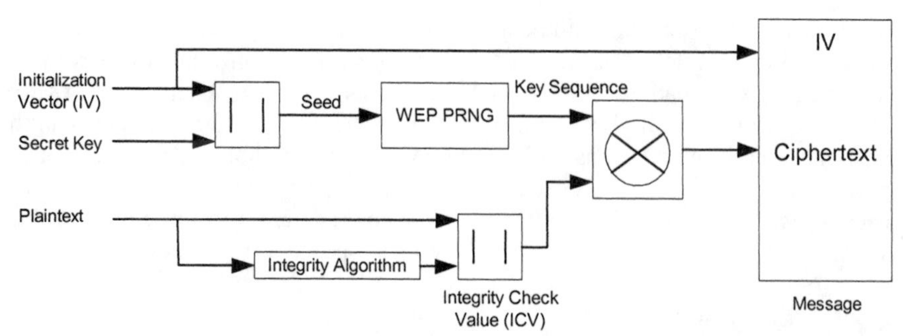

Figure 3.18 WEP encipherment block diagram [1] (reprinted with permission from IEEE 802.11-1999).

Referring to Figure 3.19 and viewing from left to right, decipherment begins with the arrival of a message. The IV of the incoming message shall be used to generate the key sequence necessary to decipher the incoming message.

Combining the ciphertext with the proper key sequence yields the original plaintext and ICV. Correct decipherment shall be verified by performing the integrity check algorithm on the recovered plaintext and comparing the output ICV′ to the ICV transmitted with the message. If ICV′ is not equal to ICV, the received MPDU is in error and an error indication is sent to MAC management. MSDUs with erroneous MPDUs (due to inability to decrypt) shall not be passed to LLC. The WEP payload is shown in Figure 3.20.

3.5.10 Roaming

The standard also allows roaming between APs either in the same channel or a different channel. The standard does not define the exact procedure for this purpose [1, 2].

3.5.11 Synchronization

Synchronization in IEEE 802.11 is done by the timing synchronization function (TSF) of the beacon. It is used for the following [1, 2]:

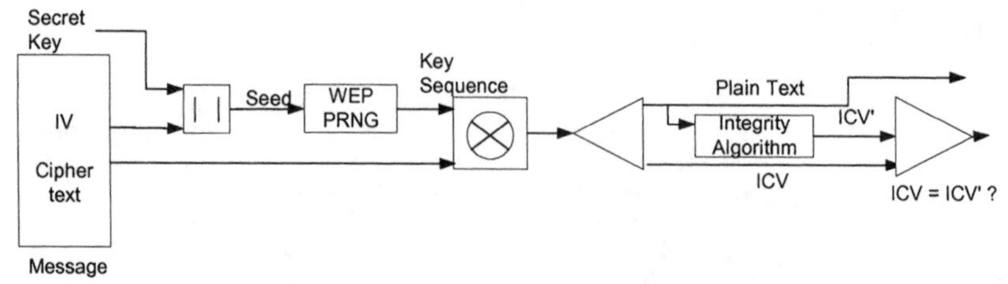

Figure 3.19 WEP decipherment block diagram [1] (reprinted with permission from IEEE 802.11-1999).

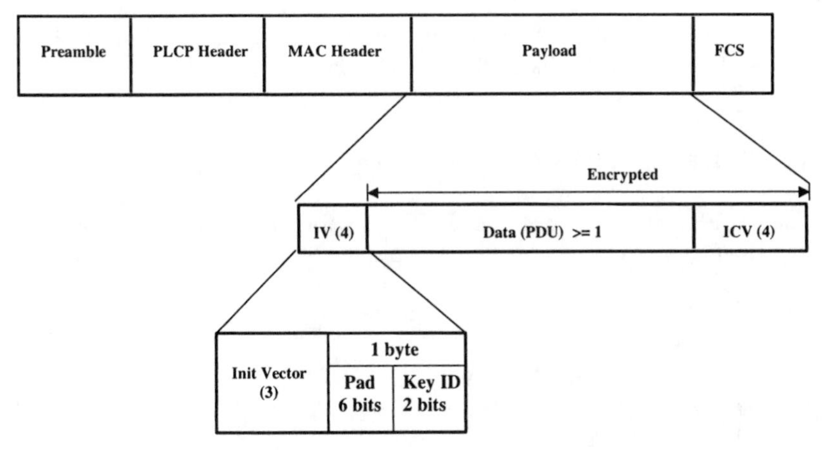

Figure 3.20 WEP payload [1] (reprinted with permission from IEEE 802.11-1999).

- Power management:
  - Beacons sent at well known intervals;
  - All station timers in BSS are synchronized.
- Acquisition:
  - Stations scan for beacons to find networks.
- Superframe timing:
  - TSF Timer used to predict start of contention-free burst.
- Hop timing for FH PHY

3.5.12 Power Management

Wireless LANs are typically related to mobile applications. In this type of application, battery power is a scare resource. This is the reason why the 802.11 standard directly addresses the issue of power saving and defines an entire mechanism that enables stations to go into sleep mode for long periods of time without losing information. The main idea behind the power saving mechanism is that the AP maintains a continually updated record of the stations currently working in power saving mode, and buffers the packets addressed to these stations until either the stations specifically request the packets by sending a polling request, or until they change their operation mode.

As part of its beacon frames, the AP also periodically transmits information about which power saving stations (PS STA) have frames buffered at the AP, so these stations wake up in order to receive the beacon frame (Figure 3.14). If there is an indication that there is a frame stored at the AP waiting for delivery, then the station stays awake and sends a polling message (PS-Poll) to the AP to get these frames. Multicasts and broadcasts are stored by the AP, and transmitted at a pre-known time (each delivery traffic indication map, DTIM), when all PS stations who wish to receive this kind of frame are awake [1, 2, 15–17].

3.6 IEEE 802.11 PHYSICAL LAYERS

The IEEE 802.11 specification calls for four different physical-layer implementations: DSSS (direct sequence spread spectrum), OFDM, FHSS, and IR.

3.6.1 DSSS

DSSS uses the 2.4-GHz ISM band (i.e., 2.4000–2.4835 GHz). The basic 802.11 DSSS [1] gives data rates of 1 and 2 Mbps. The 802.11b extension [4] added the higher rates of 5.5 and 11 Mbps to the 802.11 DSSS.

3.6.2 802.11 DSSS at 1 and 2 Mbps

This section provides a brief review of the 1 and 2 Mbps DSSS signal structure. This review is important because it shows how the 5.5-and 11-Mbps CCK signaling scheme is a natural extension of the legacy DSSS system. This harmony provides an easy extension from the lower data rates to the higher data rates. Also, interoperability is naturally provided.

The packet structure is shown in Figure 3.21. The complete packet (PPDU) is comprised of three segments. The first segment is the preamble, and it is used for signal detection and synchronization. The second segment is the header, and it contains data-rate and packet-length information. The third segment (MPDU) contains the information bits. The preamble and header are transmitted at 1 Mbps, while the data portion is sent at 1 of 4 rates.

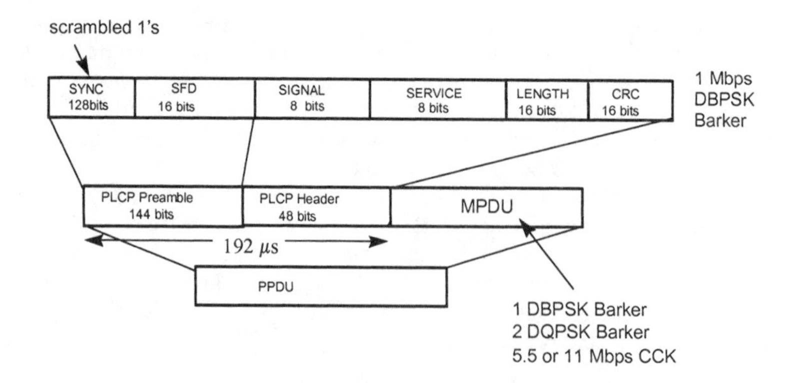

Figure 3.21 802.11 DSSS packet with a 1 Mbps header [1, 2].

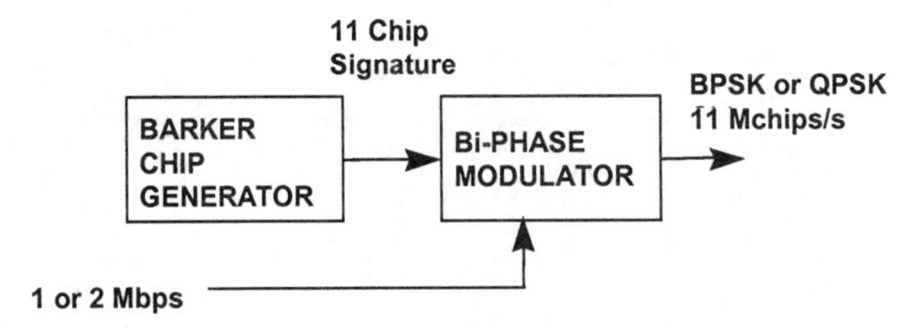

Figure 3.22 802.11 DSSS modulator for 1 and 2 Mbps.

The preamble is formed from a SYNC field and a sync field delimiter (SFD). The SYNC field is generated using 128 scrambled ones. The SYNC field is used for clear channel assessment, signal detection, timing acquisition, frequency acquisition, multipath estimation, and descrambler synchronization.

The new high-rate extension provides an optional short preamble with a preamble/header approximately half this duration.

The 1 Mbps basic rate is encoded using differential binary phase shift keying (DBPSK), and a 2 Mbps enhanced rate uses differential quadrature phase shift keying (DQPSK). The spreading is done by dividing the available bandwidth into 11 subchannels, each 11 MHz wide, and using an 11-chip Barker sequence to spread each data symbol. The maximum channel capacity is therefore (11 chips/symbol)/(11 MHz) = 1 Mbps if DBPSK is used [15].

The 1 and 2 Mbps DSSS signal is created using a fixed spreading sequence (signature) formed from an 11 chip Barker word. The Barker word has the biphase values: +1, −1, +1, +1, −1, +1, +1, +1, −1, −1, −1. This contrasts with military systems, which typically use long pseudorandom spreading sequences. For 1 Mbps the fixed spreading sequence is used to spread a 1 Mbps BPSK signal as shown in Figure 3.22. For 2 Mbps, the same spreading sequence is QPSK modulated. The modulation bits are scrambled by a seven-delay-stage shift-register (not shown). For all data rates of 1 to 11 Mbps the chip rate is 11 Mchips/s.

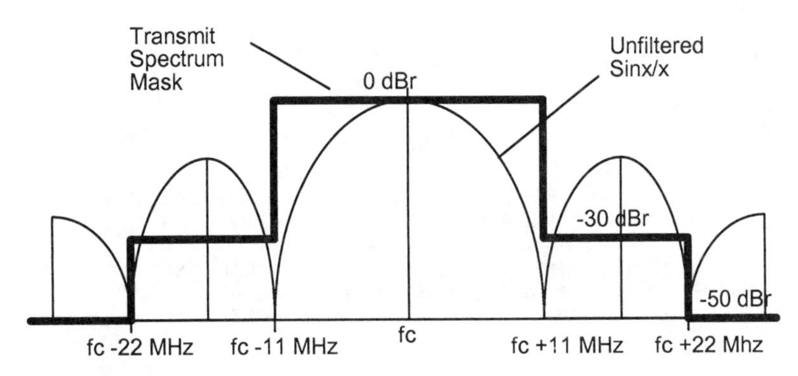

Figure 3.23 Required spectral mask for 802.11 DSSS [1].

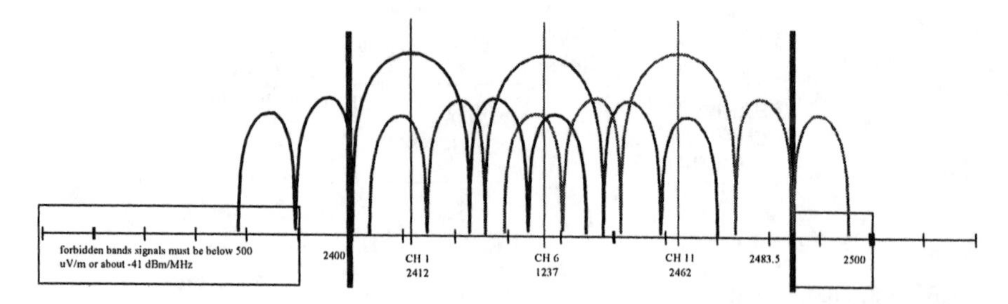

Figure 3.24 Illustration with three 802.11 DSSS channels in the 2.4GHz ISM band (based on required spectrum mask) [1] .

The 11 Mchips/s signal must conform to the spectral mask shown in Figure 3.23. An unfiltered sinc function is shown along with the mask, so the additional pulse-shaping requirements can be deduced. Signal stacking within the ISM band is shown in Figure 3.24. The amount of energy allowed in the forbidden bands is controlled by international regulatory bodies.

The recommended receiver architecture for reception of the 1- and 2- Mbps signal is shown in Figure 3.25. The channel matched filter (CMF) maximizes the detection signal-to-noise ratio by optimum combining of the different multipath signal components. This CMF filter is connected to the Barker correlator. The combination of channel matched filter and Barker correlator forms a complete signal-matched-filter detector. The channel matched filter requires that the channel impulse response be estimated for each received packet.

Channel impulse response estimation is easily accomplished by correlating for the Barker word during the SYNC portion of the preamble. The Barker word has nearly impulsive autocorrelation properties. This impulsive property makes it ideal for RAKE receiver design. Impulsive spreading patterns allow the receiver to decorrelate the undesired multipath residuals at the channel matched filter output as part of the RAKE processing.

3.7 IEEE 802.11b

802.11b gives an extension to the basic 802.11 DSSS. 802.11b is based on complementary code keying (CCK). Because the 802.11b CCK scheme is based

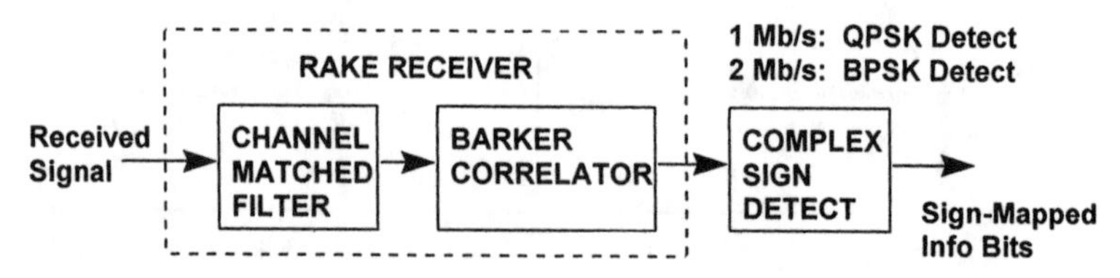

Figure 3.25 The RAKE receiver architecture used for 1 and 2 Mbps signal reception.

on the same chip rate it allows the basic 802.11 DSSS channelization scheme. Since the same PLCP header structure based on 1 Mbps is maintained the 802.11b devices have a good compatibility with the older 802.11 DSSS devices.

Complementary codes were originally conceived by M. J. E. Golay for infrared multislit spectrometry [3]. However, their properties also make them useful in radar applications and more recently for discrete multitone communications and OFDM [13]. The original publication [3] defines a complementary series as a pair of equally long sequences composed of two types of elements that have the property in which the number of pairs of like elements with any given separation in one series is equal to the number of pairs of unlike elements with the same separation in the other series. Another way to define a pair of complementary codes is to say that the sum of their aperiodic autocorrelation functions is zero for all delays except for a zero delay.

The CCK codes that were voted in at the July 1998 IEEE 802.11 conference are defined in [13]. More background information on these codes can be found in [14]. The following equation represents the 8 complex chip values for the CCK code set, with the phase variables being QPSK phases.

Basically, the three phases φ_2, φ_3, and φ_4 define 64 different codes of 8 chips, with φ_1 giving an extra phase rotation to the entire code word. Actually, the latter phase is differentially encoded across successive codewords, equivalent to the 1 and 2 Mbps DSSS differential-phase encoding. This feature allows the receiver to use differential phase decoding, eliminating a carrier-tracking PLL, if desired. Each of the four phases φ_1 to φ_4 represents 2 bits of information, so a total of 8 bits is encoded per 8-chip CCK code word.

At 5.5 Mbps, the processing is similar. Four information bits are consumed per 8 chip CCK codeword transmission. The codeword rate is still 1.375 MHz, since the chip rate is 11 Mchips/sec. Two bits select 1-of-4 CCK subcodes. The other two information bits quadriphase modulate (rotate) the whole codeword. The 4 CCK subcodes are contained in the larger 64 subcode set of 11 Mbps. At the receiver, the CCK codes can be decoded by using a modified fast Walsh transform like that described in [14].

$$\begin{aligned} c = \{ & e^{j(\varphi_1+\varphi_2+\varphi_3+\varphi_4)}, e^{j(\varphi_1+\varphi_3+\varphi_4)}, e^{j(\varphi_1+\varphi_2+\varphi_4)}, -e^{j(\varphi_1+\varphi_4)}, \\ & e^{j(\varphi_1+\varphi_2+\varphi_3)}, e^{j(\varphi_1+\varphi_3)}, -e^{j(\varphi_1+\varphi_2)}, e^{j(\varphi_1)} \} \end{aligned} \tag{3.1}$$

The 802.11b defines a packet structure where the complete packet (PPDU) comprises three segments: preamble (for signal detection and synchronization), header (containing data rate and packet length information), and payload (MPDU). The preamble and header are transmitted at 1 Mbps; the payload data is sent at 1, 2, 5.5, or 11 Mbps. Operation at 1 and 2 Mbps DSSS is based signaling with a fixed Barker-11 spreading sequence (+1, −1, +1, +1, −1, +1, +1, +1, −1, −1, −1) in combination with BPSK (1 Mbps) and QPSK (2 Mbps). With 5.5 and 11 Mbps the

802.11b CCK scheme is used, using groups of 8 bits (at 11 Mbps) and 4 bits (at 5.5 Mbps) to encode symbols composed from 8 chip intervals of 0.73 μs (8 * 1/11 MHz) (see Figure 3.26). All the 802.11b schemes apply the same chip rate; this leads to the same spectrum shaping and channelization. With 5.5 and 11 Mbps the respectively 4 and 8 bits per symbol interval (8 chip) define 2^4 and 2^8 patterns out of a set of total 4^8 different 8-chip QPSK codewords. The 2^4 and 2^8 patterns that are used have optimum properties with regard to detection distance and detection robustness with respect to channel degradation. In fact the 5.5 Mbps gives 4 different patterns but with 1 out of 4 phase (rotation symmetry positions), the 11 Mbps gives 64 different patterns with 1 out of 4 phases.

In summary the CCK high-rate extension has the following features:

- Codewords defined as 8 chip CCK. 256 CCK codewords are used at 11 Mbps; 16 CCK codewords are used at 5.5 Mbps.
- The codewords have properties that enable high-performance RAKE receivers.
- At 11 Mbps, 8 information bits per codeword are used in the following fashion: 6 information bits select one of 64 subcodes, while 2 information bits select one of 4 DQPSK phases.
- At 5.5 Mbps, 4 information bits per codeword are used in the following fashion: 2 information bits select one-of-4 subcodes, while 2 information bits select one-of-4 DQPSK phases.
- Differential-QPSK phase encoding across successive codewords is used to enable noncoherent carrier tracking loops.

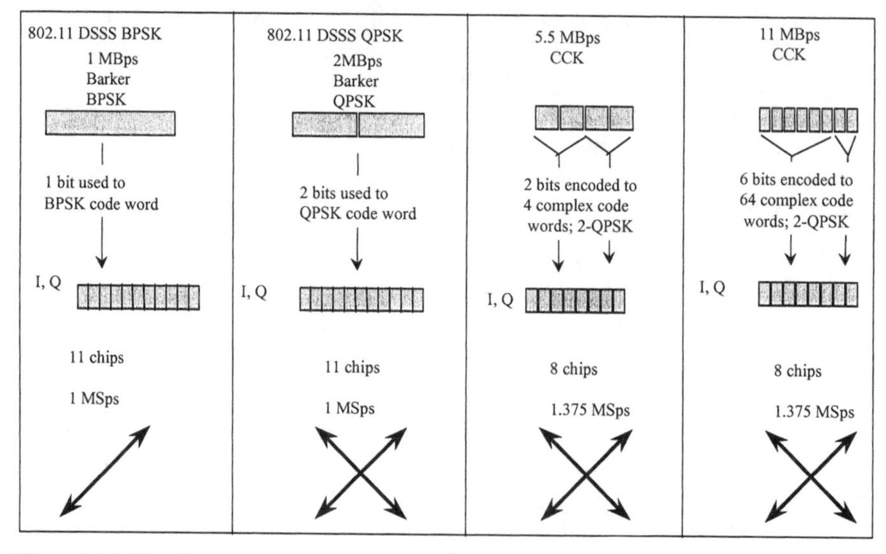

Figure 3.26 802.11b modulation scheme and bit mapping [4].

- Codeword correlation can be broken into two steps: (1) noncoherently correlate for the 64 subcodes at 11 Mbps and (2) coherently detect the codeword phase.
- At 5.5 Mbps, the 4 codewords are a subset of the 64 subcodes. This allows the same fast transform to be used for both 5.5 and 11 Mbps, if desired.

3.7.1 IEEE 802.11b Channels

Overlapping and adjacent BSSs can be accommodated by ensuring that the center frequencies of each BSS are separated by at least a multiple of 5 MHz channel positions. The basic 802.11 DSSS specified an adjacent channel rejection based on 30 MHz channel separation; with 802.11b this became more stringent because of the higher data rates and specifying 25 MHz channel separation. With the 802.11b there is room for three well isolated channels. This allows three independent channels in fully overlapping BSSs.

Table 3.2

Regulatory Domain and Channels

		NIC Regulatory Domain Variants					
		NA		ETS		FR	JP
		Regulatory Domain					
Channels		United States	Canada	Europe / CEPT-FR	Australia	France	Japan
Channel ID	Center Frequency (MHz)	IEEE 802.11 Domain ID (and code in hex)					
		FCC (10)	IC (20)	ETSI (30)	-	France (32)	MKK (40)
		Channel Mapping					
1	2412	X	X	X	X	-	X
2	2417	X	X	X	X	-	X
3	2422	X	X	X	X	-	X
4	2427	X	X	X	X	-	X
5	2432	X	X	X	X	-	X
6	2437	X	X	X	X	-	X
7	2442	X	X	X	X	-	X
8	2447	X	X	X	X	-	X
9	2452	X	X	X	X	-	X
10	2457	X	X	X	X	X	X
11	2462	X	X	X	X	X	X
12	2467			X	X	X	X
13	2472			X	X	X	X
14	2484	-	-	-	-	-	X

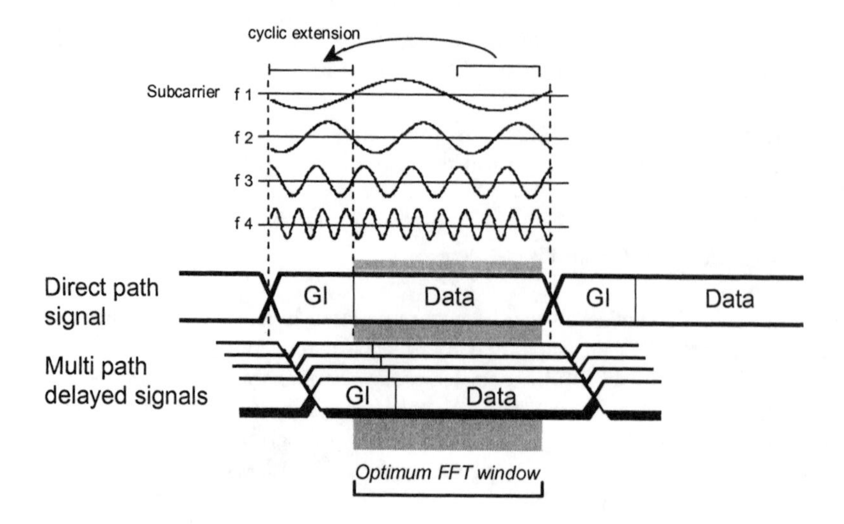

Figure 3.27 802.11a OFDM symbol with cyclic extension [5] (reprinted with permission from IEEE 802.11a-1999).

3.8 IEEE 802.11a

802.11a is based on OFDM in the 5 GHz band. The basic principle of OFDM is to split a high rate data stream into a number of lower rate streams that are transmitted simultaneously over a number of subcarriers. Since the symbol duration increases for the lower rate parallel subcarriers, the effects of time dispersion caused by multipath delay spread are decreased. Intersymbol interference is eliminated almost completely by introducing a guard time in every OFDM symbol. In the guard time, the OFDM symbol is cyclically extended to avoid intercarrier interference. Figure 3.27 shows an example of 4 subcarriers from one OFDM symbol. In practice, the most efficient way to generate the sum of a large number of subcarriers is by using the inverse fast Fourier transform (IFFT). At the receiver side, the FFT can be used to demodulate all subcarriers. It can be seen in Figure 3.28 that all subcarriers differ by an integer number of cycles within the FFT integration time, which ensures the orthogonality between the different subcarriers. This orthogonality is maintained in the presence of multipath delay spread, as illustrated by Figure 3.29. Because of multipath, the receiver sees a summation of time-shifted replicas of each OFDM symbol. As long as the delay spread is smaller than the guard time, there is no intersymbol interference nor intercarrier interference within the FFT interval of an OFDM symbol. The only remaining effect of multipath is a random phase and amplitude of each subcarrier. In order to deal with weak subcarriers in deep fades, forward error correction across the subcarriers is applied.

Table 3.3

Main Parameters of the OFDM Standard

Parameter	*Value*
Data rate	6, 9, 12, 18, 24, 36, 48, 54 Mbps
Modulation	BPSK, QPSK, 16-QAM, 64-QAM
Coding rate	½, 2/3, 3/4
Number of subcarriers	52
Number of pilots	4
OFDM symbol duration	4 μs
Guard interval	800 ns
Subcarrier spacing	312.5 kHz
-3 dB bandwidth	16.56 MHz
Channel spacing	20 MHz

3.8.1 802.11a OFDM Parameters

Table 3.3 lists the main parameters of the draft OFDM standard. A key parameter that largely determine the choice of the other parameters is the guard interval of 800 ns. This guard interval provides robustness to root-mean-squared delay spreads up to several hundreds of nanoseconds, depending on the coding rate and modulation used. In practice, this means that the modulation is robust enough to be used in any indoor environment, including large factory buildings. It can also be used in outdoor environments, although directional antennas may be needed in this case to reduce the delay spread to an acceptable amount and to increase the range.

In order to limit the relative amount of power and time spent on the guard time to 1 dB, the symbol duration was chosen to be 4 μs. This also determined the subcarrier spacing to be 312.5 kHz, which is the inverse of the symbol duration minus the guard time. By using 48 data subcarriers, uncoded data rates of 12 to 72 Mbps can be achieved by using variable modulation types from BPSK to 64-QAM. In addition to the 48 data subcarriers, each OFDM symbol contains an additional 4 pilot subcarriers, which can be used to track the residual carrier frequency offset that remains after an initial frequency correction during the training phase of the packet.

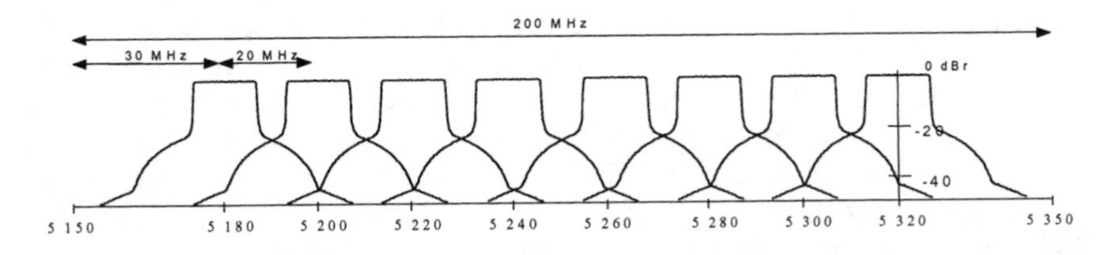

Figure 3.28 Channelization in lower and middle UNII band [5] (reprinted with permission from IEEE 802.11a-1999).

In order to correct for subcarriers in deep fades, forward error correction across the subcarriers is used with variable coding rates, giving coded data rates from 6 up to 54 Mbps. Convolutional coding is used with the industry standard rate 1/2, constraint length 7 code with generator polynomials (133,171). Higher coding rates of 2/3 and 3/4 are obtained by puncturing the rate 1/2 code.

3.8.2 802.11a Channelization

Figure 3.28 shows the channelization of 802.11a for the lower and middle UNII bands. Eight channels are available with a channel spacing of 20 MHz and guard spacings of 30 MHz at the band edges in order to meet the stringent FCC restricted band spectral density requirements. The FCC also defined an upper UNII band from 5.725 to 5.825 GHz, which carries another 4 OFDM channels. For this upper band, the guard spacing from the band edges is only 20 MHz, since the out-of-band spectral requirements for the upper band are less severe than those of the lower and middle UNII bands. In Europe and Japan the exact spectrum allocation for the new OFDM standards, 802.11a and HIPERLAN/2, is not fully established, but very likely in Europe the full range of the lower and middle UNII bands will be covered while in the Japan only the first 100 MHz (lower UNII band) will be covered. Thereby, a worldwide available range in the 5GHz band is created. In Europe also the 5.470–5.725GHz range will become available.

3.8.3 802.11a OFDM Signal Processing

The general block diagram of the baseband processing of an 802.11a OFDM transceiver is shown in Figure 3.29. In the transmitter path, binary input data is encoded by a standard rate 1/2 convolutional encoder. The coding rate may be increased to 2/3 or 3/4 by puncturing the coded output bits. After interleaving, the binary values are converted into QAM values. To facilitate coherent reception, 4 pilot values are added to each 48 data values, so a total of 52 QAM values is reached per OFDM symbol, which are modulated onto 52 subcarriers by applying the IFFT. To make the system robust to multipath propagation, a cyclic prefix is added. Further, windowing is applied to get a narrower output spectrum. After this step, the digital output signals can be converted to analog signals, which are then upconverted to the 5GHz band, amplified, and transmitted through an antenna.

The OFDM receiver basically performs the reverse operations of the transmitter, together with additional training tasks. First, the receiver has to estimate frequency offset and symbol timing, using special training symbols in the preamble. Then, it can do a fast Fourier transform for every symbol to recover the 52 QAM values of all subcarriers. The training symbols and pilot subcarriers are used to correct for the channel response as well as remaining phase drift. The QAM values are then demapped into binary values, after which a Viterbi decoder can decode the information bits.

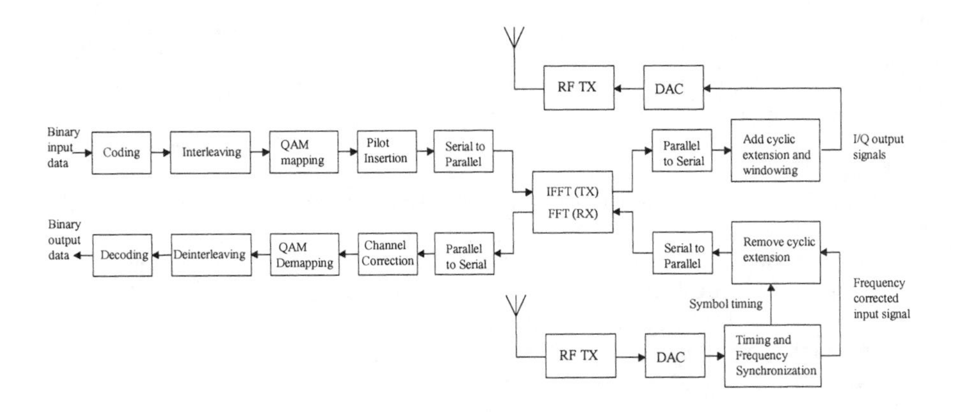

Figure 3.29 Block diagram of an OFDM transceiver [5] (reprinted with permission from IEEE 802.11a-1999).

3.8.4 Training

Figure 3.30 shows the structure of the preamble that precedes every 802.11a OFDM packet. This preamble is essential to perform start-of-packet detection, automatic gain control, symbol timing, frequency estimation, and channel estimation. All of these training tasks have to be performed before the actual data bits can be successfully decoded.

The first part of the preamble consists of 10 repetitions of a training symbol with a duration of 800 ns, which is only a quarter of the FFT duration of a normal data symbol. These short symbols are produced by using only nonzero subcarrier values for subcarrier numbers that are a multiple of 4. Hence, of all possible subcarrier numbers from −26 to +26, only the subset {−24, −20, −16, -12, -8, −4, 4, 8, 12, 16, 20, 24} is used. There are two reasons for using relatively short symbols in this part of the training; first, the short symbol period makes it possible to do a coarse frequency offset estimation with a large unambiguous range. For a repetitive signal with a duration of T, the maximum measurable unambiguous frequency offset is equal to $1/(2T)$, since higher frequency offsets result in a phase change exceeding 180 degrees from one symbol to another. Hence, by measuring the phase drift between two consecutive short symbols with a duration of 800 ns, frequency offsets up to 625 kHz can be estimated. If training symbols with a duration equal to the FFT interval of 3.2 μs were used, then the maximum frequency offset of only 156 kHz could be measured, corresponding to a relative frequency error of about 26 ppm at a carrier frequency of 5.8 GHz. The IEEE 802.11 standard specifies a maximum offset *per user* of 20 ppm, which means that the worst case offset as seen by a receiver can be up to 40 ppm, as it experiences the sum of the frequency offsets from both transmitter and receiver.

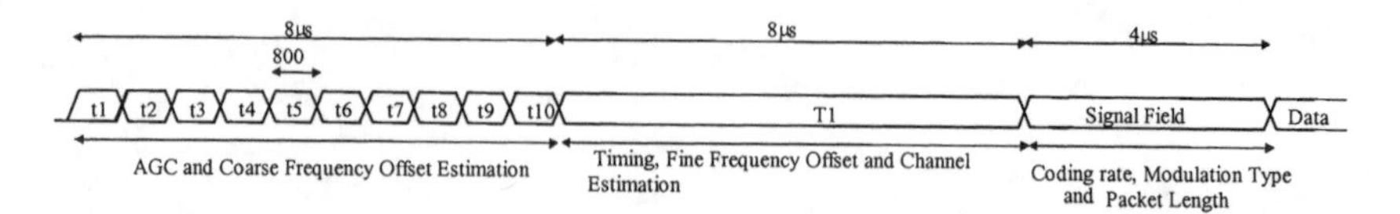

Figure 3.30 802.11a OFDM preamble [5] (reprinted with permission from IEEE 802.11-1999).

The second reason for using short symbols at the start of the training is that they provide a convenient way of performing automatic gain control (AGC) and frame detection. Detection of the presence of a packet can be done by correlating a short symbol with the next and detecting whether the correlation magnitude exceeds some threshold. After each interval equal to two short symbol durations, the receiver gain can be adjusted after which detection and gain measuring can be continued.

The short training symbols are followed by a long training symbol that contains 52 QPSK-modulated subcarriers like a normal data symbol. However, the length of this training symbol is twice that of a data symbol, which is done for two reasons; first, it makes it possible to do a precise frequency estimation on the long symbol. The long symbol is formed by cyclically extending an IFFT output signal with a duration of 3.2 μs for two and a half times. This makes it possible to do a frequency offset estimation by measuring the phase drift between samples that are 3.2 μs apart within the long training symbol. The second reason for the long symbol is to obtain reference amplitudes and phases for doing coherent demodulation. By averaging the two identical parts of the long training symbol, coherent references can be obtained with a noise level that is 3 dB lower than the noise level of data symbols.

Both the long and short symbols are designed in such a way that the peak-to-average power (PAP) ratio is approximately 3 dB, which is significantly lower than the PAP ratio of random OFDM data symbols. This guarantees the training degradation caused by nonlinear amplifier distortion to be smaller than the distortion of the data symbols.

After the preamble, there is still one training task left, which is tracking the reference phase. There will always be some remaining frequency offset, which causes a common phase drift on all subcarriers. In order to track this phase drift, 4 of the 52 subcarriers contain known pilot values. The pilots are scrambled by a length 127 pseudonoise sequence to avoid spectral lines exceeding the average power density of the OFDM spectrum.

Figure 3.31 shows the time-frequency structure of an OFDM packet, where all known training values are marked gray. It clearly illustrates how the packet starts with 10 short training symbols, using only 12 subcarriers, followed by the long training symbol and data symbols, with each data symbol containing 4 known pilot subcarriers.

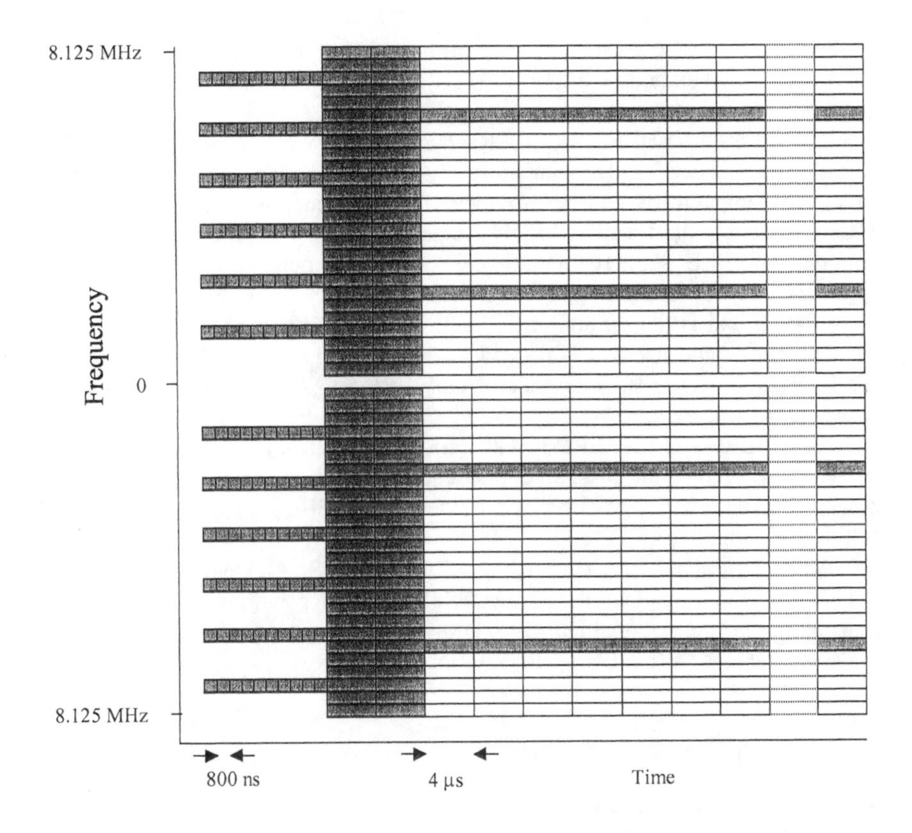

Figure 3.31 Time-frequency structure of an 802.11a OFDM packet. Gray subcarriers contain known training values [5].

In the case of the IEEE 802.11 standard, at the end of the preamble a special OFDM data symbol at the lowest 6 Mbps rate is sent, transmitting information about the length, modulation type, and coding rate of the rest of the packet. By sending this information at the lowest possible rate, it is ensured that the dynamic rate selection is at least as reliable as the most reliable data rate of 6 Mbps. Further, it makes it possible for all users to decode the duration of a certain packet, even though they may not be able to decode the data content. This is important for the IEEE 802.11 MAC protocol, which specifies that a user has to wait until the end of any packet already in the air before trying to compete for the channel.

3.9 NEW PHY: IEEE 802.11g

The standard uses OFDM and CCK, as in 802.11b, as mandatory technology. There are two optional technologies based on DSSS-OFDM, originally proposed by Intersil, and PBCC, from Texas Instruments. The previously existing PHYs in 802.11 are called extended rate PHYs (ERP). A short description of all the PHYs is given in this section and is summarized in Table 3.4 [18].

IEEE 802.11g makes use of ERP-DSSS/CCK which is the same as IEEE 802.11 and IEEE 802.11b mandatory PHY as discussed in Section 3.7; thus backward compatibility can be provided. The difference with IEEE 802.11b and ERP is that the short PLCP PPDU header format is mandatory, where it was optional for IEEE 802.11b, and that the clear channel assessment (CCA) will assess the channel for all available preambles.

ERP-OFDM is the same as IEEE 802.11a. The difference between the two is the frequency band used, 2.4- instead of 5-GHz, and slot-time, which optionally can be 9 μs when the BSS consists of only ERP STAs. In IEEE 802.11a the slot time is 20 μs; see Section 3.5.1.

As for IEEE 802.11b, ERP-PBCC is an optional PHY for IEEE 802.11g. It is a single-carrier modulation scheme that encodes payload using 256-state packet binary convolutional codes; this is an extension of PBCC in IEEE 802.11b. The achievable data rates are 22 and 33 Mbps.

Another optional PHY for IEEE 802.11g is DSSS-OFDM. This is a hybrid modulation combining a DSSS preamble and header with an OFDM payload transmission. If the optional DSSS-OFDM mode is used, the supported rates in that mode are the same as the ERP-OFDM–supported rates.

Table 3.4

IEEE 802.11g Mandatory and Optional Modes

Data Rates (Mbps)	Mandatory	Optional
1	DSSS (.11)	
2	DSSS (.11)	
5.5	CCK (.11b)	PBCC (.11b)
6	OFDM (.11a)	DSSS-OFDM
9		OFDM (.11a), DSSS-OFDM
11	CCK (.11b)	PBCC (.11b)
12	OFDM (.11a)	CCK-OFDM
18		OFDM (.11a), DSSS-OFDM
22		*PBCC*
24	OFDM (.11a)	DSSS-OFDM
33		*PBCC*
36		OFDM (.11a), DSSS-OFDM
48		OFDM (.11a), DSSS-OFDM
54		OFDM (.11a), DSSS-OFDM

3.10 SECURITY: IEEE 802.11i

Due to the security issues in WEP (see Section 4.3) IEEE 802.11 started developing MAC enhancements for security, namely IEEE 802.11i. IEEE 802.11i is not yet finalized; a short description of the draft standard is given in this section. Details of this standard are discussed in Chapter 4.

The draft standard defines a robust security network association between the AP and the STA with the support of IEEE 802.1X [19] and extensible authentication protocol (EAP) [20], also know as EAP over LAN (EAPoL). The host authentication can be done using a pre-shared key or by using authentication server (AS). In the second case a pairwise master key (PMK) is created by the STA and AS and is passed to the AP using the authentication and key management protocol (AKMP); this key can be used for mutual authentication of STA and AP. Using PMK a pairwise transient key (PTK) is derived and used for encrypting data between the AP and the STA. A group temporal key (GTK) can also be derived to protect broadcast/multicast traffic. GTK can be derived from a group master key (GMK).

Encryption and integrity check is performed using temporal key integrity protocol (TKIP) and Michael or advanced encryption standard (AES) based counter mode with cipher-block chaining message authentication code protocol (CCMP).

IEEE 802.11i also makes pre-authentication possible, while the current AP helps authenticating with new AP, thus supporting faster re-authentication and faster handover.

3.11 QOS: IEEE 802.11e

IEEE 802.11e provides MAC enhancements to support LAN applications with QoS requirements. The QoS enhancements are available to the QoS enhanced Stations (QSTAs) associated with a QoS enhanced Access Point (QAP) in a QoS enabled network. A subset of the QoS enhancements may be available for use between QSTAs. A QSTA may associate with a non-QoS AP in a non-QoS network [21, 22]. Non-QoS STAs may associate with a QAP. QoS will be covered in detail in Chapter 5.

The enhancements that distinguish the QSTAs from non-QoS STAs and the QAPs from non-QoS APs comprise an integrated set of QoS-related formats and functions that are collectively termed the QoS facility.

The IEEE 802.11e standard provides two mechanisms for the support of applications with QoS requirements. The first mechanism, designated as the enhanced distributed coordination function (EDCF), is based on the differentiating priorities at which the traffic is to be delivered. This differentiation is achieved through varying the amount of time a station would sense the channel to be idle, the length of the contention window during a backoff or the duration a station may transmit once it has the channel access.

The second mechanism allows for the reservation of transmission opportunities with the hybrid coordinator (HC). A QSTA based on its requirements requests the HC for transmission opportunities — both for its own transmissions as well as transmissions from the HC to itself. The HC, based on an admission control policy, either accepts or rejects the request. If the request is accepted, it schedules transmission opportunities for the QSTA. For transmissions from the STA, the HC polls a QSTA based on the parameters supplied by the QSTA at the time of its request. For transmissions to the QSTA, the HC queues the frames and delivers them periodically, again based on the parameters supplied by the QSTA. This mechanism is expected to be used for applications such as voice and video, which may need a periodic service from the HC. This mechanism is a hybrid of several proposals studied by the standardization committee.

3.12 IAPP: IEEE 802.11f

The IEEE 802.11f inter access point protocol (IAPP) [23] is a communication protocol, used by one AP to communicate with the other APs. It is a part of a communication system comprising the APs, the STAs, a backbone network, and the RADIUS infrastructure [24]. Details of IEEE 802.11f and mobility will be discussed in Chapter 6.

The RADIUS servers provide two functions:

- Mapping the ID of an AP to its IP address, and
- Distribution of keys to the APs to allow the encryption of the communications between the APs.

The function of the IAPP is to facilitate the creation and maintenance of the wireless network, support the mobility of the STAs, and enable the APs to enforce the requirement of a single association for each STA at a given time.

One of the services the IAPP provides is proactive caching. Proactive caching is a method that supports fast roaming by caching the context of a STA in the APs to which the STA may roam. The next APs are identified dynamically (i.e., without management pre-configuration) by learning the identities of neighboring the APs.

3.13 OTHER IEEE 802.11 ACTIVITIES

With the success of IEEE 802.11b, the IEEE 802.11 WG started working on several issues [7, 8]. In this section some of the on-going activities of IEEE 802.11 and its goals are discussed.

3.13.1 IEEE 802.11h

The IEEE 802.11h standard provides solution for dynamic frequency selection (DFS) and transmit power control (TPC). The purpose of this standard is to satisfy the regulatory requirements of the 5GHz band in Europe—that is, to provide coexistence with RADAR systems. This standard was finalized in September 2003.

3.13.2 IEEE 802.11j

The IEEE 802.11j is working on enhancements of the IEEE 802.11 standard and amendments to add channel selection for 4.9 GHz and 5 GHz in Japan to conform to the Japanese rules on operational mode, operational rate, radiated power, spurious emissions, and channel sense. The draft standard is currently going through letter ballot.

3.13.3 IEEE 802.11k

This task group is working on radio resource measurement enhancements of IEEE 802.11. Each IEEE 802.11 device will be able to measure the radio environment and respond accordingly. The group is already working on the first draft.

3.13.4 IEEE 802.11n

The IEEE 802.11n is the next step in IEEE 802.11 to achieve higher data rates. The idea is to develop a radio and a MAC that will provide greater than 100 Mbps throughput. Smart antennas and MIMO-OFDM techniques, among other things are under consideration in this group.

3.13.5 Upcoming Activities

There are several other upcoming activities in IEEE 802.11 that have not reached the stage of a task group. One of these was the fifth generation study group (5GSG), which is now working as the wireless next generation (WNG). WNG is focusing on an extension of IEEE 802.11 to new areas like interworking with mobile networks (e.g., 3GPP).

Another group is also looking at wireless home area networks (WHAN). This group is focusing on solutions for home equipment and methods which they can work together using IEEE 802.11. Yet another group is the wireless access in the vehicular environment (WAVE) group, which is looking at using some frequency that was set aside by the U.S. government for vehicle safety and using the 11a radios to support that. It used to be called dedicated short range communication (DSRC). The idea is also to provide an intelligent transportation system (ITS) solution.

There is a fast roaming group that is addressing the issues around mobility and moving around. This implies that they will be looking at what happens to security and connectivity when WLAN-connected devices move from cell to cell. Their biggest requirement is to support voice over IP (VoIP) in moving from cell to cell. The study within this group is different than IEEE 802.21 (handoff TG) in the sense that this group is looking at handover within an ESS of an administrative domain while IEEE 802.21 is working on handover between different technologies and administrative domains.

The mesh networking group is almost in the same category. This is layer 2 mesh; the group will be working on MAC layer dynamic backbone subnets. These MAC layer subnets can allow for securely moving in and out of subnets; the handoff mechanisms are taken care of just after the radio. This group is named IEEE 802.11s.

Other work is focusing on wireless device management. This work is in essence a logical next step to IEEE 802.11k activity (i.e., making use of the measurements). A more recent group is looking at wireless performance prediction (WPP). The goal of WPP is to develop a set of models and methods for predicting wireless performance metrics, including coverage and throughput, on either a point or area basis given certain information concerning the layout, usage, and devices of an IEEE 802.11 WLAN. WPP will be beneficial for planning, installing, maintaining, and developing tools for WLAN networks.

References

[1] ISO/IEC 8802-11, ANSI/IEEE Std 802.11, First Edition 1999-00-00, Information Technology — Telecommunications and information exchange between systems — Local and metropolitan area networks — Specific requirements — Part 11: Wireless LAN Medium Access Control (MAC) and Physical Layer (PHY) specifications.

[2] O'Hara, B., and A. Patrick, *The IEEE 802.11 Handbook, A Designer's Companion*, New York, NY: IEEE Press, 1999.

[3] van Nee, R., G. Awater, M. Morikura, H. Takanashi, M. Webster, and K.W. Halford, "New high-rate wireless LAN standards," *IEEE Communications Magazine*, Vol. 37, No. 12, Dec. 1999, pp. 82-88.

[4] IEEE Std 802.11b-1999, Supplement to Standard for Information Technology — Telecommunications and information exchange between systems — local and metropolitan area networks — Specific requirements — Part 11: Wireless LAN Medium Access Control (MAC) and Physical Layer (PHY) specifications: Higher speed Physical Layer (PHY) extension in the 2.4 GHz band.

[5] IEEE Std 802.11a-1999, Supplement to IEEE Standard for Information technology — Telecommunications and information exchange between systems — Local and metropolitan area networks — Specific requirements — Part 11: Wireless LAN Medium Access Control (MAC) and Physical Layer (PHY) specifications: High-speed Physical Layer in the 5GHz Band.

[6] van Nee, R.D.J., and R. Prasad, *OFDM for Wireless Multimedia Communications*, Norwood, MA: Artech House, 2000.

[7] Overview and Guide to IEEE 802 LMSC, December 2002.

[8] 802 groups websites: http://grouper.ieee.org/groups/802/.

[9] BLUETOOTH SIG website http://www.bluetooth.com.

[10] IEEE 802.18: http://grouper.ieee.org/groups/802/Regulatory/.

[11] WAPI: http://news.com.com/2100-7351_3-5122920.html.

[12] WAPI: http://news.bbc.co.uk/go/pr/fr/-/2/hi/business/3648653.stm.

[13] WECA, Web URL: http://www.wirelessethernet.com, 15 July, 2001.

[14] LaMaire, R.O. et al., "Wireless LANs and Mobile Networking: Standards and Future Directions," *IEEE Communications Magazine*, Vol. 34, No. 8, Aug. 1996, pp. 86–94.

[15] Crow, B.P., I. Widjada, J.G.Kim, P.T.Sakai, "IEEE 802.11 Wireless Local Area Networks," *IEEE Communications Magazine*, September 1997, pp.116–126.

[16] Chen, K. C., "Medium Access Control of Wireless LANs for Mobile Computing," *IEEE Network*, Vol. 8, No. 5, Sept. 1994, pp. 50–63.

[17] Weinmiller, J., H. Woesner, and A. Wolisz, "Analyzing and Improving the IEEE 802.11-MAC Protocol for Wireless LANs," *Proc. MASCOTS '96*, San Jose, CA, Feb. 1996, pp. 200–206.

[18] IEEE 802.11g, IEEE 802.11: Further High Data Rates Extension in 2.4 GHz band, D8.2, April 2003.

[19] IEEE Std 802.1X-2001 "IEEE Standard for Local and metropolitan area networks—Port-Based Network Access Control," 14 June 2001.

[20] Blunk, L., J. Vollbrecht, and Bernard Aboba, "Extensible Authentication Protocol (EAP)," October 2002. IETF pppext working group draft draft-ietf-pppext-rfc2284bis-07.txt.

[21] IEEE P802.11e/D1, March 2001, Draft Supplement to IEE Std 802.11, 1999 Edition, Draft Supplement to STANDARD FOR Telecommunications and Information Exchange Between Systems — LAN/MAN Specific

Requirements — Part 11: Wireless Medium Access Control (MAC) and physical layer (PHY) specifications: Medium Access Control (MAC) Enhancements for Quality of Service (QoS).

[22] IEEE P802.11eS/D1, March 2001, Draft Supplement to IEE Std 802.11, 1999 Edition, Draft Supplement to STANDARD FOR Telecommunications and Information Exchange Between Systems — LAN/MAN Specific Requirements — Part 11: Wireless Medium Access Control (MAC) and physical layer (PHY) specifications: Medium Access Control (MAC) Specification for Enhanced Security.

[23] IEEE Std 802.11f/D1, March 2001, Draft Supplement to IEE Std 802.11, 1999 Edition, Draft Recommended Practice for Multi-Vendor Access Point Interoperability via an Inter-Access Point Protocol Across Distribution Systems Supporting IEEE 802.11 Operation.

[24] Rigney, C., A. Rubens, W. Simpson, and S. Willens, Remote Authentication Dial In User Service (RADIUS), RFC 2138, January 1997.

Selected Bibliography

[1] FCC, "Amendment of the Commission's Rules to Provide for Operation of Unlicensed NII Devices in the 5 GHz Frequency Range," *memorandum opinion and order*, ET Docket No. 96–102, June 24, 1998.

[2] Frank, Robert L., "Polyphase Complementary Codes," *IEEE Transactions on Information Theory*, November 1980, pp. 641–647.

[3] Gerlach, K., and Kretschmer, G., "General Forms and Properties of Zero Crosscorrelation Radar Waveforms," *IEEE Transactions on Aerospace and Electronic Systems*, January 1992, pp. 98–103.

[4] Golay, Marcel J. E., "Complementary Series", *IRE Transactions on Information Theory*, April 1961, pp. 82-87.

[5] Grant, A., and R. van Nee, "Efficient maximum likelihood decoding of Q-ary modulated Reed-Muller codes," *IEEE Communications Letters*, Vol. 2, No. 5, pp. 134–136, May 1998.

[6] Green, E.R., "Call for Proposals for P802.11h – DFS TPC," *IEEE 802.11*, IEEE 802.11-00/463r1, February 2001.

[7] Halford, K., S. Halford, M. Webster, and C. Andren, "Complementary Code Keying for Rake-Based Indoor Wireless Communication," *IEEE ISCAS '99*, Orlando, Florida, Vol. 4, pp. 427–430.

[8] Halford, S., and M. Webster, "Overview of OFDM for High Rate Extension," *IEEE 802.11*, IEEE 802.11-00/389, November 2000.

[9] Heegard, C., et al., "Performance of PBCC and CCK,", *IEEE 802.11*, IEEE802.11-98/304, Sept 1998,.

[10] Heegard, C., E. Rossin, M. Shoemake, S. Coffey, and A. Batra, "Texas Instruments Proposal for IEEE 802.11g High-Rate Standard," *IEEE 802.11*, IEEE 802.11-00/384, November 2000.

[11] Kretschmer, F., and Gerlach, K., "Radar Waveforms Derived from Orthogonal Matrices," *NRL Report 9080*, Feb. 1989.

[12] Kretschmer, F. F. Jr., and B.L. Lewis, "Doppler Properties of Polyphase Coded Pulse Compression Waveforms," *IEEE Transactions on Aerospace and Electronic Systems*, July 1983, pp. 521–531.

[13] Ojha, A. and D. Koch, "Performance Analysis of Complementary Coded Radar Signals in an AWGN Environment," *IEEE Proceedings of Southeastcon*; April 8–10, 1991, pp. 842–846.

[14] Popovic, B., "Synthesis of Power Efficient Multitone Signals with Flat Amplitude Spectrum", *IEEE Transactions on Communications*, July 1991, pp. 1031-1033.

[15] Prasad, K. V. and M. Darnell, "Data Transmission Using Complementary Sequence Sets," *IEE Fifth International Conference on HF Radio Systems and Techniques*, July 22–25, pp. 222–226.

[16] Schylander, E., "5GSG Major Topics," *IEEE 802.11*, IEEE 802.11-01/072, January 2001.

[17] Sivaswamy, R., "Digital and Analog Subcomplementary Sequences for Pulse Compression," *IEEE Transactions on Aerospace and Electronic Systems*, March 1978, pp. 343–350.

[18] Sivaswamy, R., "Multiphase Complementary Codes," *IEEE Transactions on Information Theory*, September 1978, pp. 546–552.

[19] Takanashi, H., and R. van Nee, "Merged Physical Layer Specification for the 5 GHz Band," *IEEE 802.11*, IEEE P802.11-98/72-r1, March 1998.

[20] Tellado-Mourelo, J., E.K. Wesel, and J.M. Cioffi, "Adaptive DFE for GMSK in Indoor Radio Channels," *IEEE Trans. On Sel. Areas in Comm.*, Vol. 14, No. 3, April 1996, pp. 492–501.

[21] Tseng, C.-C., and C. L., Liu, "Complementary Sets of Sequences," *IEEE Transactions on Information Theory*, September 1972, pp. 644–652.

[22] van Nee, Richard, "OFDM Codes for Peak-to-Average Power Reduction and Error Correction," *IEEE Global Telecommunications Conference*, Nov. 18–22, 1996, pp. 740–744.

[23] van Nee, R., "OFDM for High Speed Wireless Networks," *IEEE 802.11*, IEEE P 802.11-97/123, November 1997.

[24] Weathers, G., and E. M. Holliday, "Group-Complementary Array Coding for Radar Clutter and Rejection," *IEEE Transactions on Aerospace and Electronic Systems*, May 1983, pp. 369–379.

[25] Webster, M., "DFE Packet Error Rate Minimization using Precursor Sliding," *IEEE 802.11*, doc.:IEEE P802.11-98/115, March 1998.

[26] Webster, M., "Multipath Issues and Architectures," *IEEE 802.11*, doc.:IEEE P802.11-98/37, January 1998.

[27] Tellambura, C., Y.J. Guo, and S.K. Barton, "Channel Estimation Using Aperiodic Binary Sequences," *IEEE Communications Letters*, May 1998, pp. 140–142.

[28] Webster, M., C. Andren, J. Boer, and R. Van Nee, "Harris/Lucent TGb Compromise CCK (11 Mbit/s) Proposal," *IEEE 802.11*, IEEE P802.11-98/246, July 1998.

Chapter 4

Security

WLANs, particularly IEEE 802.11, have gained much more ground than was expected within a very short period of time [1–3]. This growth is being hampered by security issues [4–25]; the issue is by now so well publicized that even laypeople are aware of it.

This chapter starts with a short background on security, after which some security protocols are discussed. Following that, the security issues in the IEEE 802.11 countermeasures present in the market are discussed. Finally Wi-Fi protected access (WPA) and the IEEE 802.11i standard (ratified in June 2004) are explained.

4.1 SECURITY THREATS AND GOALS

The introduction of distributed systems and the use of networks and communications facilities—wireline and now increasingly wireless—have increased the need for network security measures to protect data—both real-time and non-real-time—during transmission. To assess the security needs effectively and evaluate and choose the most effective solution a systematic definition of the security goals or requirements and understanding of the threats is a necessity [1–3, 26, 27]. In this section first the security threats and then the security goals are discussed. This section also discusses which security goals will counter a given security threat.

4.1.1 Threats

Security threats or security issues can be divided into two types: passive and active threats. Passive threats stem from individuals attempting to gain information that can be used for their benefit or maybe to perform active attacks at a later time. Active threats are those where the intruder does some modification to the data, network, or traffic in the network. In the following section the most common active and passive threats are discussed; for a comprehensive list see [1–3, 26, 27].

4.1.1.1 Passive Threats

A passive threat is a situation when an intruder does not do anything to the network or traffic under attack but collects information for personal benefit or for future attack purposes. Two basic passive threats are described as follows [1–3, 26, 27].

- Eavesdropping: This has been a common security threat to human beings for ages. In this attack the intruder listens to things he or she is not supposed to listen to. This information could contain, for example, the session key used for encrypting data during the session. This kind of attack means that the intruder can get information that is at times strictly confidential.
- Traffic analysis: This is a subtle form of passive attack. It is possible that at times for the intruder knowing the location and identity of the communicating device or user is enough. An intruder might only require information like a message has been sent, who is sending the message to whom, and at the frequency or size of the message. Such a threat is known as traffic analysis.

4.1.1.2 Active Threats

An active threat arises when an intruder directly attacks the traffic and the network and causes a modification of the network, data, etc. A list of common active attacks follows [1–3, 26, 27].

- Masquerade: This is an attack in which an intruder pretends to be a trusted user. Such an attack is possible if the intruder captures information about the user like the authentication data, simply the username and the password. Sometimes the term spoofing is used for masquerade.
- Authorization violation: An intruder or even a trusted user uses a service or resources it is not intended to use. In the case of an intruder this threat is similar to the masquerading; having entered the network the intruder can access services it is not authorized to access. On the other hand a trusted user can also try to access unauthorized services or resources; this could be done by the user performing active attacks on the network or simply by lack of security in the network/system.
- Denial of service (DoS): DoS attacks are performed to prevent or inhibit normal use of communications facilities. In the case of wireless communications it could be as simple as causing interference or it could be done by sending data to a device and overloading the central processing unit (CPU) or draining the battery. Such attacks could also be performed on a network by, for example, flooding the network with unwanted traffic.

  Sabotage is also a form of DoS attack. A DoS attack termed as sabotage could also mean the destruction of the system itself.

- Modification or forgery of information: An intruder creates new information in the name of a legitimate user or modifies or destroys the information being sent. It could also be that the intruder simply delays the information being sent. An example is an original message "Allow Neeli Prasad to read confidential Source Codes" modified to "Allow Anand Prasad to read confidential Source Codes."

4.1.2 Goals

There are five major security goals that are also known as security services and can also be used as *security requirements*. These goals are discussed as follows [1–3, 26, 27].

- Confidentiality: This is for the protection of the data from disclosure to an unauthorized person. Encryption is used to fulfill this goal. With an active attack it is possible to decrypt any form of encrypted data (given there is a good mathematician/cryptographer or a person with a powerful computer and no time limit); thus confidentiality is primarily considered a protection against passive attacks.
- Authentication: The authentication service is concerned with assuring that a communication is authentic. In the case of a single message, such as a warning or alarm signal, the function of the authentication service is to assure the recipient that the message is from the source that it claims to be from. In the case of an ongoing interaction, such as the connection of a terminal to a host, two aspects are involved. First, at the time of connection initiation, the service assures that the two entities are authentic (i.e., each is the entity that it claims to be). Second, the service must assure that the connection is not interfered with in such a way that a third party can masquerade as one of the two legitimate parties for the purposes of unauthorized transmission or reception.
- Access control: In the context of network security, access control is the ability to limit and control access to the systems, the networks, and the applications. Thus unauthorized users are kept out. Although given separately user authentication is often combined with access control purposes; this is done because a user must be first authenticated by, for example, the given server and the network so as to determine the user access rights. Implicitly access control also means authorization.
- Integrity: Prevents unauthorized changes to the data. Only authorized parties are able to modify the data. Modification includes changing status, deleting, creating, and delaying or replaying of the transmitted messages.
- Non-repudiation: Neither the originator nor the receiver of the communication should be able to deny the communication and content of the message later. Thus, when a message is sent, the receiver can prove that the message was in

fact sent by the alleged sender. Similarly, when a message is received, the sender can prove that the message was in fact received by the alleged receiver.

Besides the above mentioned security requirements there are some general requirements that play an important role in developing the security solutions; these are the following:

- Manageability: The load of the network administrator must not be unnecessarily increased by adding security, while the deployed solutions should be easy to manage and operate in the long term.
- Scalability: A network must be scalable, which requires the security scheme deployed in the network to be equally scalable while maintaining the level of security. Here the term scalability is being used in its broadest sense, scalable in terms of the number of users and in terms of an increase in network size (i.e., addition of new network elements or an extension to a new building).
- Implementability: A simple and easy way to implement a scheme is extremely important. Thus a security scheme must be devised so that it is easy to implement and still fulfills the security requirements.
- Performance: Security features must have minimum impact on the network performance. This is especially important for real-time communication where the security requirements must be met while the required quality of service is met. Performance also goes hand in hand with the resource usage of the medium; the security solutions must not, for example, cause a decrease in the overall capacity of the network.
- Availability: This goal is closure to the five goals mentioned earlier in the section. Any service or network should be available to the user. Several attacks are possible to disrupt the availability, DoS being the major one.

4.1.3 Mapping Security Threats to Goals

In this section the security threats are mapped to the security goals. Knowing which threat can be countered by which goal, the next step is to find the security solutions or mechanisms that can fulfill the security goals. A mapping of security threats and goals is given in Table 4.1 [1], where X in a cell represents that the given security goal can counter the given threat. It should be noted that a security goal can fail to counter a threat sometimes.

4.2 RELATED INFORMATION

In this section some security technologies are presented as background information to the reader. This information is also beneficial for understanding later sections.

Table 4.1

Mapping of the Security Threats to the Security Goals

Security Goals	Security Threats					
	Eaves-dropping	Traffic Analysis	Masquerade	Authorization violation	DoS	Modification
Confidentiality	X	X	X	X		
Authentication			X	X		X
Access Control			X	X		X
Integrity			X	X		X
Non-repudiation			X	X		X
Availability			X	X	X	X

IP protocol stack and security protocols from IETF in those layers are shown in Figure 4.1 as an overview.

4.2.1 IPSec

IPSec is developed by Internet Engineering Task Force (IETF). It is an IP layer protocol. IPSec provides authentication of packet source, data integrity, confidentiality by use of encryption, and anti-replay protection by using sequence numbering. Non-repudiation, access control, and key exchange mechanism are not part of the protocol [26–35].

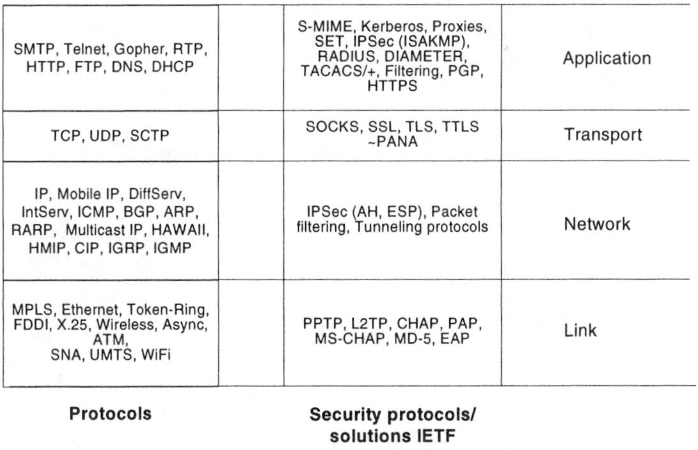

Protocols	Security protocols/ solutions IETF	
SMTP, Telnet, Gopher, RTP, HTTP, FTP, DNS, DHCP	S-MIME, Kerberos, Proxies, SET, IPSec (ISAKMP), RADIUS, DIAMETER, TACACS/+, Filtering, PGP, HTTPS	Application
TCP, UDP, SCTP	SOCKS, SSL, TLS, TTLS ~PANA	Transport
IP, Mobile IP, DiffServ, IntServ, ICMP, BGP, ARP, RARP, Multicast IP, HAWAII, HMIP, CIP, IGRP, IGMP	IPSec (AH, ESP), Packet filtering, Tunneling protocols	Network
MPLS, Ethernet, Token-Ring, FDDI, X.25, Wireless, Async, ATM, SNA, UMTS, WiFi	PPTP, L2TP, CHAP, PAP, MS-CHAP, MD-5, EAP	Link

Figure 4.1 IP protocol stacks and security protocols.

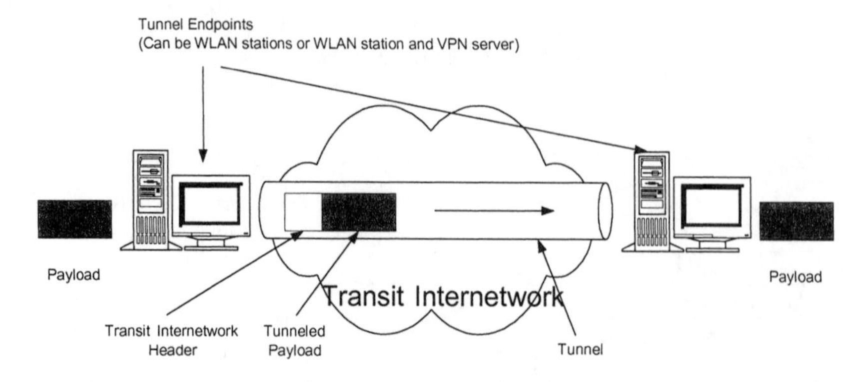

Figure 4.2 Tunneling.

IPSec is defined in several RFC's also given in the reference [29–35]:

- Security Architecture for the Internet Protocol (RFC 2401);
- IP Security Document Road Map (RFC 2411);
- IP Authentication Header (AH) (RFC 2402);
- IP Encapsulating Security Payload (ESP) (RFC 2406).

In this section various features of IPSec are discussed briefly.

4.2.1.1 AH and ESP

Protocols within IPSec used to provide the security are AH and ESP. Both AH and ESP can work in transport and tunnel mode (Figure 4.2). In tunnel mode a new IP header is added to the packet.

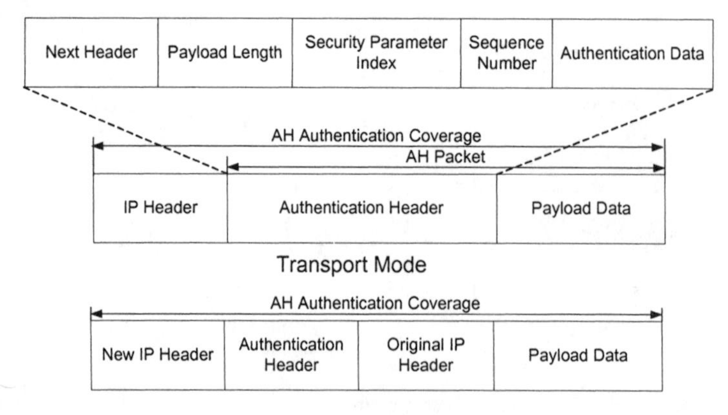

Figure 4.3 IPSec AH transport and tunnel modes packet format for IPv4.

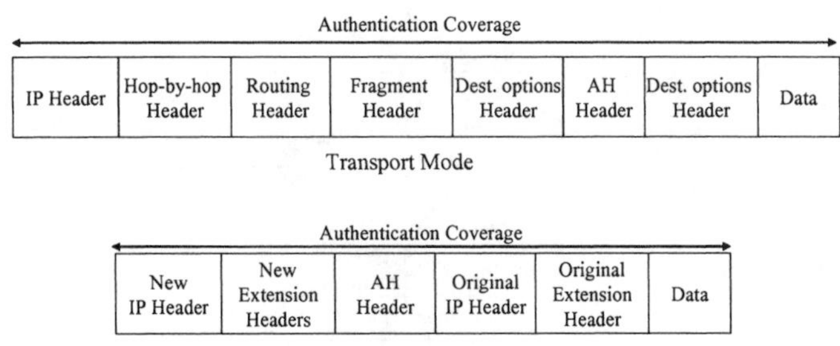

Figure 4.4 IPSec AH transport and tunnel modes packet format for IPv6.

AH provides data origin authentication of the whole IP packet; fields that would change in transit are not used for calculation of authentication data. It provides connectionless data integrity and replay protection using sequence number and may also provide non-repudiation. In transport-mode AH, the IP payload and selected header fields are included in the authentication calculation. In tunnel-mode AH, the entire original IP datagram is included in the authentication calculation. The result is placed within a new IP datagram. Selected header fields of this new IP datagram are also included in the authentication calculation. The AH transport and tunnel modes for IPv4 and IPv6 are shown in Figure 4.3 and Figure 4.4, respectively.

ESP provides confidentiality and integrity protection for IP packets. ESP, without authentication, encrypts the IP payload and with authentication it encrypts and authenticates the payload but not the IP header as in AH. Protection against traffic analysis is not provided by this mechanism. Unencrypted synchronization data (if required) is carried in the beginning of the payload data field. Possible encryption algorithms are DES, 3DES, AES, RC5, IDEA, 3-key triple IDEA, CAST, and Blowfish. In transport-mode ESP, the IP payload (which contains a transport-layer packet) is placed in the encrypted portion of the ESP frame and that entire ESP frame is placed in the payload of the original IP datagram, the IP headers of which are not encrypted. In tunnel-mode ESP, the entire original IP datagram is placed in the encrypted portion of the ESP and that entire ESP frame is placed within a new IP datagram having unencrypted IP headers. Encapsulating the protected data is necessary when confidentiality protection is required for the entire original datagram. Figure 4.5 shows the transport and tunnel modes of ESP for IPv4. For IPv6 the extension headers follow the IP header.

4.2.1.2 Security Policy and Security Association

The Security Policy (SP) specifies what service is offered to a specific packet based on its selectors: source and destination addresses, protocol, and upper layer ports (TCP and UDP). The SP is kept in the security policy database (SPD). For outbound traffic the SPD is searched to see if the packet should be protected. For

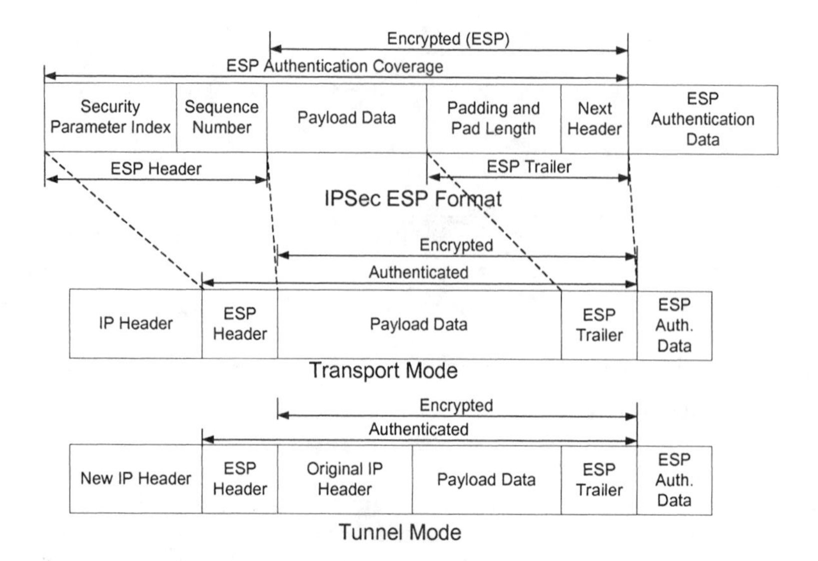

Figure 4.5 IPSec ESP packet format for transport and tunnel modes in IPv4. For IPv6 the extension header follows the IP header.

inbound traffic the SPD is searched to check if the decrypted/authenticated packet was supposed to be protected.

A security association (SA) is the method IPSec uses to track a given communication session. It defines how the communicating systems will use security services, including information about the traffic security protocol, the authentication algorithm, and the encryption algorithm to be used. SAs also contain information on dataflow, lifetime, and lifedata as well as sequence numbering for anti-replay. SAs are negotiated between two IPSec systems—in tunnel mode, between the two endpoints of the tunnel. The two IPSec systems can also negotiate the level of authorization for a range of addresses, protocols, and ports that will be protected by the SA. An SA is unidirectional; that is, for each pair of communicating systems there are at least two security connections—one from A to B and one from B to A. A given SA can use ESP or AH, not both. If a connection needs both protocols, it needs to establish two SAs for each direction; four for a bidirectional connection. SAs are identified with their security policy identifier (SPI). For one peer, an SA is identified with a unique nonambiguous SPI/<remote peer IP> pair. The SA's main pieces of information are the algorithm to use to protect data (e.g., DES), algorithm-specific attributes (e.g., keys), mode (tunnel/transport), tunnel destination (peer), and proxy identity (selector). A security association is identified by a combination of a security parameter index (SPI), which is a randomly chosen unique number; the destination IP address of the packet; and the traffic security protocol to be used (AH or ESP). Two databases are required for SAs: a security policy database (SPD), specifies the security services that will be provided for IP packets, and a security association database (SAD), in which each entry defines the parameters associated with one SA.

4.2.1.3 Key Management

As mentioned earlier, IPSec does not provide any key management mechanism and thus permits several different key management mechanisms to be used, including manual configuration (pre-established keys). This method also allows separate development and modification of the key management in one hand and AH and ESP protocols on the other hand. The only coupling between the key management protocol and the security protocol is with the SPI.

The key management mechanism is used to negotiate a number of parameters for each SA, including not only the encryption keys but also other information (e.g., the authentication algorithm and its mode) used by the communicating parties. Internet key exchange (IKE) is a family of protocols that is based on the Internet security association and key management protocol (ISAKMP), which specifies a framework for key management, parts of Oakley, a key exchange protocol; parts of the secure key exchange mechanism (SKEME), a key exchange protocol; and parts of station-to-station (STS), a key exchange protocol.

There are 6 variations of an IKE negotiation, two modes (aggressive mode and main mode), and three authentication methods (pre-shared, public key encryption, and public key signature). IKE uses two modes from ISAKMP phase 1, main mode and aggressive mode, and there is a single phase 2 mode: quick mode.

During phase 1, IKE negotiates the following:

- How to protect phase 1 (crypto and hash algorithms);
- Hardness of the keys (D-H group);
- How to authenticate with the remote peer (pre-shared, public-key encryption, and digital signature);
- Keying material for phase 2;
- Overall, IKE ensures that it is talking to the right peer.

During phase 2, IKE negotiates the following:

- A protection suite (e.g., ESP and AH);
- Algorithms in the protection suite (e.g., DES and SHA);
- Whom we are protecting (proxy identities);
- Optional keying material for negotiated protocols;
- Phase 2 creates a pair of actual IPSec SAs.

A new version of IKE, also known as son of IKE (SOI) or IKEv2 [72] is also being standardized by IETF. IKEv2 can establish IPSec SA in 2 request/response pairs as compared to 6 or 3–4 steps in IKE. It is also less complex and thus more secure. On the other hand the mobike working group is working on enhancements of IKEv2 for multihoming and mobility.

4.2.1.4 IP Address Configuration

Let us now look at the issue of IP address configuration in IPSec. Using IPSec as a Virtual Private Network (VPN) solution (see Section 4.4.3) is only beneficial if one can have the user in an external network as part of the internal network in a secure way. This means that there is a necessity of assigning an internal LAN IP address to the device the user is using while away from the LAN. This is possible roughly based on the following steps:

1. The user device connects to the VPN server and creates a IPSec tunnel.
2. The DHCP address provides the device an internal LAN IP address, which is used as source address by the device. Here SA is needed with the DHCP server.
3. The device, in IPSec tunnel mode uses the address from the LAN as the internal IP packet source address and the address assigned by the network where it is located as the external IP packet source address.

4.2.2 Network Address Translation

Due to lack of IPv4 addresses there are some non-routable private addresses [36]; these are Class A from 10.0.0.0 until 10.255.255.255, Class B from 172.16.0.0 until 172.16.255.255, and Class C 192.168.0.0 until 192.168.255.255. When communication between devices with the non-routable or private address is needed with a device in a public network (like the Internet) then address translation is required. This is where Network Address Translator (NAT) comes in. So NAT routers (or NATificators) are located at borders of public and private networks. The payload of the packet must also be considered during the translation process. NAT must also regenerate the User Datagram Protocol (UDP), or Transmission Control Protocol (TCP), the checksum (optional in UDP) as it is computed from a pseudo-header containing source, and the destination IP address. Thus IP checksum must also be recalculated.

Basically a pool of public IP addresses is shared by an entire private IP subnet (static or dynamic NAT). Edge devices that run dynamic NAT create bindings "on the fly" (i.e., for each private address a public address is mapped thus building a NAT table). After the connection is terminated (or a timeout is reached which is usually short), the binding expires, and the address is returned to the pool for reuse. A variation of dynamic NAT known as Network Address Port Translation (NAPT) may be used to allow many hosts to share a single IP address by multiplexing streams differentiated by TCP/UDP port number.

4.2.3 IPSec and NAT

IPSec when used together with NAT faces some issues. In this section these issues are discussed together with possible solutions.

4.2.3.1 Issues

The first thing that we notice is that there is a change in IP address when using NAT; this means that the integrity in AH will fail. A further change in IP address means recalculation of TCP checksum, which is encrypted in the case of ESP. This is a non-issue for ESP tunnel mode.

Another issue is with IKE and SA setup and endpoint authentication. IKE is based on the IP address as identifier, which must change when NAT is used but this is always hashed or encrypted. Even in the case where IP addresses are not used in IKE payloads and an IKE negotiation could occur uninterrupted, there is difficulty with retaining the private-to-external address mapping on NAT from the time IKE completed negotiation to the time IPSec uses the key on an application [37]. There are other issues related to SA time-out and IP address time-out.

4.2.3.2 Solutions

There are a few solutions discussed in IETF [38–44]. The most prominent one is the NAT traversal [38], which is discussed here. The main technology behind this solution is UDP encapsulation, wherein the IPSec packet is wrapped inside a UDP/IP header, allowing NAT devices to change IP or port addresses without modifying the IPSec packet.

For NAT traversal to work properly, two things must occur. First, the communicating VPN devices must support the same method of UDP encapsulation. Second, all NAT devices along the communication path must be identified.

Usually, NAT assignments last for a short period of time and are then released. For IPSec to work properly, the same NAT assignment needs to remain intact for the duration of the VPN tunnel. NAT traversal accomplishes this by requiring any end point communicating through a NAT device to send a "keep alive" packet, which is a one-byte UDP packet sent periodically to prevent NAT endpoints from being remapped midsession.

All NAT Traversal communications occur over UDP port 500. This works well because port 500 is already open for IKE communications in IPSec VPNs, so new holes do not need to be opened in the corporate firewall. This solution does add a bit of overhead to IPSec communications; namely, 200 bytes is added for the phase 1 IKE negotiation and each IPSec packet has about an additional 20 bytes.

4.2.4 Secure Socket Layer

The Secure Sockets Layer (SSL) mechanism was developed by Netscape. SSL v3.0 was standardized by IETF as Transport Layer Security (TLS). SSL/TLS uses TCP as the underlying transport protocol for reliable data transport; it is independent of the higher level applications. SSL/TLS provides the following [27, 45]:

- Server and client authentication are based on asymmetric encryption (RSA and DSS) and certificates (X.509 format) issued by a CA;
- Data integrity and authenticity are based on a keyed one-way hash function (MD5 and SHA-1); a sequence number (a message counter kept by the sender) is used in the hash calculation, to protect against replay;
- Data confidentiality is based on symmetric encryption (DES, DES EDE3, RC2, RC4, and IDEA).

Various key agreement mechanisms are supported by SSL/TLS: RSA, Diffie-Hellman, and Fortezza-KEA (key exchange algorithm).

SSL describes four protocols (Figure 4.6). At the most basic layer is the SSL record protocol, which places application data into an encrypted SSL payload. The SSL Handshake protocol negotiates session-specific parameters (such as the cipher suite to be used and the pre-master secret) and handles the authentication of the remote computer. The last two protocols, the change cipher spec protocol and the alert protocol, are used during SSL session operation.

With the packet prepared, the client can now send the data to the receiving machine, assuming that an SSL session exists. The process of building a session, authenticating each machine, negotiating an encryption algorithm, and exchanging information used to derive cryptographic keys is the domain of the Handshake protocol (see Figure 4.7).

The current SSL/TLS standard only provides explicit support for HTTP, NNTP, POP3, LDAP, and SMTP application layer protocols (for these protocols, port numbers have been assigned by IANA). SSL is also used by universal access method (UAM) proposed by the WiFi alliance in its WISP roaming recommendation.

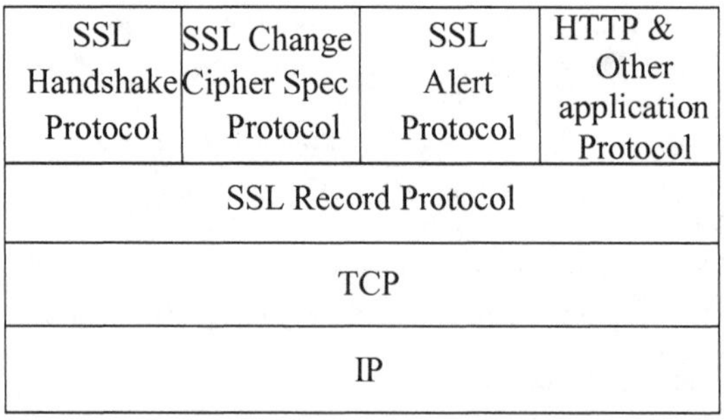

SSL Handshake Protocol	SSL Change Cipher Spec Protocol	SSL Alert Protocol	HTTP & Other application Protocol
SSL Record Protocol			
TCP			
IP			

Figure 4.6 SSL protocol stack.

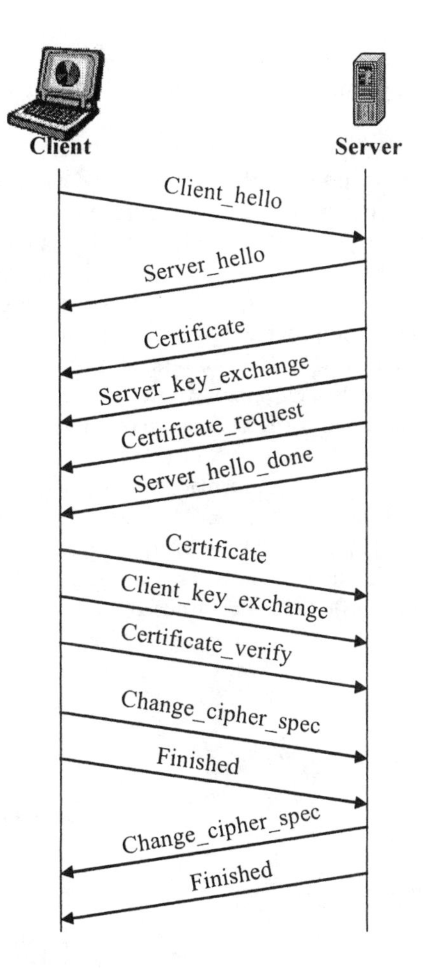

Figure 4.7 SSL/TLS handshake procedure.

4.2.5 Kerberos

Kerberos provides a means of verifying the identities of principals, (e.g., a workstation user or a network server) on an open (unprotected) network [46–49]. This is accomplished without relying on authentication by the host operating system, without basing trust on host addresses, without requiring physical security of all the hosts on the network, and under the assumption that packets traveling along the network can be read, modified, and inserted at will. Many applications use Kerberos' functions only upon the initiation of a stream-based network connection, and assume the absence of any "hijackers" who might subvert such a connection. Such use implicitly trusts the host addresses involved. Kerberos performs authentication under these conditions as a trusted third-party authentication service by using conventional cryptography (i.e., the shared secret key).

Kerberos is a secure protocol and fulfills all the security goals such as confidentiality, authentication, access control, integrity, non-repudiation, availability, scalability, and manageability if implemented and used correctly. As Kerberos is also implemented in Microsoft Windows, implementation is not an issue. In Kerberos v4, the user ID can be spoofed and password can be guessed but this issue is solved in Kerberos v5. The further data encryption system (DES) used for Kerberos v4 is considered broken. The scalability and availability of Kerberos can be a problem because it requires a centralized trusted server, which constitutes a single point of failure. If the centralized trusted server is broken then everything else is broken. Other limitations of Kerberos are discussed in [48].

Kerberos protocols rely on encryption keys, known only to the appropriate parties in a transaction, to protect information sent across an open network. The example in Figure 4.8 shows the sequence that takes place when users first log in to the system and gain access to their files. Only the first exchange requires a user's password; subsequent requests rely on the session key shared by the user and the ticket granting service (TGS). A Kerberos-based solution for WLAN is discussed in [49].

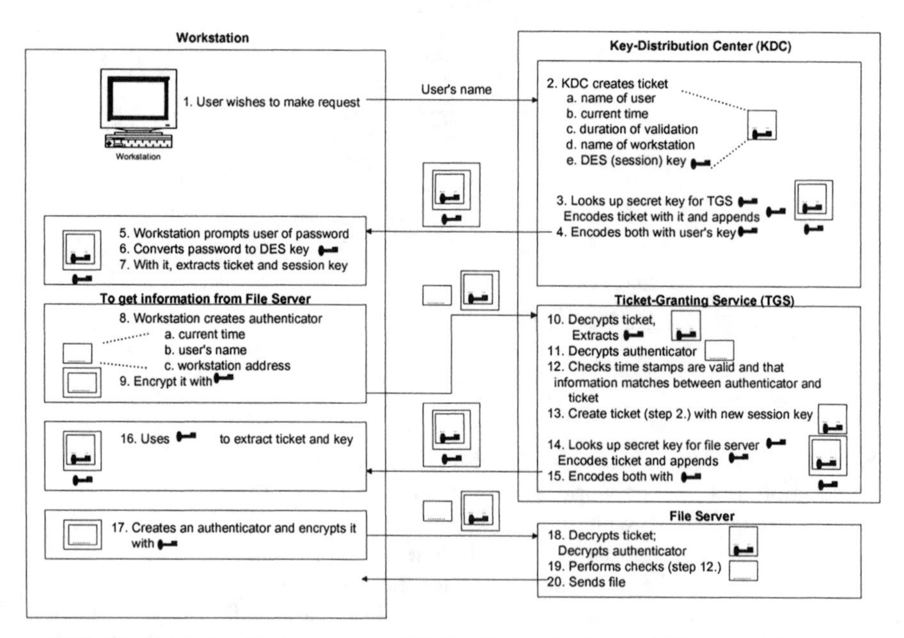

KDC: Key Distribution Center
TGS: Ticket Granting Service
DES: Data Encryption Standard (Kerberos v4.0)

Figure 4.8 Kerberos example.

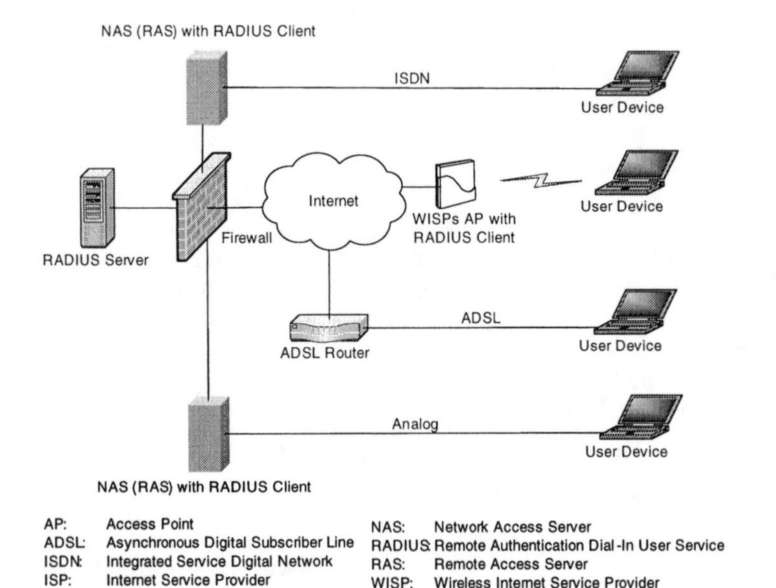

Figure 4.9 RADIUS network configuration.

4.2.6 RADIUS and DIAMETER

Remote Authentication Dial In User Service (RADIUS) is the industry standard protocol for authenticating remote users [49–51]. Today it is widely deployed in remote access servers, routers, and firewalls. RADIUS servers are strategically placed on the network to provide authentication services to all users through a common security protocol. In addition to authenticating and authorizing users, RADIUS enables accounting for the network services. A network configuration of RADIUS is given in Figure 4.9. Key features of RADIUS are listed as follows:

1. Client/server model: A network access server (NAS) operates as a client of RADIUS. The client is responsible for passing user information to designated RADIUS servers, and then acting on the response that returned. RADIUS servers are responsible for receiving the user connection requests, authenticating the user, and then returning all configuration information necessary for the client to deliver the service to the user. A RADIUS server can act as a proxy client to other RADIUS servers or other kinds of authentication servers.
2. Network security: Transactions between the client and RADIUS server are authenticated through the use of a shared secret, which is never sent over the network. In addition, any user passwords are sent encrypted

between the client and RADIUS server to eliminate the possibility that someone snooping on an insecure network could determine a user's password.

DIAMETER [52] can be considered as a next generation RADIUS protocol. It has been developed to address RADIUS flaws in inter-domain roaming support and to provide a much more scalable architecture. Its framework consists of a base protocol and a set of protocol extensions (e.g., end-to-end security, PPP, Mobile IP, and accounting). The base protocol provides all the basic functionalities that must be provided to all the services supported in DIAMETER, while application-specific functionalities are provided through extension mechanisms. The most important difference between DIAMETER and RADIUS is that DIAMETER is based on a peer-to-peer architecture, instead of client/server model. This easily allows service providers to cross-authenticate their users and to support mobility between many different domains.

4.2.7 IEEE 802.1x

IEEE 802.1x or Port-based network access control was designed to provide higher layer authentication mechanisms to layer 2 [53]. Basically IEEE 802.1x has three entities (see Figure 4.10):

- Supplicant: The device desiring to join the network. In our case the IEEE 802.11 station;
- Authenticator: The device that controls the access; in a WLAN network it can be the IEEE 802.11 AP or the access router (AR);
- Authentication Server: This makes the authentication decision, for example, the RADIUS server.

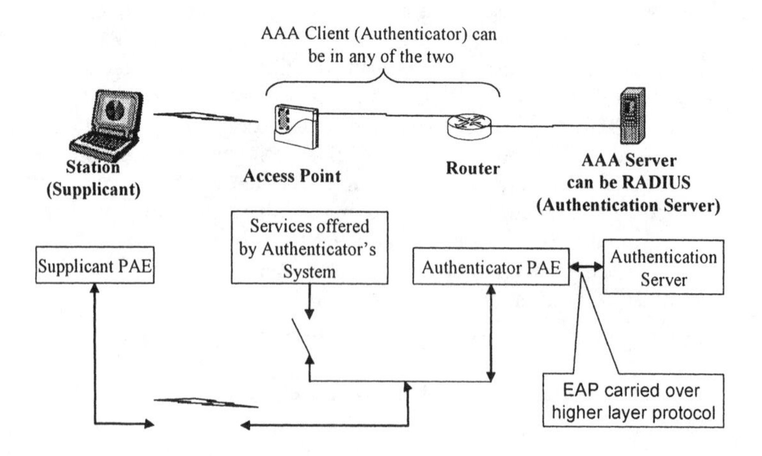

Figure 4.10 Roles of supplicant, authenticator, and authentication server in IEEE 802.1x.

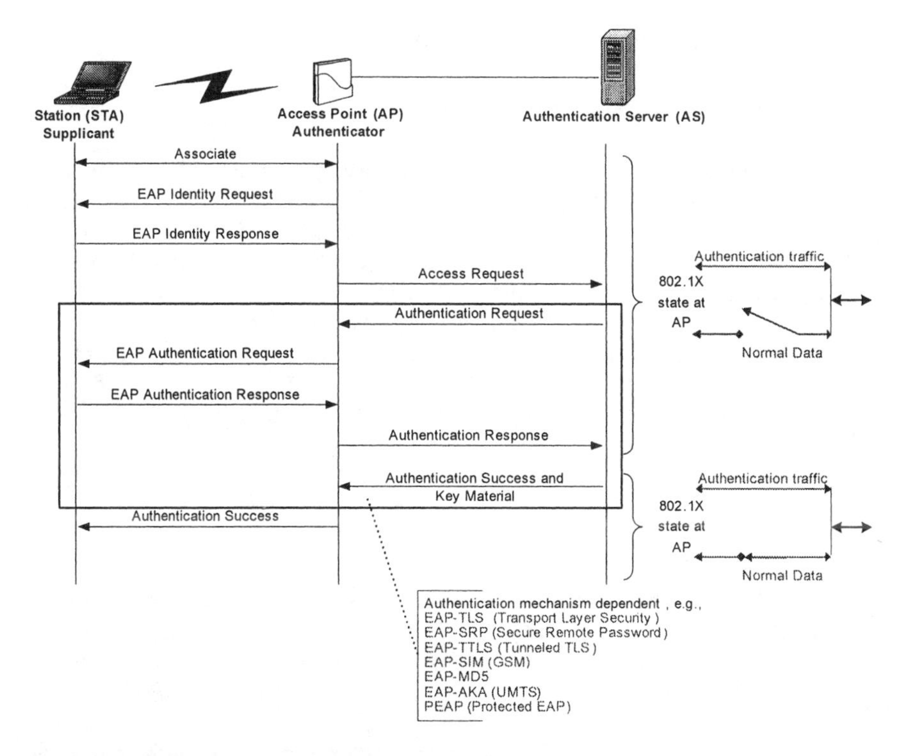

Figure 4.11 IEEE 802.1x message sequence.

The point where the supplicant connects to a network via the authenticator is called the port or port access entity (PAE), thus the designation "port based ...". Basically there are two ports controlled by the authenticator. When a supplicant connects it goes through the authenticator port to the authentication server; once the authentication is successful for the supplicant the service's port is made available. Now the supplicant can access the services through the authenticator.

The protocol that can be used for communication is extensible authentication protocol (EAP). In case the authentication server is at a remote location RADIUS may be used. In any case EAP works between the supplicant and the authenticator. As EAP is part of the point-to-point (PPP) protocol the protocol used is called EAP over LAN (EAPOL) [53]. The EAPOL messages of concern for IEEE 802.11 are listed as follows (the message sequence is shown in Figure 4.11):

- EAPOL-Start: Determines if there is an authenticator. Used by sending this message to a special group multicast to MAC address reserved for 802.1x authenticator. Response is an EAPOL-Identity Request in EAPOL-Packet.
- EAPOL-Key: Authenticator sends encryption keys to the supplicant.
- EAPOL-Packet: A container for transferring EAP messages on LAN.

- EAPOL-Logoff: Disconnection message.

4.2.8 Extensible Authentication Protocol

EAP was designed to solve a major problem, the assignment of an IP address after authentication in an IP network [54]. IPSec and SSL run on an IP layer with knowledge of the IP address. Today EAP has become an important part for WLAN. EAP can be used over layer-2, over IP, or any other higher layer; it was designed as an extension of point-to-point protocol (PPP).

EAP does not provide authentication; it is only a wrapper that gives flexibility in usage of any kind of authentication protocol. Thus an AP does not need to know all the kinds of authentication protocols. Due to a lack of space, a detailed explanation of the discussed EAP protocols is not possible; the authors hope that the message sequence charts (MSCs) help the readers to understand the protocols. The protocol stack of EAP is shown in Figure 4.12.

4.2.8.1 EAP-TLS

The EAP-TLS procedure is basically the SSL/TLS procedure as shown in Figure 4.7 wrapped in an EAPOL packet [55]. Thus, Figure 4.7 combined with Figure 4.11 gives the EAP-TLS message sequence. The only difference is that in the authentication success message from the authentication server the session key is sent to the authenticator (AP in the case of Figure 4.11). In EAP–TLS mutual authentication is achieved by the mandatory certificate used at both the client and server side.

The communication between the authenticator (AP) and the authenticator server (e.g., RADIUS server) can be encrypted by the AP-RADIUS key. On authentication success the message is also encrypted by a master key which is only known by the station (supplicant) and the authenticator server. With this success message the session key is sent by the authentication server to the AP.

The use of a client certificate that is not understood by the end user, lack of user identity protection, and unprotected EAP-success/fail messages are drawbacks or weaknesses of TLS. This has led to development of EAP-tunneled TLS (TTLS) and protected EAP (PEAP). These two are explained below.

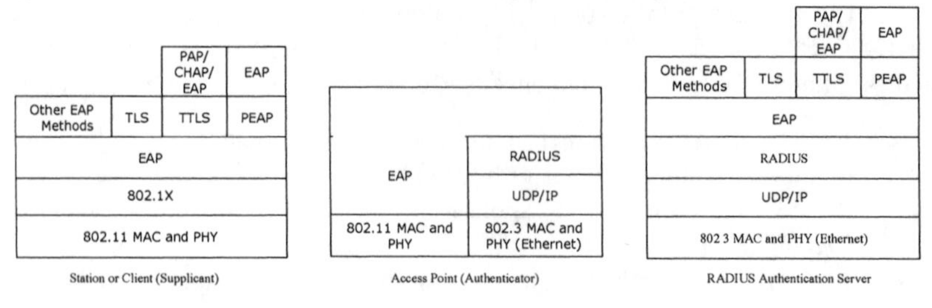

Figure 4.12 EAP protocol stack.

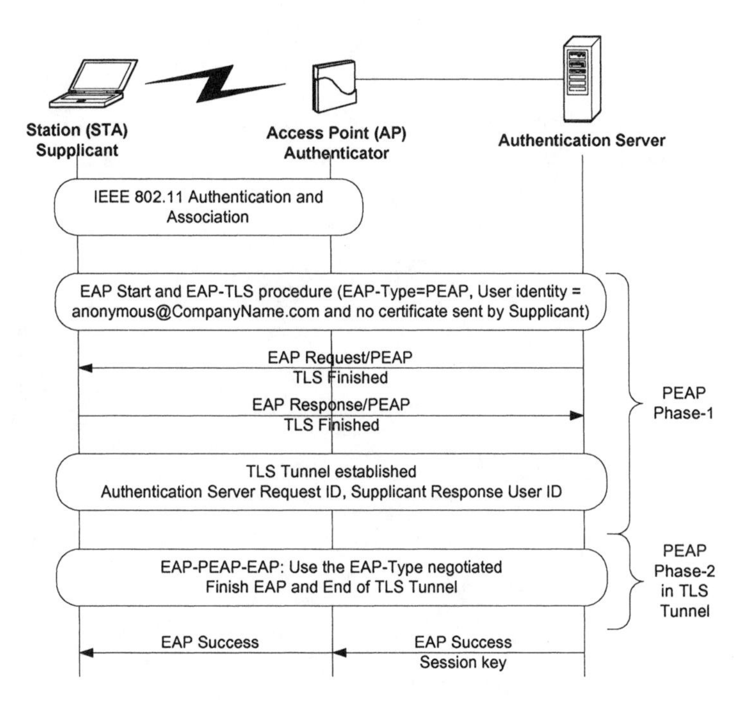

Figure 4.13 Example of PEAP MSC.

4.2.8.2 PEAP

The lack of privacy in EAP-TLS by sending the identity of the user in the open is the issue that PEAP first tries to achieve in phase 1. In the first phase only the server is authenticated. Once privacy is achieved PEAP performs mutual authentication, phase 2. In the following the two phases are discussed. The MSC is shown in Figure 4.13 [56].

In phase 1 the normal TLS is used except that the user does not send a username; instead of that it sends an arbitrary name. Usually this name will contain information to identify the backend authentication server; thus a normal network access identifier (NAI) is used (e.g., anonymous@companyname.com) [57]. The server sends its certificate in this phase but the client does not have to do so or else the privacy issue remains, and the procedure simply becomes the same as EAP-TLS.

After phase 1 the protocol automatically starts phase 2. In phase 2 the protocol restarts with the user identity part. Also in phase 2 any upper layer protocol can be used. Note that the user identity in phase 2 is the real one and is not compared with that in phase 1. In this phase the communication is encrypted by using the keys created during phase 1.

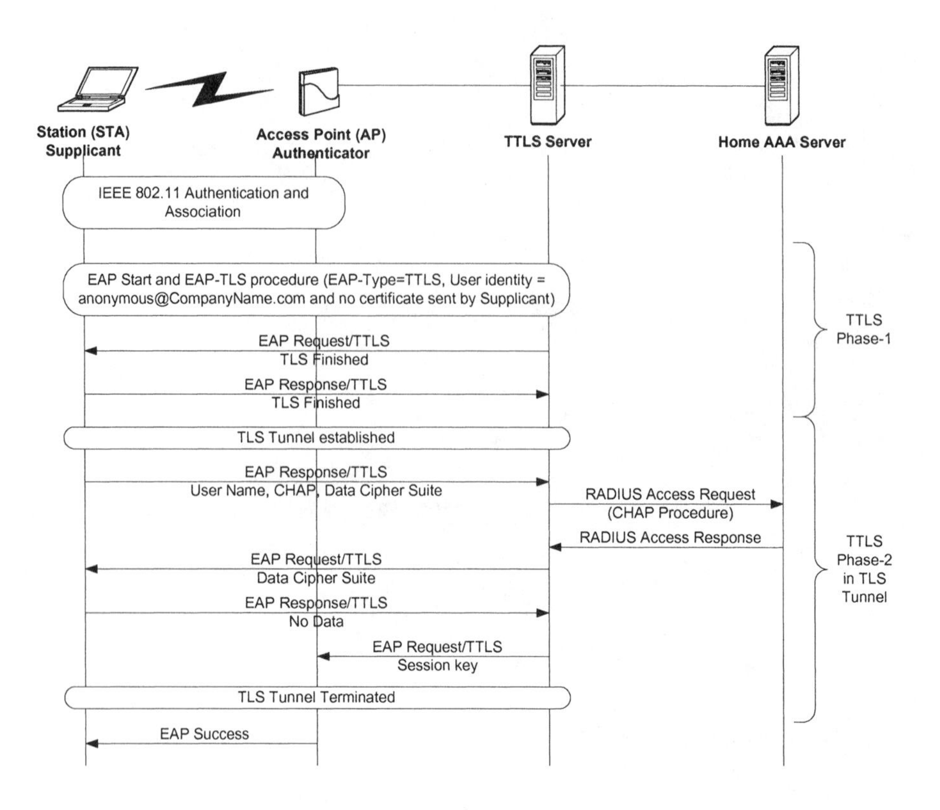

Figure 4.14 Example of EAP-TTLS MSC.

PEAP provides mutual authentication, supplicant identity protection, and key generation.

4.2.8.3 EAP-TTLS

Development of EAP-TTLS started with the thought of leaving the legacy systems untouched and still providing the required level of security. The solution was to introduce a TTLS server that lies at the hotspot network or at an Internet service provider (ISP) while the AAA server is in the home network. Besides that, backward compatibility is achieved by using attribute-value Pairs (AVPs) [58], which are compatible with both RADIUS and Diameter. EAP-TTLS also consists of two phases as explained below. The MSC is given in Figure 4.14.

In the first phase, similar to PEAP, an EAP-TLS secure channel is created. The client side certificate and thus client authentication is optional.

When in phase 2 there is a tunnel that exists until the TTLS server. Separate protection must be provided between the TTLS server and the home AAA server. The client sends the AVPs to the TTLS server, which checks if the sequence of

AVPs includes authentication information and forwards the information to the home AAA server.

TTLS provides mutual authentication, supplicant identity protection, key generation, and data cipher suite negotiation.

4.2.8.4 EAP-FAST

EAP-flexible authentication via secure tunneling (FAST) is a relatively new proposal in IETF. The basic idea of this protocol is to avoid the usage of certificates. EAP-FAST tunnel establishment relies on a protected access credential (PAC) that can be managed dynamically by EAP-FAST through the AAA server. This method also has two phases with an optional phase 0 [59].

The optional phase 0 is used infrequently. In this phase the user credential is securely generated between the user and the network. This credential, known as PAC, is used in phase 1.

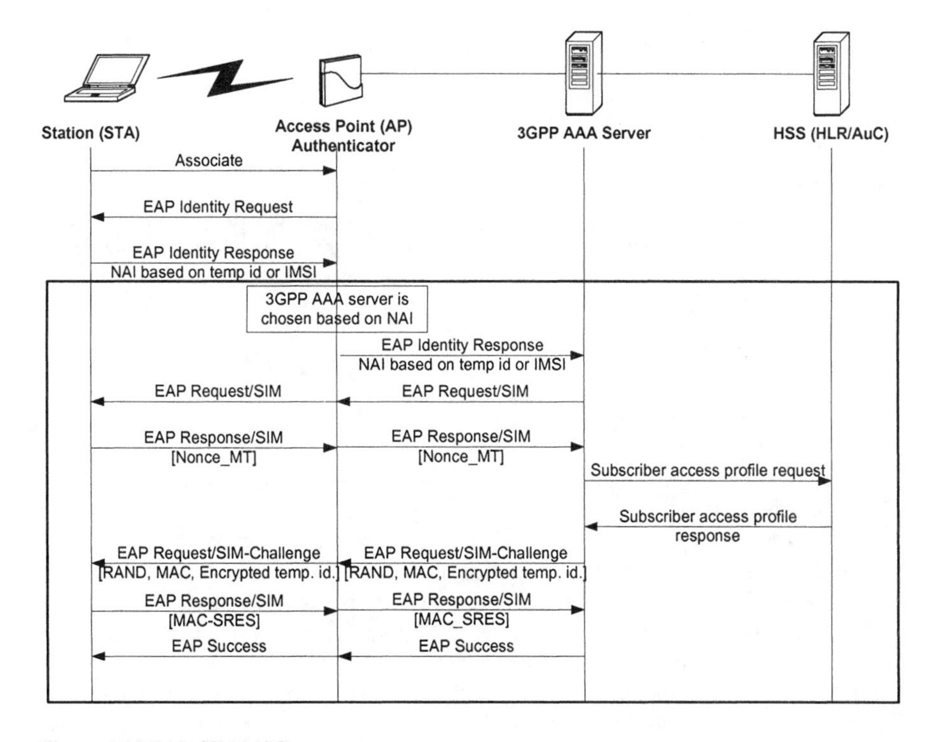

Figure 4.15 EAP-SIM MSC.

The phase 1 establishes a tunnel between the station and the AAA server. PAC is used for authentication purposes.

The tunnel created in phase 1 is used in phase 2 to securely perform client authentication. The user sends the username and password in this phase.

EAP-FAST is said to provide protection against MitM, weak IV attack, replay attack, and dictionary attacks.

4.2.8.5 EAP-SIM

A number of mobile operators are already providing WLAN access. The method for them to do so is to reuse their current infrastructure. The current infrastructure of operators makes use of subscriber identity module (SIM)–based authentication [60]. The MSC is given in Figure 4.15.

There are certain security issues related to the EAP-SIM method, the first being that an attack is possible in the public network area that is between the station and the authenticator and the authenticator and the 3GPP AAA server. The other two issues are related to GSM; there is no mutual authentication possible and the GSM encryption method has been cracked.

4.2.8.6 EAP-AKA

The authentication and key agreement (AKA) protocol is used by the third generation (3G) standard developed by the 3G partnership project (3GPP). It is based on symmetric keys and runs on universal mobile telecommunications systems (UMTS) subscriber identity module (SIM) or USIM. EAP-AKA was developed so that WLAN users could be authenticated by a 3GPP network. The MSC is given in Figure 4.16 [61].

4.2.8.7 Other EAP Methods

There are other EAP methods that are not discussed above [62]. These are discussed briefly below.

EAP-message digest 5 (MD5) provides only user authentication by using user ID and password. It is vulnerable to dictionary attacks and MitM.

Another EAP method is EAP-secure remote password (SRP). It uses the Diffie-Hellman method to authenticate both sides. The methods provides mutual authentication and uses user ID and password.

EAP-SecureID uses one time password (OTP) so as to authenticate the client. There is no authentication of the server in this method; the proposal is to use some sort of tunneling. Some issues can occur in this method, the main one being MitM.

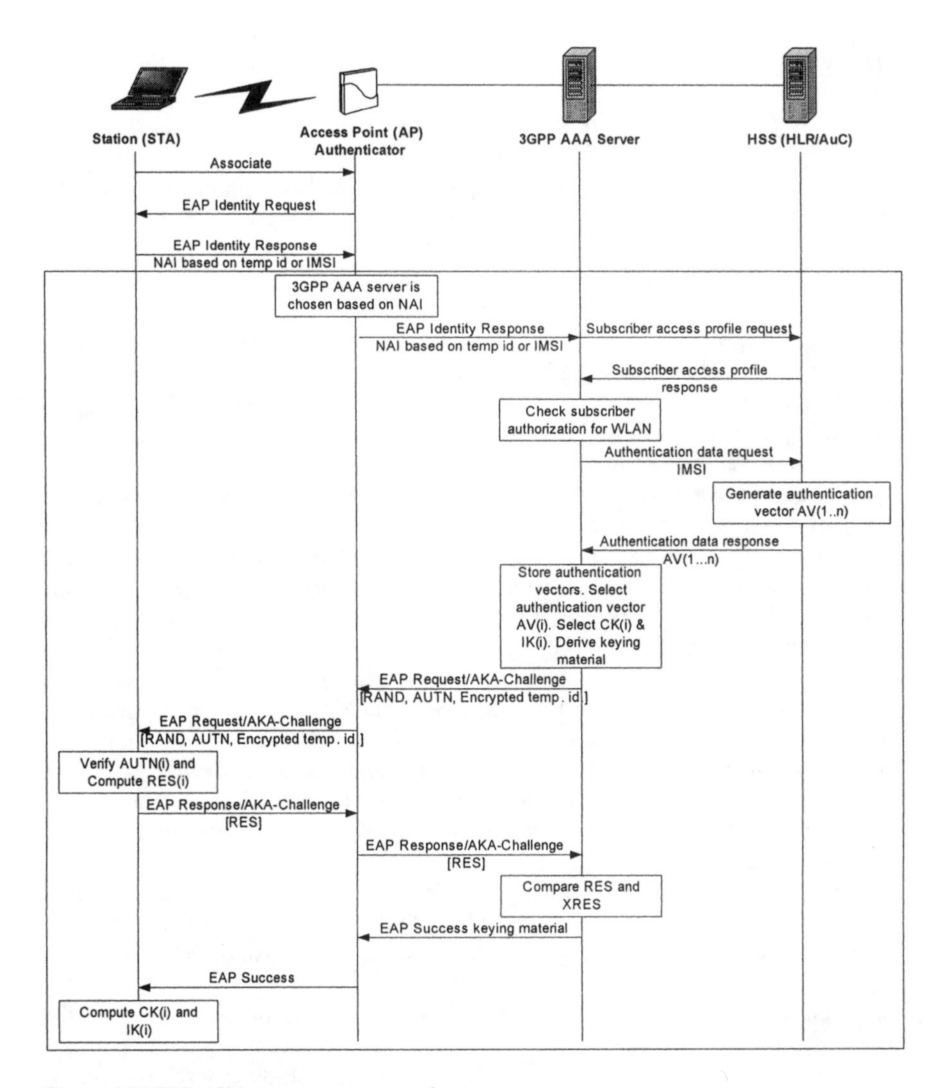

Figure 4.16 EAP-AKA message sequence chart.

4.3 IEEE 802.11 SECURITY ISSUES

Although the IEEE 802.11 standard–based WLAN has evolved into a major wireless technology, its growth is being hampered by security flaws. In this section the security issues in the current IEEE 802.11 standard are discussed; the security solution itself is discussed in Section 3.4. The authors have tried to present the security issues from the point of view of security goals so as to map them with the discussion in Section 4.2. Availability and non-repudiation are not considered because availability is more related to the network and not to the protocol while non-repudiation is related to the service. Note that there are several security issues concerning IEEE 802.11 that appear every day; thus the list might not be complete. The WEP algorithm has been shown to have several weaknesses [1, 5–15].

4.3.1 Authentication

IEEE 802.11 provides two types of authentication mechanisms (see Section 3.4.11); these are open system and shared key–based. The first method is obviously a "non-security" solution; everyone is let in. The second method is dependent on "shared key"; that is, each station of the network has the same key and so does the AP thus the AP authenticates the station.

There are flaws in key generation and key management, discussed in a later section that affect the authentication. Note that this is a very basic security architecture flaw; one should separate the authentication and encryption (confidentiality) keys. The flaw from the authentication point of view is that the shared key is used for a very long time as renewing the key in each station means a lot of overhead for the IT staff [1].

Further the use of challenge and response makes it easier for an attacker to break in. As discussed in Section 3.4.12, the data is XORed with the key sequence to produce an encrypted message. This means the attacker just has to capture the challenge or plaintext and the encrypted response and then XOR the two giving the key sequence. Now the attacker can simply request authentication, encrypt the challenge with the key sequence, and send it to the AP. The AP will simply decrypt and authenticate the station [1].

Another authentication issue is that it is usually not per station–based, although key per MAC address is possible; neither is the key different per AP. Once again this issue is related not only to the standard but also to the IT overhead [1].

Even if one could provide the station–based authentication, it remains at the air interface level (i.e., the AP and the station). The point is to provide per user authentication, which is not possible with the current IEEE 802.11 solution and is understandable because it is a layer-2 solution. Thus it is important to integrate a higher layer authentication mechanism [1].

An important issue is related to mutual authentication. IEEE 802.11 does not provide mutual authentication; thus an AP can verify a station but not the other way round [1]. This is an important issue because the station can easily attach itself to a rogue AP or network.

With the shared key mechanism, let us not forget the issue of stolen stations and time taken to know that a station is stolen to informing the IT and renewing the secret key [1].

4.3.2 Confidentiality

Usually one would say confidentiality, which is provided by encryption, is dependent on authentication (i.e., once authentication is successful confidentiality can be provided). This does not mean that the keys should be the same for confidentiality and authentication. For a WLAN network knowing that authentication has its share of weakness one could simply skip the authentication of station at the air interface level and use confidentiality to provide user authentication at the network level. Thus one is talking about confidentiality and then authentication.

As confidentiality is provided by encryption let us now look at the issues related to key generation and key management as alluded to in the previous section.

4.3.2.1 Key Management

There is no real key management in WEP, but two methods of using WEP keys are provided [1]. The AP and the stations share the usage of the four (default) keys. The compromising of each of the nodes means a compromise of the wireless network. A key mappings table is used at the AP. In this method, each unique MAC address can have a separate key. The size of a key mappings table should be at least 10 entries according to the 802.11 specification; however, it is likely chip-set dependent. The use of a separate key for each user mitigates the known cryptographic attacks but requires more efforts on the manual key management. Since key distribution is not defined in the WEP and can be done only manually, many of the organizations deploying wireless networks use either a permanent fixed cryptographic variable, or key, or no encryption at all.

4.3.2.2 Small IV Space and IV Renewal [1, 4, 5, 7, 11]

The IEEE 802.11 standard specifies the use of 24 bits for initialization vector (IV). The IV and the secret key are used as initialization vector for the pseudorandom number generator (PRNG), which is in turn used to generate a key sequence and encrypt the message. If the IV was not used then the seed for the PRNG would be the same until the secret key was changed thus leading to the same key sequence all the time. But if the IV is not changed often one gets the same key sequence anyway; thus this does not really solve the problem. Many vendors initially did not change the IV. It is recommended to change the IV every frame until all the IVs are used (2^{24}). Choosing a random value is not recommended because it leads to about a 50% probability of IV collision.

So the point is that if the attacker knows the plaintext and the encrypted text then the key sequence can be known by simply XORing the two. An attacker just needs to know the plaintext; the encrypted data is available to everyone as it is sent in the wireless. For an earnest attacker it is not hard to find the plaintext; one might even use an active attack. Other issues also exist; within an encrypted frame there are several known fields from such areas as the IP layer and IP header; thus the attacker can find the partial key sequence and on further analysis maybe can find the whole sequence.

4.3.2.3 RC4 Keys' Weakness

RC4 PRNG contains an array (8*8 S-box = 2^8 rows of 8 bits each) of 256 bytes. The entries of these are permutations of 0 through 255. The S-box is first initialized with another 256–byte array containing the key [8]. Now with the key as input, the random number key sequence is generated and used for encryption. There are certain weak IVs that do not rearrange the array or barely rearranges it.

In IEEE 802 networks the first byte is known (0xAA, i.e., the LLC header) thus the pseudorandom bytes can be easily recovered. With packets using weak IVs it is possible to guess the true key byte with 5% accuracy. According to [9] based on [8] it is possible to recover the correct key byte from 256 packets. One way to fight back is to implement WEP, which filters the weak IVs.

4.3.3 Integrity

Integrity in 802.11 is provided by using the 32-bit cyclic redundancy check (CRC) known as integrity check value (ICV), which is a linear function of the message. Further in WEP the original message is encrypted as is (i.e., the message is not mixed so location of the original bits remains the same). Due to linear function it is possible to change a bit in the message part that will have a known effect in the CRC part. So one can flip one bit in the message and on flipping some bits (this can be known) in the CRC part one will get a correct frame [4].

An attacker can use this behavior of CRC to get the whole set of bit sequences. The basic idea of such an attack is that certain fields are known and thus their key sequence can be easily found (XOR the encrypted with the known value). After this the attacker only needs to send some information to the network that will cause a known response (e.g., a ping). This ping can be with the known field and a CRC that is computed until a response comes. The method is repeated by incrementing the frame size until the full key sequence is known. This is then repeated for the next IV thus producing a 25 GB [2^{24} possible IVs and 1,500 bytes of key sequence, (i.e., IP packet size in Ethernet)] of key sequence dictionary; now the attacker can decrypt any data.

4.3.4 Access Control

Let us look at this goal in two parts, the first being the access control itself and the second being authorization.

4.3.4.1 Access Control

Although IEEE 802.11 does not define access control, it can be provided by using MAC address. The AP can have a list [access control list (ACL)] of MAC addresses that are allowed to access the network. It is relatively easy to forge MAC addresses so this is not really a secure solution; still it is used by several organizations.

4.3.4.2 Replay Attack (Authorization)

IEEE 802.11 does not provide any form of replay protection, which also helps in other security attacks. Any intercepted message can be simply resent. There is no concept of timestamp and packet numbering in the security mechanism. This lack of authentication can permit attackers to launch potential man-in-the-middle (MitM) attacks or DoS attacks.

4.3.5 Other Issues

In this section some other security issues related to IEEE 802.11 are discussed.

4.3.5.1 Password Protection

Most APs being sold today have a password to access the management functionalities. Very often it is noted that these passwords are not changed due to ease of use. Thus an attacker can simply try to use the default product password of various well known vendors and modify the settings of the AP for its benefit. Included in this is the issue of the factory default state or reset method in APs. These problems are not only for WLANs but also for network elements in mobile networks or in fact any system using default passwords.

4.3.5.2 Location of APs

Location of APs should be such that they are not stolen easily, or else there should be an alarm on the network or the device. A stolen AP can be used as well as a stolen station.

4.3.5.3 DoS Attack

A DoS attack can be performed in a wireless medium very simply by creating a lot of interference but let us look at some specific situations. Within IEEE 802.11 there is no integrity protection for management frames; this means that an attacker can simply send a dissociation or deauthentication to the station or the AP—not only that a broadcast MAC address can also be used to disconnect all the STAs connected to the AP.

An attacker can also use the network allocation vector (NAV) field as discussed in Chapter 2. The NAV field is used to indicate other devices in the network the period during which a device will transmit so that other devices do not try to access the channel in that period. Setting a long NAV time will do the trick.

4.3.5.4 MitM Attack

MitM attack is also possible using some of the thoughts presented in the previous section. The attacker can simply disconnect a user and act as a fake AP. In this case the station will try to connect to the fake AP (lack of mutual authentication); the attacker will then simply pass the connection information to the AP and thus will be in a situation to do several things including modification of frames and accessing the network.

It is also possible to give fake a response to an address resolution protocol (ARP) request and masquerade as a user. Thus the traffic of a user will go to the attacker.

4.3.5.5 Dynamic Host Configuration Protocol

The Dynamic Host Configuration Protocol (DHCP) provides an automatic IP address to the stations from an address pool. There is usually no DHCP authentication; thus a stolen device or cracked network will give direct access to the network.

4.3.5.6 Management

The Simple Network Management Protocol (SNMP) is often used for network management and monitoring. It should be noted that SNMPv1 and SNMPv2 do not have much security support and this makes a path open for the attacker. It is normally advised to use SNMPv3 with better security support.

4.3.6 Tools

There are several tools available on the Internet today that can be used for attack purposes but there is also a positive side. These tools can also be used to find security issues in the WLAN network and even sometimes for network planning purposes. Several such tools are listed in [16].

One of the most famous tools for war driving is NetStumbler [17]. War driving basically maps the APs that are available in a given region with no security. It also comes as MinioStumbler for PDAs. Another tool that works on Linux is Kismet with all the functionalities of NetStumbler [18].

For just sniffing purposes one could use the tools like Ethereal [19], which works for both Windows and Linux. There are several others, some of which are listed under [16].

Tools for forging MAC addresses are SMAC for Windows and MAC change for Linux [20, 21]. The tools that can be used for cracking WEP are AirSnort and WEPCrack; both are for Linux [22, 23].

If one is not satisfied with just cracking the WEP then there are tools to perform MitM also. The most known one is AirJack [24] but another one for Internet Explorer is SSL MitM attack and sslsniff [25].

4.3.7 Security Issues in Other Solutions

In this section security issues related to IEEE 802.1x, PEAP, and TTLS are discussed briefly.

It was found in [15] that IEEE 802.1x has a major flaw. Basically MitM is possible when IEEE 802.1x is used. The attacker waits until the station is authenticated and then sends a disassociation or deauthentication message to the station. For the AP the session is still alive; thus the attacker can now use the session until reauthentication. The MitM method presented in Section 4.3.5.4 can be used by the attacker too.

In the case of PEAP or TTLS there is the possibility of the attacker masquerading as a station to the AP and as an AP to the station. The attacker creates a tunnel with the AP and the station; anyway the first phase is simple as anyone can get in as an anonymous person. Next the attacker simply passes the station's responses to the AP and on achieving the connection with the AP simply disconnects the station.

4.4 COUNTERMEASURES

There are several countermeasures for the security issues in WLANs. In this section let us look at them briefly. The following section will give further information on a few higher layer protocols that can be used.

Some of the methods to provide countermeasure are using personal firewall, intrusion detection systems (IDS), virtual private networks (VPNs), public key infrastructure (PKI), and possibly biometrics. Note that these are in addition to the correct methods of configuring APs and WLAN networks.

4.4.1 Personal Firewalls

It is advisable to install personal firewalls on user devices whether they are from a company IT department or personal. Most of the firewalls, like those from Norton or OutPost, are very simple to install. With firewalls one can set a different level of security. In general a setting should be such that only the traffic that is needed is allowed to enter and only required data are allowed to leave the system, but this can have some problems. Some applications need to access the network automatically; thus pop-ups might appear often asking permission. Another point is that certain protocols need to make a connection to a user device so as to function [e.g., file transfer protocol (FTP) server].

4.4.2 Biometrics

Biometric is a method of identifying a user based on users' physical information, for example, fingerprints or reading the retina, etc. [63]. Although biometrics is an option it should be noted that it is not a perfect science. Human physical condition changes with time; further, solutions are also not yet available that can work independently. It is better to use biometrics, in its current stage, with another security solution (e.g., biometrics can combine with VPN solutions).

4.4.3 Virtual Private Networks

A VPN connects the components and resources of one network over another network by allowing the user to tunnel through the Internet or another public network, giving the participants the same security and features as those available in private networks [64]. VPNs allow telecommuters, remote employees, or even branch offices to connect in a secure fashion to a corporate server using the routing infrastructure provided by a public internetwork (such as the Internet). From the user's perspective, the VPN is a point-to-point connection between the user's computer and a corporate server (see Figure 4.17).

The secure connection across the internetwork appears to the user as a private network communication—despite the fact that this communication occurs over a public internetwork—hence the name.

Some of the common uses of VPN are listed as follows:

- Remote user access over the Internet: VPNs provide remote access to corporate resources over the public Internet, while maintaining privacy of information.
- Connecting networks over the Internet: The VPN software uses the connection to the local ISP to create a virtual private network between the branch office router and the corporate hub router across the Internet.
- Connecting computers over the Internet: VPNs allow the department's LAN to be physically connected to the corporate internetwork but separated by a VPN server. The network administrator can ensure that only those users on the corporate internetwork who have appropriate credentials can establish a connection with the VPN server and gain access to the protected resources of the department. All communication across the VPN can be encrypted for data confidentiality.

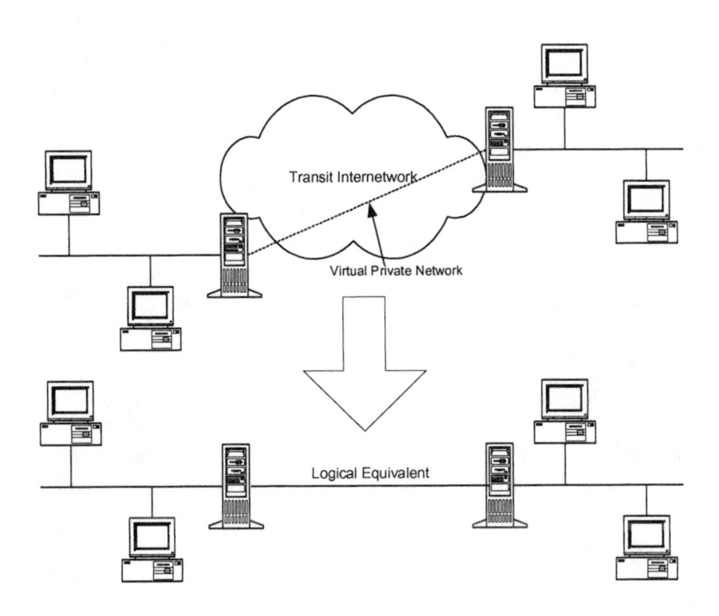

Figure 4.17 VPN and its logical equivalent.

The requirements for VPN are the following:

- User authentication: The solution must verify a user's identity and restrict VPN access to authorized users. In addition, the solution must provide audit and accounting records to show who accessed what information and when.
- Address management: The solution must assign a client's address on the private net, and must ensure that private addresses are kept private.
- Data encryption: Data carried on the public network must be rendered unreadable to unauthorized clients on the network.
- Key management: The solution must generate and refresh encryption keys for the client and server.

VPN fulfills the requirements by making use of tunneling. Tunneling is a method of using an internetwork infrastructure to transfer data from one network over another network (see Figure 4.2). The data to be transferred (or payload) can be the frames (or packets) of another protocol. Instead of sending a frame as it is produced by the originating node, the tunneling protocol encapsulates the frame in an additional header. The additional header provides routing information so that the encapsulated payload can traverse the intermediate internetwork. The encapsulated packets are then routed between tunnel endpoints over the internetwork. The logical path through which the encapsulated packets travel through the internetwork is called a tunnel. Once the encapsulated frames reach their destination on the internetwork, the frame is unencapsulated and forwarded to its final destination. Tunneling includes encapsulation, transmission, and unencapsulation of packets.

The VPN tunnel can provide a secure connection to users of WLANs. Today VPN is often used by corporate users. The device does not need any special hardware; a VPN software is enough. User authentication to the VPN gateway can occur using RADIUS or onetime passwords (OTPs). The VPN gateway may or may not have an integral firewall to restrict traffic to certain locations within the enterprise network. Today, most VPN devices have integrated firewalls that work together to protect both the network from unauthorized access and the user data going over the network. The VPN gateway may or may not have the ability to create an audit journal of all activities. An audit trail is a chronological record of system activities that is sufficient to enable the reconstruction and examination of the sequence of environments and activities. A security manager may be able to use an audit trail on the VPN gateway to monitor compliance with a security policy and to gain an understanding of whether only authorized persons have gained access to the wireless network.

It should be noted that issues like authentication and authorization to enterprise applications are not addressed with VPN. Some VPN devices can use user-specific policies to require authentication before accessing enterprise applications.

There are several VPN solutions available on the market although IP security (IPSec) is the most popular one. Protocol choices for VPN include the following:

- HTTPS: secure HTTP (SSL/TLS);
- SSH: secure shell;
- SSL/TLS: secure socket layer/transport layer security;
- IPSec: IP security;
- L2TP: layer 2 tunneling protocol;
- PPTP: point to point tunneling protocol.

VPN organizations and product information can be found at [64], the VPN Consortium (VPNC).

4.4.4 Public Key Infrastructure

PKI basically means certificates, a commonly known term. PKI provides the framework and services for the generation, production, distribution, control, and accounting of public key certificates. It provides applications with secure encryption and authentication of network transactions as well as data integrity and non-repudiation, using public key certificates [65]. Both user and device authentication can be provided using this technique, together with signing and encrypting messages. All these benefits come with complexity and the cost of implementing and administering a PKI. On the other hand PKI has grown a lot in recent times and has come as far as becoming integrated in enterprise networks, thus spurring possible growth in WLAN networks.

WLANs can integrate PKI for authentication and secure network transactions. Third-party manufacturers, for instance, provide wireless PKI, handsets, and smart cards that integrate with WLANs. Smart cards provide even greater utility since the certificates are integrated into the card. Smart cards serve both as a token and a secure (tamper-resistant) means for storing cryptographic credentials.

CableLabs has paved the path towards developing a PKI solution for a complete industry (http://www.cablelabs.com/). All manufacturers and thus their products have certificates and so there is certification for the Internet service providers (ISPs). This method when used by WLANs can allow user, device, and network authentication along with the other benefits of using certificates/PKI.

4.4.5 Intrusion Detection System

The intrusion detection system (IDS) can be used as a host-based or network-based system or as a hybrid of the two systems. A host-based agent is installed in individual systems while a network-based IDS monitors the network traffic. The IDS solutions are wired specific and need to be modified to work in wireless medium. Some of the enhancements required are [66] determination of location of devices, detecting a rogue AP, and detection of attacks at the wireless medium between stations.

4.5 WPA AND IEEE 802.11I RSN

While the security issues of IEEE 802.11 were widespread and the standard was still working on security enhancements or IEEE 802.11i, the WiFi Alliance came up with an intermediary solution known as WiFi protected access (WPA). For most existing products a firmware (FW) upgrade can be used to provide WPA but IEEE 802.11i requires a hardware upgrade. WPA makes use of temporal key integrity protocol (TKIP) and it is an optional mode in IEEE 802.11i [1, 3, 67–70].

IEEE 802.11i defines a robust security network (RSN) that requires a number of capabilities at the station and the AP [70]. The standard also defines a transitional security network (TSN) in which both WEP and RSN can work. Unlike WPA, RSN makes use of an advanced encryption standard (AES).

The biggest differences between the original IEEE 802.11 security and WPA and RSN are the possibility of using higher layer protocols for authentication and the possibility of key exchange. Both of them make use of IEEE 802.1x, EAP, and other methods explained in Sections 4.2.5–4.2.8.

A variety of keys are used in both WPA and RSN; this is first explained in this section. After the hierarchy is clarified the TKIP and AES methods are explained. However, before all that the benefits of WPA and RSN are given.

4.5.1 IEEE 802.11i Services

IEEE 802.11i provides the following services:

1. Association and reassociation: An IEEE 802.1x port maps to an association. Once the STA is authenticated by a higher layer protocol and authorized to access the network the AP allows data traffic for the STA.
2. Access control: This is provided by use of IEEE 802.1x and higher layer protocols.
3. Authentication and deauthentication: As explained in Chapter 3, IEEE 802.11 provides link layer authentication using open system and shared key authentication. In IEEE 802.11i authentication is provided by using IEEE 802.1x with EAP or by pre-shared key (PSK). When using IEEE 802.11i the STA must first use open system authentication to perform authentication with the AP (note that shared key authentication is not allowed). After open system authentication the association takes place and then the IEEE 802.1x process starts.

   As authentication (open system) is done before association, a deauthentication will also terminate any association between the AP and the STA. Any authenticated party can send the deauthentication message and this message cannot be refused as it is not a request but a notification.
4. Confidentiality: Confidentiality is provided by using WEP, TKIP, and counter mode with the CBC-MAC protocol (CCMP). CCMP makes use of AES and the rest use RC4. The default confidentiality state is to send data in clear (i.e., not use any protection).
5. Key Management: All the new services in IEEE 802.11 require key management which uses 4-way handshake and group key handshake mechanisms to provide fresh keys.
6. Data origin authenticity: Data origin authenticity is provided by TKIP and CCMP and is applicable to unicast frames only.
7. Replay protection: Again, TKIP and CCMP provide replay protection.

4.5.2 RSN Information Elements

The RSN information element (IE) is sent in the beacon and probe response from the AP. The IE contains security information like the kind of authentication supported, the cipher suite supported, and the key exchange mechanism supported. This is to be negotiated between the two parties.

On receiving the IE from the AP the STA also sends an IE, which contains only a single choice of the different types of algorithms. This choice made by STA is used for further communication.

The IE is also sent during 4-way handshake so as to prevent bidding attack. The contents of IE can be changed during the handshake.

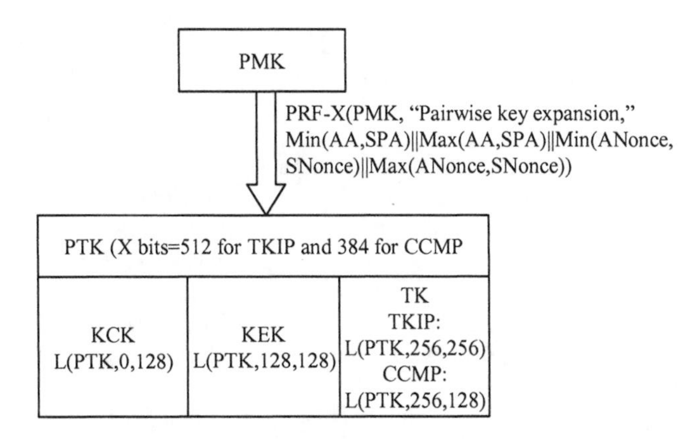

Figure 4.18 Pairwise key hierarchy (reprinted with permission from IEEE 802.11i-2004).

4.5.3 Key Hierarchy

IEEE 802.11i defines a pairwise key hierarchy for unicast traffic and a group key hierarchy for multicast and broadcast traffic. The basic element in the hierarchy is the pairwise master key (PMK). In the following a short discussion is given on PMK after which the two hierarchies are discussed.

4.5.3.1 PMK

PMK is the top of the key hierarchy. A PSK can also be used as a PMK. In case PSK is not available, upper layer authentications methods (with the help of IEEE 802.1x and EAP) are used to create PMK, i.e., PMK is created from the AAA key, which is also sometimes known as the master key (MK). The AAA key is jointly negotiated between the STA and the AS. This key information is transported via a secure channel from the AS to the authenticator. The PMK is computed as the first 256 bits of the AAA key.

Once PMK is created between the STA and the AS by using IEEE 802.1x, it has to be transferred to the AP. IEEE 802.11i does not define any secure method for doing so but it is defined in WPA.

4.5.3.2 Pairwise Key Hierarchy

Before starting a discussion of the key hierarchy, there are two functions used by IEEE 802.11i. These are the following:

- L(Str,F,L): from Str starting from left to right, extract bits F through F+L−1.

- PRF-n: pseudorandom function (PRF) producing n bits of output. PRF is a function that hashes various inputs to derive a pseudorandom value (the key).

The PMK is used to create pairwise transient keys (PTKs). Transient keys are used for confidentiality algorithms and their maximum lifetime is PMK lifetime. PTKs are created for each association. The PTK consists of EAPOL-key confirmation key (KCK), the EAPOL-key encryption key (KEK), and temporal keys (TKs) for TKIP and CCMP. In the following these keys and their use are explained (see Figure 4.18, where the AA is the authenticator address and the SPA is the supplicant address). The nonces are explained in Section 4.5.3.4.

- KCK: It is 128 bits and is used by IEEE 802.1x in a 4-way handshake (see Section 4.5.3.4) for data origin authenticity. One can also call this key the integrity key.
- KEK: This is also 128 bits long and is used by handshake to provide confidentiality.
- TKs for TKIP: TKIP makes use of RC4, which only had the possibility for encryption. Thus TKs in TKIP consist of the integrity and encryption keys of 128 bits each. Bits 0–127 of TKs are input to the TKIP phase 1 and 2 mixing functions (see Section 4.5.8), (i.e., for encryption). Bits 128–191 of TK are used as the Michael key [i.e., integrity key for MAC service data units (MSDUs) from the authenticator (AP) to the supplicant (STA)] while bits 192–255 are used as the Michael key for MSDUs from the STA to the AP. Note that a MSDU is a packet of data between the software and the MAC in contrast to the MAC Protocol Data Units (MPDUs) which are the MAC layer packets. Thus a MPDU can be a portion of the MSDU if the MSDU is bigger than MPDU.
- TKs for CCMP: In case of CCMP both encryption and integrity are incorporated in a single calculation. Thus there is one key of length 128 bits.

KeyID 0 is used when sending a pairwise key.

4.5.3.3 Group Key Hierarchy

A 256-bits group master key (GMK) is created. From this GMK the group transient key (GTK) is created out of which the group in the pairwise connection is established and the GTK is sent to the STA of which the acknowledgment is checked. Similar to pairwise key hierarchy, the TKIP has two keys each of 128 bits for encryption and integrity while CCMP has one key for both purposes. The group key hierarchy is shown in Figure 4.19.

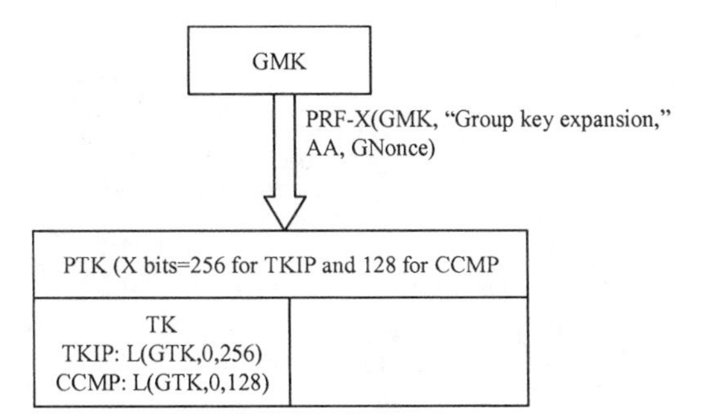

Figure 4.19 Group key hierarchy (reprinted with permission from IEEE 802.11i-2004).

Group keys use the key rotation method. If a given group key is using KeyID1 then the new key is stored at KeyID 2. The new key is used as soon as keys in all STAs are updated.

A STA shall use bits 0–127 of TK as the input to the TKIP phase 1 and 2 mixing functions and bits 128–191 as the Michael key for MSDUs from the AP to the STA. Bits 192–255 of TK are used as the Michael key for MSDUs from the STA to the AP. Bits 0–39 and bits 0–103 are used as WEP-40 and WEP-104 keys, respectively. For CCMP the TK is used as the key.

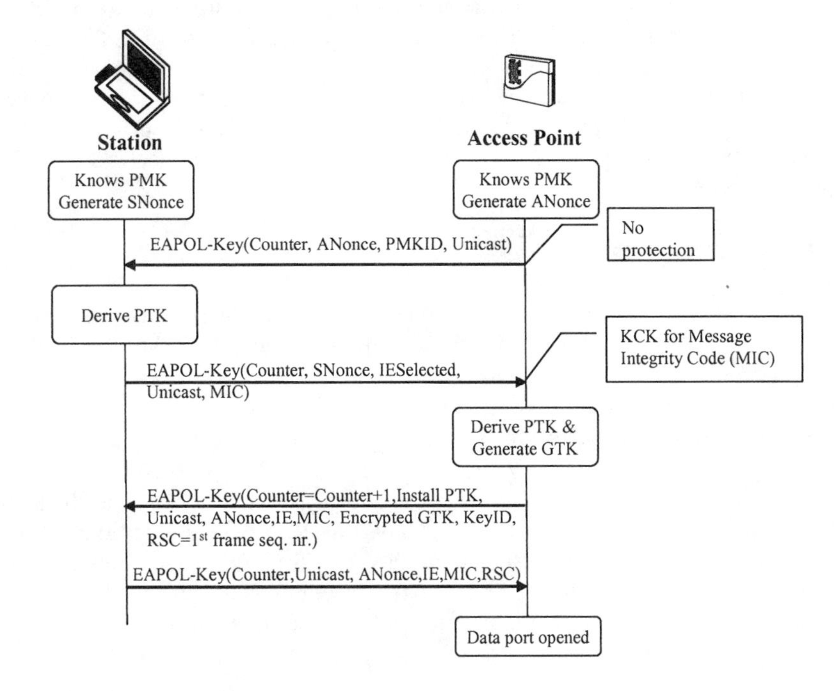

Figure 4.20 4-way handshake.

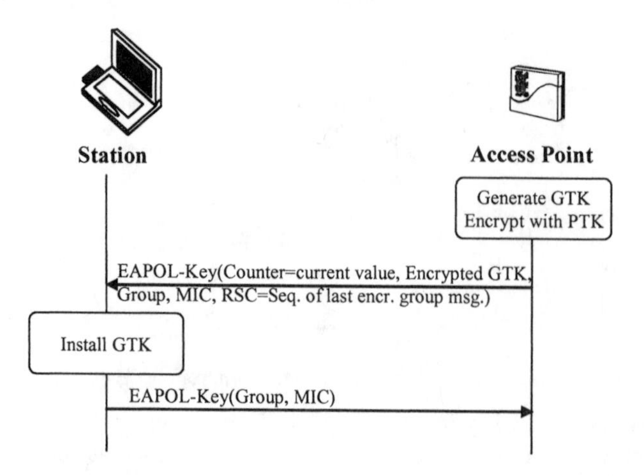

Figure 4.21 2-way handshake.

4.5.3.4 Liveness

As per IEEE 802.11i, liveness is a method to demonstrate that the peer is actually participating in this instance of communication. Thus its purpose is to prevent a replay of the same message in different sessions. Liveness is added to the PRF in the form of a nonce as one of the inputs. A nonce is guaranteed never to be reused.

Each device generates a nonce and sends it to the other device. Both nonces are taken together with the two MAC addresses and the PMK to produce PTKs. The nonce from the authenticator (AP in our case) is called ANonce and from the supplicant (STA) is SNonce.

4.5.4 Handshake Protocols

After the PMK is transferred by the AS to the AP the 4-way handshake takes place to create the PTKs and 2-way handshake takes place to create the GTKs. Both these procedures are shown Figures 4.20 and 4.21.

4.5.5 SAs in RSN Association

Within RSN the STAs, the AP, and the AS create an association known as the RSN association (RSNA) by using IEEE 802.1x. Within RSNA secure communication takes place by using security associations (SAs). A SA is a relation between two communicating ends, which defines the method in which the secure communication will take place; it is stored at both ends and contains an ID. There are four types of SAs defined in RSNA; these are explained below.

The result of a successful IEEE 802.1x authentication leads to a PMKSA between an AP (Authenticator) and a STA derived from EAP authentication and authorization parameters. This SA is bidirectional. The PMKSA is used to create the PTKSA. PMKSAs are cached for up to their lifetimes. The PMKSA consists of the PMKID, which identifies the SA, AP MAC address, PMK (or PSK), lifetime, authentication and key management protocol (AKMP), and authorization parameters specified by AS or by local configuration at the AP.

The PTKSA is a result of the 4-way handshake and is bidirectional. PTKSAs are cached for the life of the PMKSA. There is only one PTKSA with the same supplicant and authenticator MAC addresses. The PTKSA consists of the PTK, pairwise cipher suite selector, STA MAC address, and AP MAC address.

The GTKSA results from a successful 4-way or 2-way handshake, and is unidirectional. A GTKSA is used for encrypting and decrypting broadcast and multicast messages. A GTKSA consists of the direction vector (whether the GTK is used for transmit or receive), group cipher suite selector, GTK, AP MAC address, and authorization parameter specified by local configurations in the AP GTKSA.

The STAKeySA is a result of the STAKey handshake. This security association is unidirectional from the initiator to the peer. There is only one STAKeySA with the same initiator and peer MAC addresses. The STAKeySA consists of the STAKey, pairwise cipher suite selector, initiator MAC address, and the peer MAC address.

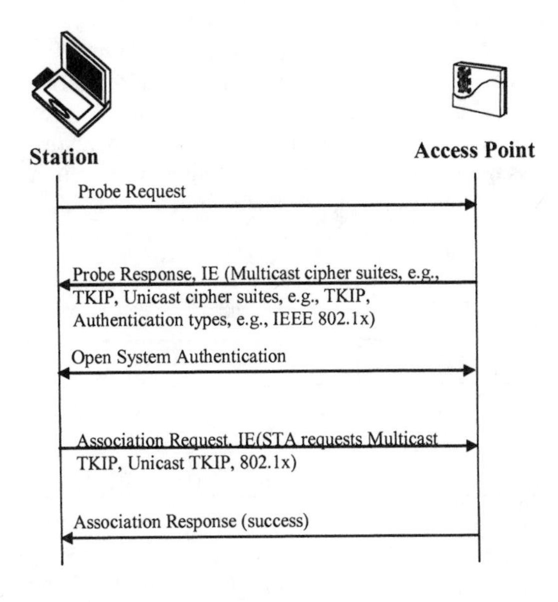

Figure 4.22 Discovery process.

4.5.6 Discovery Process

The most important thing of course is that the STA recognizes the AP and connects to it. This is what we call here the discovery process. Each AP advertises its capabilities in the beacon, and probe response. The detailed process is shown in Figure 4.22. After discovery the STA is ready to perform authentication. Once authentication is done the keys are generated after which the port is opened for data transfer.

4.5.7 Pre-Authentication

In pre-authentication the STA can be authenticated with multiple APs at a time. These APs may or may not be in the radio range of the STA. The result of pre-authentication may be a PMKSA, if the IEEE 802.1X authentication completes successfully. If pre-authentication produces a PMKSA, then, when the Supplicant (STA) associates with the preauthenticated AP, the STA can use the PMKSA with the 4-way handshake. The PMKSA is inserted into the PMKSA cache. If the STA and AP lose synchronization with respect to the PMKSA, the 4-way handshake will fail. Even if a STA has pre-authenticated, it is still possible that it may have to undergo a full IEEE 802.1X authentication, as the AP may have purged its PMKSA due to, for example unavailability of resources or delay in the STA associating.

4.5.8 TKIP

TKIP was developed to provide an intermediary solution until the AES solution was available and to be usable with the existing hardware [1, 4, 68, 69]. Thus TKIP can counter most of the security issues in WEP. The mechanisms that TKIP makes use of are the message integrity check (MIC) value called Michael, the extended IV as TKIP sequence counter (TSC), and encryption using RC4.

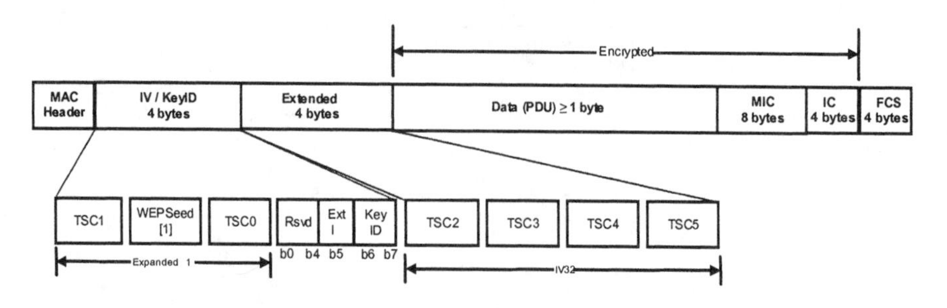

Figure 4.23 The TKIP MPDU (reprinted with permission from IEEE 802.11i-2004).

4.5.8.1 Michael

Michael is a performance-friendly MIC providing 20 bits of security and thus is vulnerable to brute force attack. 20 bits of security means a one out of 2^{20} chance of forgery of MIC. To fight this vulnerability the countermeasures are used. Whenever there is a MIC validation error a rekeying is done with a maximum of one per minute and the network operator is informed of the attack. Michael is calculated on a MSDU in contrast to encryption, which is done at the MPDU level. The MIC is added to one or more MPDUs if MSDU is fragmented. Michael uses a 64-bit key, and partitions packets into 32-bit blocks. Michael then uses shifts, XORs, and additions to process each 32-bit block into two 32-bit registers that will represent the final output, a 64-bit authentication tag. Michael is calculated on the actual data and the source and destination addresses (SA and DA).

4.5.8.2 IV and TSC

In TKIP the IV is extended to 48 bits and is used as a sequence counter (TSC) that starts from 0 and is incremented by 1 for each MPDU. In reality 32 bits are added to the 24 IV bits of WEP but 8 bits (WEPSeed [1]) are not used so as to get rid of weak keys (see Figure 4.23). TSC1 and TSC0 or lower IVs are the sequence number and are used in TKIP phase 2 mixing while TSC2–TSC5 or upper IVs increment by one after lower IVs rotate and are used in phase 1; phases are discussed in Section 4.5.8.3. This solution allows the usage of different keys for each frame while taking care of weak IVs and gives sequence numbering. The solution also allows the use of the original WEP 24 bits IVs.

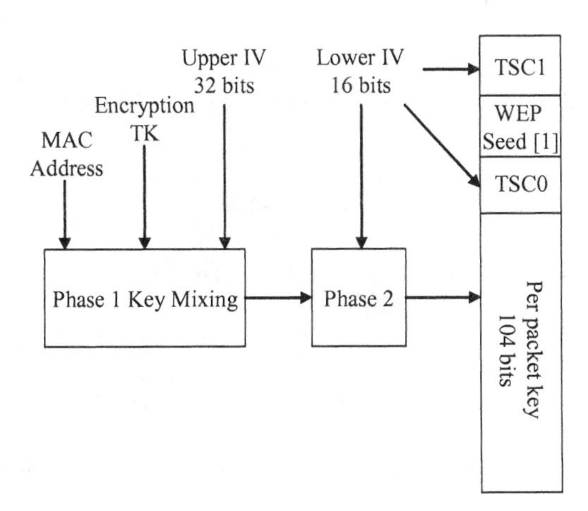

Figure 4.24 Per-packet mixing in TKIP.

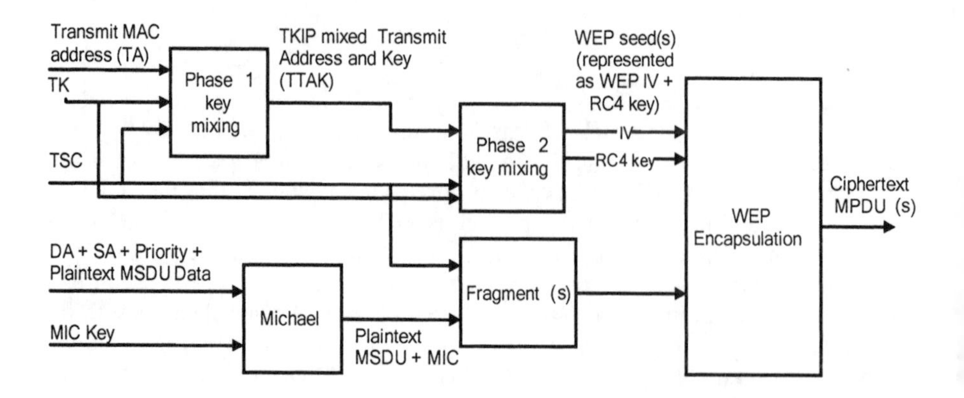

Figure 4.25 TKIP encapsulation procedure (reprinted with permission from IEEE 802.11i-2004).

So as to accommodate burst-acks (send 16 frames in quick succession and allow one ACK for all 16) and retransmissions TKIP uses the concept of windowing. A replay window of 16 is used. This means that any packet with a sequence number one greater than that accepted is also accepted; if the sequence number is smaller than the accepted packet then the packet is rejected while the packet is stored if the sequence number is within the window size.

4.5.8.3 Per-Packet Key Mixing

Unlike WEP, TKIP does not concatenate IV to the key; instead a key mixing procedure is used in two phases. In the first phase the host MAC address, encryption TK, and upper 32 bits of IV are used. The MAC address gives a unique value per host. This value is static for 2^{16} packets.

In phase 2 the lower 16 bits of IV are used together with the phase 1 key to produce a per packet key of 104 bits. The mixing procedure is shown in Figure 4.24.

4.5.8.4 TKIP Encapsulation and Decapsulation

The encapsulation and decapsulation processes in TKIP are given in Figures 4.25 and Figure 4.26, respectively. At the receiver the MPDU is decrypted and passed to a higher layer where the MIC check is performed on the MSDU.

4.5.9 CCMP

CCM merges two known techniques: the counter (CTR) mode for encryption and the cipher block chaining message authentication code (CBC-MAC) for integrity protection. Thus CCM stands for CTR CBC-MAC and CCMP stands for CCM Protocol [69]. It was specially designed for IEEE 802.11i and is proposed to NIST

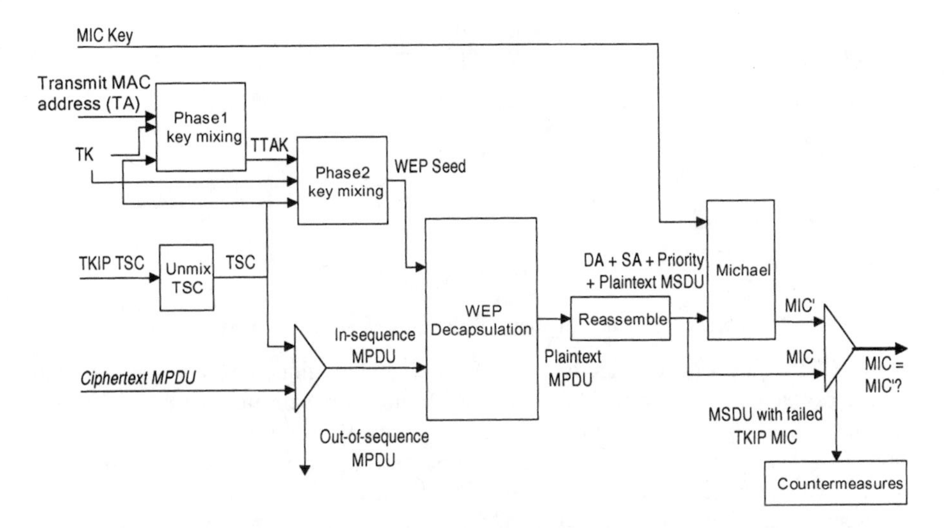

Figure 4.26 TKIP decapsulation procedure (reprinted with permission from IEEE 802.11i-2004).

for consideration as a federal information processing standard. AES was chosen by the IEEE 802.11i committee as the encryption algorithm especially because of its tough selection procedure [70].

CCM was designed to meet the security requirements of IEEE 802.11 [71] and requirements in general set by the wireless medium (e.g., throughput, overhead, etc.). CCM protects the integrity of both the MPDU data field and selected portions of the IEEE 802.11 MPDU header [70].

In CCM there are two parameters of concern, M, which indicates the MIC size and L, which indicates the length of a MPDU in bytes. M = 8 bytes and L = 2 bytes have been chosen for IEEE 802.11i.

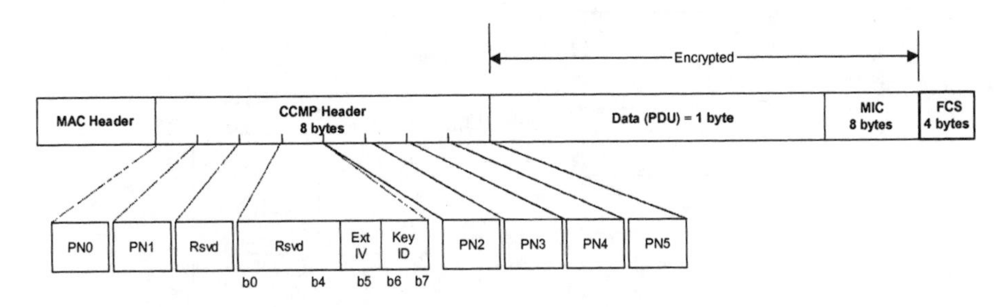

Figure 4.27 CCMP MPDU (reprinted with permission from IEEE 802.11i-2004).

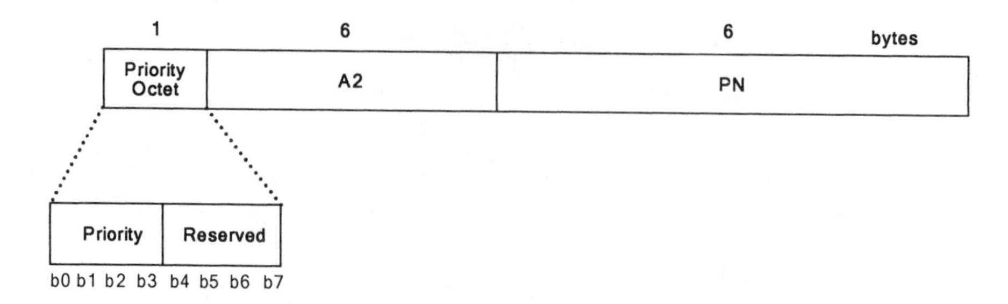

Figure 4.28 The nonce construction (reprinted with permission from IEEE 802.11i-2004).

CCM requires a fresh TK for every session and a unique nonce value for each frame protected by a given TK; CCMP uses a 48-bit packet number (PN) for this purpose. As discussed earlier CCM uses the same key for both confidentiality and integrity. Using the same key for encryption and integrity can be dangerous but CCM avoids the danger by guaranteeing that the space for the counter mode never overlaps with that used by the CBC-MAC initialization vector (i.e., they are independent).

The CCMP MPDU is given in Figure 4.27. 16 bytes are added in CCMP, 8 for MIC, and another 8 for CCMP header. Similar to TKIP, CCMP employs a 48-bit IV, ensuring that the lifetime of the AES key is longer than any possible association. In this way, key management can be confined to the beginning of an association and ignored for its lifetime. CCMP uses a 48-bit IV as a PN to provide replay detection; as in TKIP, this value is incremented for each frame.

As AES is used there is no need for per-packet keys, so CCMP has no per-packet key derivation function. CCMP uses the same AES key to provide encryption and integrity protection for all of the packets in an association. The 8-byte MIC is significantly stronger than Michael and the encrypted ICV is no longer required.

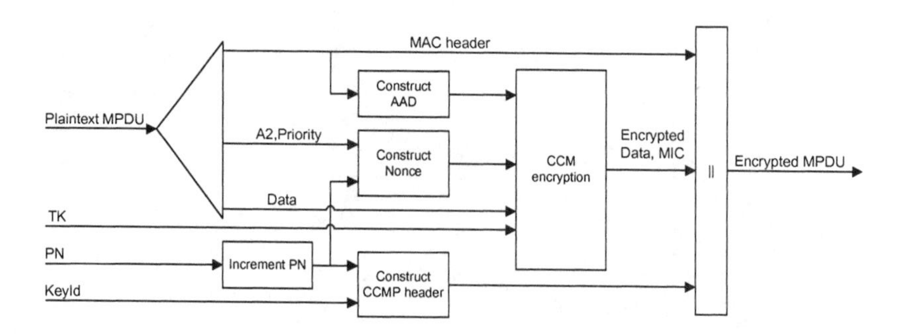

Figure 4.29 CCMP encapsulation (reprinted with permission from IEEE 802.11i-2004).

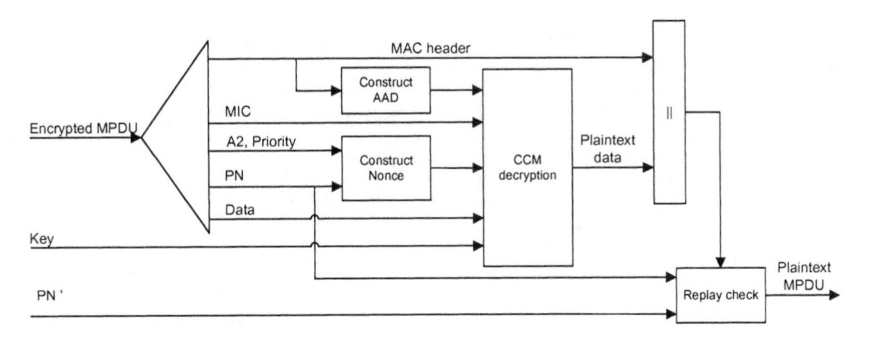

Figure 4.30 CCMP decapsulation (reprinted with permission from IEEE 802.11i-2004).

Integrity in CCM is provided over MPDU, unlike TKIP where it is provided over a MSDU. Further CCM provides integrity over the MPDU header by using additional authentication data (AAD). Within AAD the static MPDU header fields are used the rest is masked to 0.

The nonce is also created from the PN, address 2 (A2) of MPDU (see Figure 3.6) and priority of the MPDU. The nonce construction is shown in Figure 4.28. The encapsulation and decapsulation methods of CCMP are given in Figures 4.29 and 4.30, respectively.

4.5.10 IBSS

IEEE 802.11i security in independent basic service set (IBSS) is discussed briefly below. Note that in IBSS each STA can act as both a supplicant and authenticator.

1. Share a key, PMK, is decided and distributed among all members. It can be verbal.
2. Start IBSS and use standard IBSS procedure (i.e., the beacon is sent by a STA at a time). The beacon sending procedure is based on the backoff method. The STA of which the backoff ends first sends the beacon. This continues until the network is alive.
3. PTK is created with a 4-way handshake between STAs desiring to communicate. The STA with the smallest MAC becomes the IEEE 802.1x Supplicant while the other STA becomes the Authenticator.
4. GTK is created with all the STAs in the IBSS using the 2-way handshake.

This procedure does provide security but at the cost of complexity in terms of number of keys to be stored.

4.6 COMPARISON

Now that we have gone through different security solutions of IEEE 802.11, let us look at a comparison. The comparison is given in Table 4.2.

Table 4.2
Comparison of WEP, WPA, and IEEE 802.11i

	WEP	**WPA (Optional in 802.11i)**	**IEEE 802.11i**
Cipher algorithm	RC4	RC4 with TKIP	AES with CCMP
Key length	Encryption: 40 or 104 bits	Encryption: 128 bits, Authentication (message integrity): 64 bits	Encryption & Authentication (message integrity): 128 bits
Key per-packet	Concatenated IV and Secret key	Per-packet mixing	Not needed
Key uniqueness	Secret key per network. IV change per packet	Key per session, packet, user	Key per session, packet, user
Data integrity	CRC-32	Michael	CCM
Header integrity	-	Michael	CCM
Sequence counter	-	IV	IV
Key management and hierarchy	-	IEEE 802.1x EAP	IEEE 802.1x EAP
IBSS	-	-	Yes
Pre-authentication	-	-	Yes
Authentication mechanism	Open System and Shared Key	Open System and Upper Layer	Open System and Upper Layer
Security-related negotiation	-	Yes with RSN IE	Yes with RSN IE

References

[1] Prasad, A. R., *WLANs: Protocols, Security and Deployment*, Ph.D. Thesis, Delft University Press (DUP), Delft, The Netherlands, December 2003.

[2] Prasad, N. R., *Adaptive Security in Heterogeneous Networks*, Ph.D. Thesis, University of Roma "Tor Vergata," Rome, Italy, April 2004.

[3] Prasad, N. R., and A. R. Prasad (eds.), *WLAN Systems and Wireless IP for Next Generation Communications*, Norwood, MA: Artech House, January 2002.

[4] Edney, J., and W.A. Arbaugh, *Real 802.11 Security: Wi-Fi Protected Access and 802.11i*, Reading, MA: Addison-Wesley, 2004.

[5] Arbaugh, W.A., website: http://www.cs.umd.edu/~waa/wireless.html, 2 June 2004.

[6] Borisov, N., I. Goldberg, and D. Wagner, "Intercepting Mobile Communications: The Insecurity of 802.11," *7th Annual International Conference on Mobile Computing and Networking*, July 2001.

[7] Walker, J.R., "Unsafe at any key size: An analysis of the WEP encapsulation." IEEE 802.11/00-362, October 2000.

[8] Fluhrer, S., I. Martin, and A. Shamir, "Weaknesses in the Key Scheduling Algorithm of RC4," *8th Annual Workshop on Selected Areas in Cryptography*, August 2001.

[9] Stubblefield, A., J. Ioannidis, and A. D. Rubin, "Using the Fluhrer, Mantin and Shamir Attack to break WEP," *Network and Distributed Systems Security Symposium '02*, 2002.

[10] Burrows, M., M. Abadi, and R. Needham, "A Logic of Authentication," *ACM Transactions on Computer Systems*, Vol. 8:1, February 1990.

[11] Arbaugh, W.A., N. Shankar, and J. Wang, "Your 802.11 Network has no Clothes," *1st IEEE International Conference on Wireless LANs and Home Networks*, Singapore, December 2001, pp. 131–144.

[12] Cam-Winget, N., R. Housley, D. Wagner, and J. Walker, "Security Flaws in 802.11 Data Link Protocols," *Communications of the ACM*, Vol. 46, issue 5, May 2003, pp. 34–39.

[13] Housley, R., and W. A. Arbaugh, "Security Problems in 802.11-based Networks," *Communications of the ACM*, Vol. 46, Issue 5, May 2003, pp. 31–34.

[14] Arbaugh, W.A., "An Inductive chosen Plaintext Attack against WEP/WEP2," IEEE Document 802.11-02/230, May 2001.

[15] Petroni Jr., N.L., and W. A. Arbaugh, "The Dangers of Mitigating Security Design Flaws: A Wireless Case Study," *IEEE Security and Privacy Magazine*, January-February 2003 (Vol. 1, No. 1), pp. 28–36.

[16] UNINETT: http://www.uninett.no/wlan/wlanthreat.html#02, 2 June 2004.

[17] NetStumbler: http://www.netstumbler.com/download.php?op=viewdownload&cid=1&orderby=hitsD, 2 June 2004.

[18] Kismet: http://www.kismetwireless.net/, 2 June 2004.

[19] Ethereal: http://www.ethereal.com/http://www.uninett.no/wlan/wlanthreat.html#02, 2 June 2004.

[20] SMAC: http://www.klcconsulting.net/smac/, 2 June 2004.

[21] GNU MAC Changer: http://www.alobbs.com/modules.php?op=modload&name=macc&file=index, 2 June 2004.

[22] Airsnort: http://airsnort.shmoo.com/, 2 June 2004.

[23] WEPCrack: http://wepcrack.sourceforge.net/, 2 June 2004.

[24] AirJack: http://www.11ninja.net/, 2 June 2004.

[25] SSLSniff: http://www.thoughtcrime.org/ie.html, 2 June 2004.

[26] Black, U., *Internet Security Protocols: Protecting IP Traffic*, Upper Saddle River, NJ: Prentice Hall, 2000.

[27] Stallings, W., *Cryptography and Network Security: Principles and Practice*, Upper Saddle River, NJ: Prentice Hall, July 1998.

[28] Frankel, S., *Demystifying IPSec Puzzle*, Norwood MA: Artech House, April 2001.

[29] Thayer, R., N. Doraswamy, and R. Glenn, "IP Security Document Roadmap," RFC 2411, November 1998.

[30] Kent, S., and R. Atkinson, "Security Architecture for the Internet Protocol," RFC 2401, November 1998.

[31] Kent, S., and R. Atkinson, "IP Authentication Header," RFC 2402, November 1998.

[32] Kent, S., and R. Atkinson, "IP Encapsulating Security Payload (ESP)," RFC 2406, November 1998.

[33] Maughan et al., "Internet Security Association and Key Management Protocol (ISAKMP)," RFC 2408, November 1998.

[34] Harkins, D.C., "The Internet Key Exchange (IKE)," RFC 2409, November 1998.

[35] Kaufmann, C., ed., "Internet Key Exchange (IKEv2) Protocol," draft-ietf-ipsec-ikev2-13.txt, March 2004.

[36] Phifer, L., "Trouble with NAT," Cisco IP Journal, pp. 2–13, Vol. 3, nr. 4, December 2000.

[37] Holdrege, M., and P. Srisuresh, "Protocol Complications with IP Network Address Translation," RFC 3027, January 2001.

[38] Rosenberg, J., J. Weinberger, C. Huitema, and R. Mahy, "STUN—Simple Traversal of User Datagram Protocol (UDP) Through Network Address Translators (NATs)," RFC 3489, March 2003.

[39] Kivinen, T., B. Swander, A. Huttunen, and V. Volpe, "Negotiation of NAT Traversal in the IKE," draft-ietf-ipsec-nat-t-ike-05.txt, January 2003.

[40] Carpenter, B., and K. Moore, "Connection of IPv6 Domains via IPv4 Clouds," RFC 3056, February 2001.

[41] Borella, M., J. Lo, D. Grabelsky, and G. Montenegro, "Realm Specific IP: A Framework," RFC 3102, October 2001.

[42] Borella, M., D. Grabelsky, J. Lo, and K. Taniguchi "Realm Specific IP: Protocol Specification," RFC 3103, October 2001.

[43] Montenegro, G., and M. Borella, "RSIP Support for End-to-End IPSec," RFC 3104, October 2001.

[44] Aboba, B., and W. Dixon, "IPSec NAT Compatibility Requirements," March 2004, RFC 3715.

[45] Dierks, T., and C. Allen, "The TLS Protocol Version 1.0.," RFC 2246, January 1999.

[46] Borman, D., "Telnet Authentication: Kerberos Version 4," RFC 1411, January 1993.

[47] Kohl, J., and C. Neuman, "The Kerberos Network Authentication Service (V5)," RFC 1510, September 1993.

[48] Bellovin, S.M., and M. Merritt, "Limitations of the Kerberos Authentication System," *ACM SIGCOMM Computer Comm. Review*, Vol. 20, iss. 5, Oct. 1990, pp. 119–132.

[49] Prasad, A.R., H. Moelard, and J. Kruys, "Security Architecture for Wireless LANs: Corporate & Public Environment," VTC 2000 Spring, May 2000, pp. 283–287.

[50] Rubens, A., C. Rigney, W. Simpson, and S. Willens, "Remote Authentication Dial In User Service (RADIUS)," RFC 2138, April 1997.

[51] Rigney, C., W. Willats, and P. Calhoun, "RADIUS Extensions," RFC 2869, June 2000.

[52] Calhoun, P., J. Loughney, E. Guttman, G. Zorn, and J. Arkko, "Diameter Base Protocol, " RFC 3588, September 2003.

[53] IEEE 802.1X, IEEE Standard for Local and metropolitan area networks—Port-Based Network Access Control, July 2001.

[54] Blunk, L., and J. Vollbrecht, "PPP Extensible Authentication Protocol (EAP)," RFC 2284, March 1998.

[55] Aboba, B., and D. Simon, "PPP EAP TLS Authentication Protocol," RFC 2716, October 1999.

[56] Palekar, A., D. Simon, G. Zorn, J. Salowey, H. Zhou, and S. Josefsson, "Protected EAP protocol (PEAP) version 2," draft-josefssonpppext-eap-tls-eap-07, IETF, October 2003.

[57] Aboba, B., and M. Beadles, "The network access identifier (NAI)," RFC 2486 (Standards Track), January 1999.

[58] Funk, P., and S. Blake-Wilson "EAP tunneled TLS authentication protocol (EAP-TTLS)," draft-ietf-pppext-eap-ttls-03, IETF, August 2003.

[59] Cam-Winget, N., D. McGrew, J. Salowey, H. Zhou, "EAP Flexible Authentication via Secure Tunneling (EAP-FAST)," draft-cam-winget-eap-fast-00.txt, February, 2004.

[60] Haverinen, H., and J. Salowey, (eds.), "Extensible Authentication Protocol Method for GSM Subscriber Identity," draft-haverinen-pppext-eap-sim-13.txt, April 5, 2004.

[61] Arkko, J., and H. Haverinen, "Extensible Authentication Protocol Method for UMTS Authentication and Key Agreement (EAP-AKA)," draft-arkko-pppext-eap-aka-12.txt, April 5, 2004.

[62] EAP charter: http://www.ietf.org/html.charters/eap-charter.html, 2 June 2004.

[63] Danielyan. E., "The Lure of Biometrics," *Cisco IP Journal*, pp. 15–34, Vol. 7, nr. 1, March 2004.

[64] VPNC: http://www.vpnc.org/, 2 June 2004.

[65] PKI: http://www.pki-page.org/, 2 June 2004.

[66] IDS: http://www.intrusion-detection-system-group.co.uk/, 2 June 2004.

[67] Prasad, A., and A. Raji, "A Proposal for IEEE 802.11e Security," IEEE 802.11e, 00/178, July 2000.

[68] Wi-fi protected access: http://www.wi-fi.org/opensection/protected_access.asp, 2 June 2004.

[69] Whiting, D., R. Housley, and N. Ferguson, "Counter with CBC-MAC (CCM)," RFC 3610, September 2003.

[70] IEEE 802.11i, IEEE 802.11: Specification for Robust Security, D10.0, April 2004.

[71] Godfrey, T., and J. Walker, "Adopted TGe Functional Requirements," IEEE 802.11-00/245r1, July 2000.

[72] Eronen, P., and H. Tschofenig, "Extension for EAP Authentication in IKEv2," draft-eronen-ipsec-ikev2-eap-auth-01.txt.

Chapter 5

Quality of Service

Quality of service (QoS) is a classic topic in telecommunications. QoS cannot be covered in one chapter. This is a wide topic requiring discussion of all the protocol layers and queuing methods, signaling, bandwidth and resource management, service level agreement (SLA), and policies. In this chapter the discussion will be about QoS in IEEE 802.11 Medium Access Control (MAC), signaling protocols in the IP layer, internetwork communication protocols, and transport protocols [1–27].

5.1 INTRODUCTION

QoS is becoming an increasingly important element of any communications system. In the simplest sense, QoS means providing consistent, predictable data delivery service, in other words, satisfying the customer application requirements. Providing QoS means providing real-time (e.g., voice) as well as non-real-time services. Voice communication is the primary form of service required by humankind. So the focus of this chapter will be on voice.

Support for voice communications using the Internet Protocol (IP), which is usually just called "*Voice over IP*" or VoIP. VoIP has become especially attractive given the low-cost, flat rate pricing of the public Internet. VoIP can be defined as the ability to make telephone calls (i.e., to do everything that can be done today with the (public switched telecommunications/telephone network) PSTN) over the IP-based data networks with a suitable QoS. This is desirable because of a much superior cost-benefit ratio as compared to the PSTN. Equipment producers see VoIP as a new opportunity to innovate and compete. Their challenge of turning the vision to reality is already happening; VoIP-enabled equipment is available and providing toll quality service. For Internet service providers (ISPs) the possibility of introducing usage-based pricing or distinguishing themselves from competition and increasing their traffic volumes is very attractive. On the other hand the users are seeking integrated voice and data applications as well as cost benefits.

As WLANs are an extension of the IP to the wireless, it is necessary to have a voice over WLAN (VoWLAN) protocol that fulfills the requirement. A simplified system for voice over WLAN, IP to POTS is depicted in Figure 5.1. End-to-end voice delivery will require (1) QoS provision in WLAN, which means QoS enabled MAC including admission control and radio resource management (RRM); (2) signaling protocols [e.g., H.323 and Session Initiation Protocol (SIP)]; (3) routing protocols [e.g., Routing Information Protocol (RIP), Open Shortest Path First (OSPF), and Border Gateway protocol (BG)]; (4) communication between the IP network and the PSTN network using gateway protocols [e.g., Media Gateway Control Protocol (MGCP), and H.323]; and (5) transport protocols that include Real-Time Transport Protocol (RTP) with Real-Time Transport Control Protocol (RTCP), Differentiated Service (DiffServ), and Resource Reservation Protocol (RSVP). RSVP could also be put under signaling protocols. Note that this is only a partial list of things required to provide QoS. In this chapter we will not discuss the routing protocols.

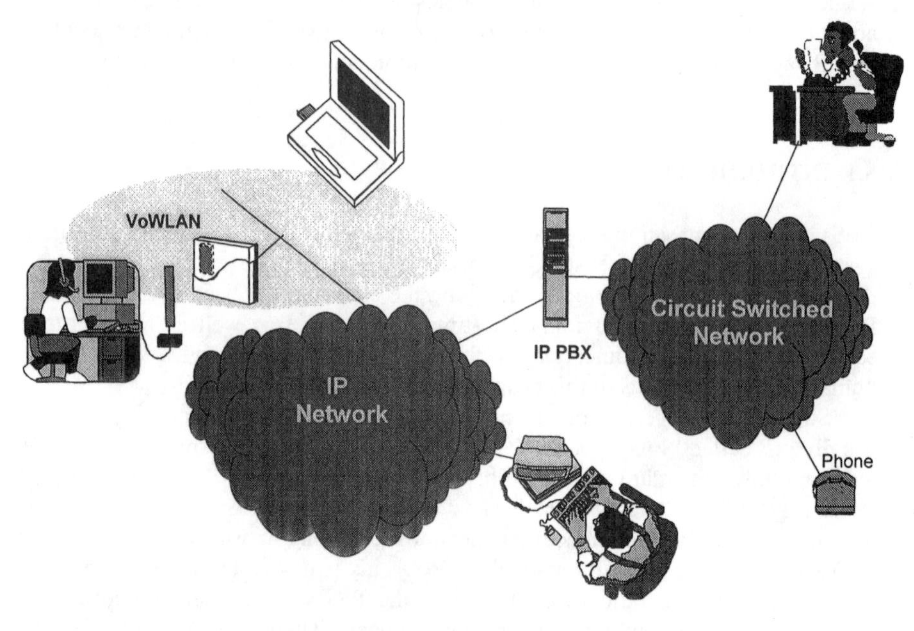

VoWLAN:	Voice over Wireless Local Area Network
IP:	Internet Protocol
PBX:	Private Branch eXchange

Figure 5.1 VoWLAN systems interconnected with POTS.

5.2 VOICE COMMUNICATION REQUIREMENT

Voice is a real-time communication, which means it has severe delay constraints [4–6]. Dedicated systems for voice do not face any problems considering delay but in a system made for asynchronous data transmission it will become a big issue. In the following subsections the challenges that will be faced during VoWLAN product development and the requirements of voice transmission are given.

5.2.1 Voice over Wireless Challenges

The goal is relatively simple: add telephone calling capabilities (both voice transfer and signaling) to WLANs with backbone IP-based networks and interconnect these to the public telephone network and to the private voice networks in such a way as to maintain current voice quality standards and preserve the features everyone expects from the telephone. The challenges for the product developer arise in five specific areas:

- Voice quality should be comparable to what is available using the PSTN, even over networks having variable levels of QoS.
- The underlying network must meet strict performance criteria including minimizing call refusals, network latency, packet loss, and disconnects. This is required even during congestion conditions or when multiple users must share the network resources.
- Call control (signaling) must make the telephone calling process transparent so that the callers need not know what technology is actually implementing the service.
- PSTN/VoIP/VoWLAN service interworking (and equipment inter-operability) involves gateways between the voice and the wireline data network environments and the wireline and the wireless data networks.
- System management, security, addressing (directories, dial plans) and accounting must be provided, preferably consolidated with the PSTN operation support systems (OSSs).

5.2.2 Voice Quality and Characteristics

Providing a level of quality that equals that of the PSTN (this is usually referred to as “toll quality voice”) is viewed as a basic requirement. It has been found that there are three factors that can profoundly impact the quality of the service:

1. Delay: Talker overlap (the problem of one caller stepping on the other talker’s voice) becomes significant if the one-way delay becomes greater than 250 milliseconds. The end-to-end delay budget is therefore the major constraint and driving requirement for reducing the delay through a packet network.

2. Jitter (delay variability): Jitter is the variation in inter-packet arrival time as introduced by the variable transmission delay over the network. Removing the jitter requires collecting packets and holding them long enough to allow the slowest packets to arrive in time to be played in the correct sequence, which causes additional delay. The jitter buffers add delay, which is used to remove the packet delay variation that each packet is subjected to as it transits the packet network.
3. Packet loss: Wireless and IP networks cannot provide a guarantee that packets will be delivered at all, much less in order. The packets will be dropped under peak loads.

The above three parameters can be used for objective voice quality measurements. The subjective voice quality measurements are given in terms of mean opinion score (MOS). MOS is the average score given by a large number of users (about 100) with similar listening ability listening to the same segment of voice under the same conditions [11].

5.3 LIMITATIONS OF LEGACY 802.11 MAC

The legacy IEEE 802.11 MAC is discussed in Chapter 3. In this section the limitations in terms of the QoS provision of legacy MAC will be discussed. First the limitations of the distributed coordination function (DCF) are discussed after which the limitations of point coordination function are discussed (PCF).

Referring to Section 5.2.2, packet loss basically is an issue of the channel. In the case of IEEE 802.11, packet loss can be decreased by decreasing the data rate. As forward error correction (FEC) is not used by the IEEE 802.11b there is no other way to decrease packet loss [3, 6]. In the case of IEEE 802.11a,g FEC is used but in combination with modulation it is used to vary data rate [6, 13].

Jitter can be taken care of by the AP. For this purpose the AP will have to implement a scheduling scheme that can distinguish between different service classes. The service classes should not be distinguished by unwrapping the received packet until the application layer is reached but there should be mechanisms to identify the service class from the packet header.

5.3.1 Distributed Coordination Function

The advantage of DCF is that it promotes fairness among stations, but its weakness is that it cannot support time-bounded services [5]. Fairness is maintained because each station must recontend for the channel after every transmission of a packet. All stations have an equal probability of gaining access to the channel after each DIFS interval. Time-bounded services typically support applications such as packetized voice or video that must be maintained with a specified minimum delay. With DCF, there is no mechanism to guarantee minimum delay to stations supporting time-bounded services. It might be possible to transmit voice over DCF in an isolated cell with few users but in normal conditions (data and voice traffic in a network) performance for voice communication will be relatively bad [6].

As DCF is the basic access method of the IEEE 802.11 WLANs, implementation is a non-issue; it has to be implemented and is already implemented in all the IEEE 802.11 products. The issue with the DCF, like any other QoS solution for WLAN, will be the implementation of scheduling and prioritized queue in the AP and also in the STAs. This could be easily done for example, by delaying the nonreal-time traffic from the STA and giving priority to real-time traffic and in the AP by implementing a round-robin method for transmission of real-time packets to the stations while giving low priority to non-real-time traffic.

As the basic MAC mechanism of IEEE 802.11 is dependent on the DCF, compatibility with the standard is a nonissue.

Scalability, in terms of the number of people that can use an AP is not a problem for the DCF. The standard does not define the number of users per channel using the DCF either. As the channel access is based on the CSMA/CA mechanism the overall delay will increase if a large number of users joins a channel and has data to transmit at the same time. Scalability in terms of a large number of APs that use the DCF is again not a problem; such deployments are common. Neighboring APs will use different channels and if there is some effect of overlapping, the CSMA/CA characteristics will take care of it as they take care of channel access of the STAs to an AP.

5.3.2 Point Coordination Function

There are three issues related to real-time services and PCF [5]; the first issue is that the centralized mode cannot be operated simultaneously in the neighboring cells; this happens because there are very few independent or non overlapping channels (three channels) defined by the IEEE 802.11 standard [6–7]. With so few independent channels there is a very high probability of interference from the neighboring cells. Even if the neighboring cell is not using an overlapping channel or the same channel a far away cell using the same channel will cause interference. Further, the IEEE 802.11 APs are not synchronized; thus the contention free period (CFP) in different cells can start at the same time and thereby cause collision. Thus the overall quality will be degraded. Not only that, a STA in an AP with the same frequency can also cause interference and thus degradation of quality. This issue is often known as the overlapping cell issue.

Second, as a result of the CSMA/CA protocol, the PC might be unable to gain control of the channel at the nominal beginning of the CFP. If, for instance, a station starts a transmission during the DCF period that lasts longer than the remaining time between the start of the transmission and the nominal start of the next CFP, the PC has to defer the start of its transmission until the medium has been free for a PIFS. This decreases the CFP period and thus the QoS will be degraded.

Third, PCF is central–control-based but a study done in [1] show that high overhead introduced by the IEEE 802.11 WLAN standard results in a low number of possible voice conversations.

Besides the above issues, PCF is an optional feature of the IEEE 802.11 standard and is not implemented by most vendors. To the best of the author's knowledge it was only implemented by one vendor and that for test purposes only. Implementation of a fully functional PCF might not be very difficult but a good implementation could be very complicated because it would require resolving all the issues discussed in this section.

PCF is defined by the standard; thus its compatibility is not an issue but a good implementation might actually make it incompatible with the IEEE 802.11 standard.

As already presented in this section, the PCF is not scalable.

5.4 QOS SUPPORT MECHANISM OF 802.11e

The legacy 802.11 MAC protocol has no means of differentiating traffic streams or sources. All data flows are treated equally (i.e., same priority to access the medium) in both DCF and PCF mechanisms. This means that no special considerations can be given to traffic on the channel for services with critical requirements in bandwidth, delay, jitter, and packet losses for applications such as voice and video. For example a low-priority bursty traffic can choke out a running video stream and thereby destroy the user's experience. In this chapter stations operating under 802.11e protocols are referred to as 802.11e stations. Stations that operate as central coordinator for all other stations within the same QoS supporting BSS (QBSS) are called the hybrid coordinator (HC). Similar to the PC, HC also resides within an 802.11e AP. A BSS that includes an 802.11e compliant HC is referred to as a QBSS. WLAN 802.11e enhanced the original legacy 802.11 MAC to support applications with QoS requirements. The QoS facility includes an additional coordination function called the hybrid coordination function (HCF), which is only usable in QoS network (QBSS) configuration. The HCF combines functions from the DCF and PCF with some enhanced, QoS-specific mechanisms and frame subtypes to allow a uniform set of frame exchange sequences to be used for QoS data transfers during both the CP and CFP. As shown in Figure 5.2 the HCF uses both a contention-based channel access method, called the enhanced distributed channel access (EDCA) mechanism for contention based transfer and a controlled channel access, referred to as the HCF controlled channel access (HCCA) mechanism, for contention-free transfer. HCF defined in 802.11e supports up to eight priority traffic classes that map directly to the differentiated services code point (DSCP).

Most of this section is reprinted with permission from IEEE P802.11e-2004.

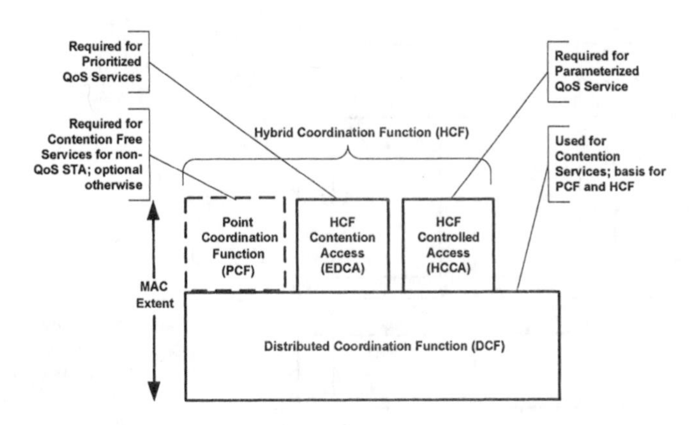

Figure 5.2 Enhanced 802.11e MAC Architecture [3].

The Wi-Fi Alliance is developing Wi-Fi Multimedia (WMM) interoperability certification for IEEE 802.11e. A white paper is available from the Wi-Fi Website.

5.4.1 Enhanced Distributed Channel Access

EDCA also known as HCF contention-based channel access provides differentiated, distributed access to the WM for QoS stations (QSTAs). EDCA defines the access category (AC) mechanism that provides support for the priorities at the stations. Each station may have up to four ACs to support eight user priorities (UPs). One or more UPs are assigned to one AC. A station accesses the medium based on the AC of the frame to be transmitted. The mapping from priorities to ACs is defined in Table 5.1. Each AC is an enhanced variant of the DCF. It contends for transmission opportunities (TXOPs) using a set of EDCA parameters. TXOP is a time interval when a particular station has the right to initiate transmissions onto the WM. An AC with higher priority is assigned a shorter CW in order to ensure that a higher priority AC will be able to transmit before lower priority ones. This is done by setting the CW limits CWmin[AC] and CWmax[AC], from which CW[AC] is computed, to different values for different ACs. For further differentiation, different inter-frame space (IFS) is introduced according to ACs. Instead of DIFS, an arbitration IFS (AIFS) is used. The AIFS is at least DIFS, and can be enlarged individually for each AC. Similar to DCF, if the medium is sensed to be idle in the EDCA mechanism, a transmission can begin immediately. Otherwise, the station defers until the end of current transmission on the WM. After deferral, the station waits for a period of AIFS (AC) to start a backoff procedure. The backoff interval is now a random number drawn from the interval [1, CW(AC)+1]. Each AC within a single station behaves like a virtual station. It contends for access to the WM and independently starts its backoff time after sensing the medium is idle for at least AIFS. Collision between ACs

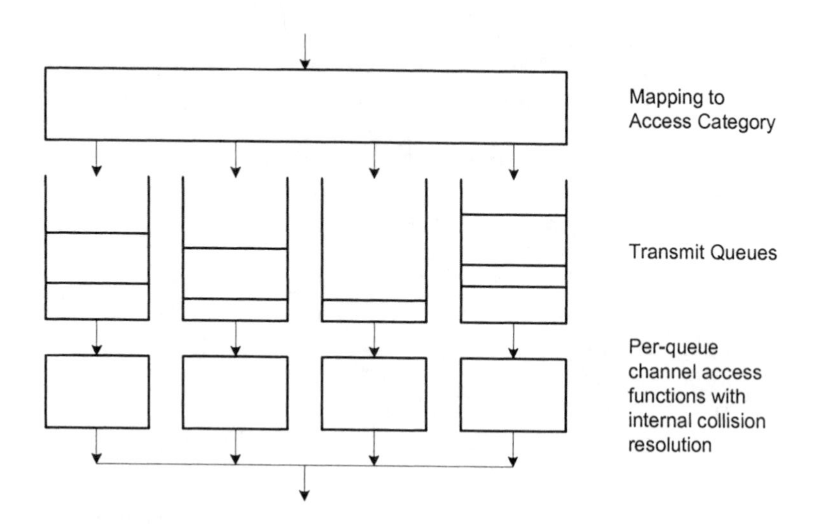

Figure 5.3 Reference implementation model [3].

within a single station are resolved within the station such that the data frames from higher valued AC receive the TXOP, and the data frames from lower valued colliding ACs behave as if there were an external collision on the WM. The timing relationship for an EDCA is shown in Figure 5.4.

The prioritized medium access of the EDCA in 802.11e is realized by assigning different CWs and different AIFS to different ACs. Data units are now delivered through multiple backoff instances within one station. Each backoff instance is parameterized with TC-specific parameters. The typical values of CW limits and AIFSs for different ACs in the QoS parameters set are shown in Table 5.2. A model of the reference implementation is shown in Figure 5.3. It illustrates a mapping from frame type or priority to ACs, the four queues, and four independent channel access functions, one for each queue.

There are two modes of EDCA TXOP, the initiation of the EDCA TXOP and the EDCA TXOP continuation. An initiation of the TXOP occurs when the EDCA rules permit access to the medium. A TXOP continuation occurs when a channel access function retains the right to access the medium following the completion of a frame exchange sequence, such as on receipt of an ACK frame. The TXOP limit duration values are advertised by the QoS access point (QAP) in the EDCA parameter set information element in beacons and probe response frames transmitted by the QAP. A TXOP limit value of 0 indicates that a single MSDU or MMPDU, in addition to a possible RTS/CTS exchange or CTS to itself, may be transmitted at any rate for each TXOP. Non-AP QSTAs shall ensure that the duration of TXOPs obtained using the EDCA rules do not exceed the TXOP limit. The duration of a TXOP is the duration during which the TXOP holder maintains uninterrupted control of the medium, and it includes the time required to transmit frames sent as an immediate response to the TXOP holder's transmissions. A QSTA shall fragment a unicast MSDU so that the transmission of the first MPDU of the TXOP does not cause the TXOP limit to be exceeded at the PHY rate selected for the initial transmission attempt of that MPDU. The TXOP limit may be exceeded, when using a lower PHY rate than selected for the initial transmission attempt of the first MPDU, for a retransmission of an MPDU, for the initial transmission of an MPDU if any previous MPDU in the current MSDU has been retransmitted, or for broadcast/multicast MSDUs. When the TXOP limit is exceeded due to the retransmission of a MPDU at a reduced PHY rate, the STA shall not transmit more than one MPDU in the TXOP. It should be noted, when transmitting frames using acknowledgment mechanisms other than immediate ACK a protective mechanism should be used (such as RTS/CTS or the protection mechanism). In an infrastructure QBSS, for broadcast/multicast frames belonging to QoSLocalMulticast service either one frame may be sent at a time and perform backoff or multiple frames may be sent by sending first the RTS frame, with the duration/ID covering the entire burst, to the QAP and upon receiving the corresponding CTS from the QAP. A QAP may send broadcast/multicast frames without using any protection mechanism. In a QoS independent basic service set (QIBSS), for broadcast/multicast frames belonging to QoSLocalMulticast service shall be sent one at a time and perform backoff.

5.4.2 HCF-Controlled Channel Access (HCCA)

The HCF-controlled channel access mechanism uses a QoS-aware centralized coordinator, called a hybrid coordinator (HC), and operates under rules that are different from the point coordinator (PC) of the PCF. The HC is collocated with

Table 5.1

Priority to Access Category Mapping

	User Priority (Same as 802.1D Priority)	**Access Category (AC)**	**Designation (Informative)**
Lowest	1	0	Best Effort
	2	0	Best Effort
	0	0	Best Effort
	3	1	Video Probe
	4	2	Video
	5	2	Video
↓	6	3	Voice
Highest	7	3	Voice

Table 5.2

Typical QoS Parameters

AC	CWMIN	CWMAX	AIFS
0	Cwmin	Cwmax	2
1	Cwmin	Cwmax	1
2	[(CWmin + 1)/2] – 1	Cwmin	1
3	[(CWmin + 1)/4] – 1	[(CWmin + 1)/2] - 1	1

the QAP of the QBSS and uses the HC's higher priority of access to the WM to initiate frame exchange sequences and to allocate TXOPs to itself and other QSTAs, so as to provide limited-duration controlled access phase (CAP) for contention-free transfer of QoS data.

The HC traffic delivery and TXOP allocation may be scheduled during the CP and any locally generated CFP (generated optionally by the HC) to meet the QoS requirements of a particular TC or TS. TXOP allocations and contention-free transfers of QoS traffic can be based on the HC's QBSS-wide knowledge of the amounts of pending traffic belonging to different TS and/or TCs and is subject to QBSS-specific QoS policies. A QAP may indicate availability of CF-Polls to non-QoS STAs, thereby providing non-QoS contention-free transfers during the CFP. This provisioning of contention-free transfers during the CFP to non-QoS STAs, however, is not recommended. QAP that provides non-QoS CF-polling to adhere to frame sequence restrictions considerably more complex than, and less efficient than, those specified for either PCF or HCF. In addition, the achievable service quality is likely to be degraded when non-QoS STAs are associated and being polled. The HCF protects the transmissions during each CAP using the virtual carrier sense mechanism. A QSTA may initiate multiple frame exchange sequences during a polled TXOP of sufficient duration to perform more than one such sequence. The use of virtual carrier sense by the HC provides improved protection of the CFP.

5.4.3 Coexistence of DCF, PCF, and HCF

The DCF and a centralized coordination function (either PCF or HCF) coexist in a manner that permits both to operate concurrently within the same BSS. When a PC is operating in a BSS, the PCF and DCF access methods alternate, with a contention-free period (CFP) followed by a contention period (CP). When an HC is operating in a QBSS, it may generate an alternation of CFP and CP in the same way as a PC, using the DCF access method only during the CP. The HCF access methods (controlled and contention-based) operate sequentially when the channel is in CP. Sequential operation allows the polled and contention-based access methods to alternate, within intervals as short as the time to transmit a frame exchange sequence.

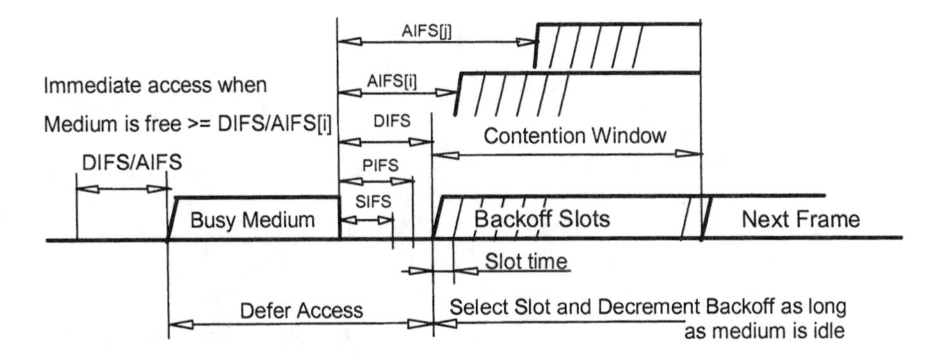

Figure 5.4 IEEE 802.11e inter frame space [3].

Five different IFSs are defined to provide priority levels for access to the wireless media; in the following they are listed in order, from the shortest to the longest except for AIFS; Figure 5.4 shows some of these relationships.

1. SIFS = Short inter frame space
2. PIFS = PCF inter frame space
3. DIFS = DCF inter frame space
4. AIFS = Arbitration inter frame space (used by the QoS facility)
5. EIFS = Extended inter frame space

The different IFSs are independent of the STA bit rate. The IFS timings are defined as time gaps on the medium, and those except for AIFS are fixed for each PHY (even in multirate-capable PHYs). The IFS values are determined from attributes specified by the PHY.

5.4.4 Interpretation of Priority Parameters in MAC Service Primitives

The MAC frame format is shown in Figure 5.5. The QoS facility supports eight priority value UPs. The values a UP may take are the integer values from 0 to 7 and are identical to the IEEE 802.1D priority tags. An MSDU with a particular UP is said to belong to a traffic category (TC) with that UP. The UP is provided with each MSDU at the medium access control service access point (MAC_SAP), either directly, in the UP parameter, or indirectly, in a traffic specification (TSPEC) designated by the UP parameter.

Octets: 2	2	6	6	6	2	6	2	n	4
Frame Control	**Duration / ID**	**Address 1**	**Address 2**	**Address 3**	**Sequence Control**	**Address 4**	**QoS Control**	**Frame Body**	**FCS**
MAC Header									

Figure 5.5 MAC frame format [3].

The received unicast frames at the QAP may be either of the following:

1. Non-QoS subtypes, in which case the QAP assigns a priority of contention if they are received during the CP, or CF, if they are received during the CFP.
2. QoS subtypes, in which case the QAP assumes the UP value from the TID in the QoS control field directly for TID values between 0 and 7. For TID values between 8 and 15, the QAP extract the UP value in the user priority subfield of the TSInfo field in the associated TSPEC or from the user priority field in the associated TCLAS element, as applicable.

QAPs deliver the UP with the received MSDUs to the differentiated services (DS) field. The value of the priority parameter in the MAC service primitives may be a non-integer value of either contention or contention-free, or may be any integer value in the range 0 through 15. When the priority parameter has an integer value, it is used in the traffic identifier (TID) fields that appear in certain frames that are used to deliver, and to control the delivery of QoS data across the WM. Priority parameter and TID field values 0 through 7 are interpreted as UPs for the MSDUs. Outgoing MSDUs with UP values 0 through 7 are handled by MAC entities at QSTAs in accordance with the UP. Priority parameter and TID field values 8 through 15 specify TIDs that are also traffic stream identifiers (TSIDs), and select the TSPEC for the TS designated by the TID. Outgoing MSDUs with priority parameter values 8 through 15 are handled by MAC entities at QSTAs in accordance with the UP value determined from the user priority subfield as well as other parameter values in the selected TSPEC. Selection of a TSPEC for which the MAC sub layer management entity (MLME) has not provided QoS parameter values is equivalent to using a TSPEC with a user priority subfield value equal to the TID value minus 8 and the current, local default values for all other QoS parameters. The non integer values of the priority parameter are allowed at all non-QoS STAs. The integer values of the priority parameter (i.e., TID) are supported only at QSTAs that are either associated in a QBSS or members of a QIBSS. A range of 0 through 15 is supported by QSTAs associated in a QBSS; whereas a range of 0 through 7 is supported by QSTAs that are members of a QIBSS. If a QSTA is associated in a non-QoS BSS (nQBSS), the QSTA is functioning as a non-QoS STA, so the priority value is always contention or contention-free. At QSTAs associated in a QBSS, MSDUs with priority of contention are considered equivalent to MSDUs with TID 0, and those with priority of contention-free are delivered using the contention-free delivery, if a PC is present in the QAP. If a PC is not present, a QSTA attempts to deliver traffic with CF priority using a polled service to initiate a TSPEC with default parameters and a UP of zero and access policy set to HCCA. If a TSPEC cannot be setup by the QSTA, traffic marked CF will be delivered using an UP of 0. At QSTAs associated in an nQBSS, all MSDUs with an integer priority are considered equivalent to MSDUs with priority contention. If a QSTA is associated in a QBSS, the MSDUs it receives in QoS data frames are reported with the TID value contained in the MAC header of that frame. The MSDUs such a QSTA receives in non-QoS data frames are reported to logical link control with priority of contention if they are received during the CP, or CF, or if they are received during the CFP.

5.4.5 Admission Control at the HC

An 802.11e network may use admission control (AC) to administer policy or regulate the available bandwidth resources. Admission control is also required when a QSTA desires guarantees on the amount of time that it can access the channel. The HC, which is in the QAP, is used to administer admission control in the network. As the QoS facility supports two access mechanisms, there are two distinct admission control mechanisms — one for contention-based access and another for the controlled-access mechanism. Admission control, in general, depends on vendors' implementation of the scheduler, available channel capacity, link conditions, retransmission limits, and the scheduling requirements of a given stream. All of these criteria affect the admissibility of a given stream; any stream may be rejected. If the HC has admitted no streams that require polling, it may not find it necessary to perform the scheduler or related HC functions.

5.4.5.1 Contention-Based Admission Control Procedures

A non-AP QSTA supports admission control procedures. QAPs support admission control procedures, at least to the minimal extent of advertising that admission is not mandatory on their ACs. The QAP uses the admission control mandatory (ACM) subfields advertised in the EDCA parameter set element to indicate whether admission control is required for each of the ACs. While the CWmin, CWmax, AIFS, TXOP limit parameters may be adjusted over time by the QAP, the ACM bit is static for the duration of the lifetime of the BSS. An ADD traffic stream (ADDTS) request shall be transmitted by a non-AP QSTA to the HC in order to request admission of traffic in any direction (uplink, downlink, direct, or bidirectional) employing an AC that requires admission control. The ADDTS request contains the user priority associated with the traffic and indicates EDCA as the access policy. The QAP associates the received user priority of the ADDTS request with the appropriate AC as per the UP to AC mappings. The non-AP QSTA may transmit unadmitted traffic for those ACs for which the QAP does not require admission control. If a QSTA desires to send data without admission control using an AC that mandates admission control, the QSTA uses EDCA parameters that correspond to a lower priority and do not require admission control. All ACs with priority higher than that of an AC with ACM flag equal to 1 should have ACM flag set to 1.

5.4.5.2 Controlled-Access Admission Control

In this section the schedule management of the admitted HCCA streams by the HC is discussed. When the HC provides controlled channel access to non-AP QSTAs, it is responsible for granting or denying polling service to a TS based on the parameters in the associated TSPEC. If the TS is admitted, the HC is responsible for scheduling channel access to this TS based on the negotiated TSPEC parameters. The HCs do not initiate a modification of TSPEC parameters of an admitted TS unless requested by the STA. The HCs do not tear down a TS unless explicitly requested by the STA or at the expiry of the inactivity timer. The polling service based on admitted TS provides a "guaranteed channel access" from the scheduler in order to have its QoS requirements met. This is an achievable goal when the WM operates free of external interference (such as operation within the channel by other technologies and co channel overlapping BSS interference). The nature of wireless communications precludes absolute guarantees to satisfy QoS requirements. However, in a controlled environment (e.g., no interference), the behavior of the scheduler is compliant to meet the service schedule. The normative behavior of the scheduler is as follows:

- The scheduler is implemented such that, under controlled operating conditions, all stations with admitted TS are offered TXOPs that satisfy the service schedule.
- Specifically, if a TS is admitted by the HC, then the scheduler shall service the non-AP QSTA during a service period (SP). An SP is a contiguous time during which a set of one or more downlink frames and/or one or more polled TXOPs are granted to the QSTA (even if the QSTA is not using automatic power save delivery). An SP starts at fixed intervals of time specified in service interval field. The first SP starts when the lower order 4 bytes of the TSF timer equals the value specified in the service start time field. A SP ends after (1) expiration of the time duration specified in the max service duration field or (2) if the non-AP QSTA receives a frame with EOSP field set to 1 in QoS control field during an SP. Additionally, the minimum TXOP duration is at least the time to transmit one maximum MSDU Size successfully at the minimum PHY rate specified in the TSPEC. If maximum MSDU Size is not specified in the TSPEC then the minimum TXOP duration is at least the time to transmit one nominal MSDU size successfully at the minimum PHY rate.

The HC may aggregate admitted HCCA TS for a single non-AP QSTA and establishes a service schedule for the non-AP QSTA. When the HC aggregates the admitted TS, it sets the aggregation field in the granted TSPEC to 1. A QAP shall schedule the transmissions in HCCA TXOPs and communicate the service schedule to the non-AP QSTA. The HC shall provide an aggregate service schedule if the non-AP QSTA sets the aggregation field in its TSPEC request. If the QAP establishes an aggregate service schedule for a non-AP QSTA, it aggregates all HCCA streams for the QSTA. The service schedule is communicated to the non-AP QSTA in a schedule element contained in an ADDTS response. In the ADDTS response frame the modified service start time does not exceed the requested service start time, if specified in the ADDTS request frame, by more than one minimum service interval. The service schedule could be subsequently updated by a QAP as long as it meets TSPEC requirements.

The HC may update the service schedule at any time by sending a schedule element in a schedule frame. The updated schedule is in effect when the HC receives the acknowledgment frame for the schedule frame. The service start time in the schedule element in the schedule frame does not exceed the beginning of the immediately previous SP by more than the maximum service interval. The service start time shall not precede the beginning of the immediately previous SP by more than the minimum service interval.

5.5 OTHER QOS-RELATED IEEE 802.11 STANDARDS

There are several other IEEE 802.11 activities besides IEEE 802.11e that are looking towards better QoS provisioning either directly or indirectly. IEEE 802.11k provides radio resource measurement enhancements of IEEE 802.11. Each IEEE 802.11 device will be able to measure the radio environment and respond accordingly. Another recent group is looking into the use of these measurements. A new group is also working on fast handover with VoIP application in mind.

5.6 QOS REQUIREMENTS FOR HETEROGENEOUS TRAFFIC

The convergence of data communications (packet-switched) and voice- and video-based communications (traditionally circuit-switched) into a single flexible IP-based architecture would mean that the network should be capable of handling traffic ranging from constant bit rate to bursty in nature with the required QoS characteristics (see Table 5.3).

The term QoS may mean different things depending on the type of application or users expectations (e.g., low latency for IP telephony, guaranteed bandwidth for streaming audio and video, predictable delivery for collaborative interactive service, protection of network control for different grades of services from competing users, and minimum guarantee on service quality to meet user expectations). This all leads to definition of QoS, which can be classified into two categories:

1. Application-level QoS is basically the user-perceived QoS and how well users expectations are satisfied, (e.g., clear voice, and jitter-free video). Application-level QoS mechanisms are used to improve QoS without making changes to the network infrastructure by modifying the application implementations in order to make them more adaptive to the variations in delay and packet losses, (e.g. reconstruction methods at the audio receiver such as FEC and using transport protocols such as RTP and RTCP). These mechanisms are discussed in detail in Section 5.9.
2. Network-level QoS refers to tangible measurements of factors such as delay, bandwidth, and packet loss. This approach tries to modify the network implementation in order to support different grades of services with some performance guarantees. Issues that play a role in the network-level QoS implementations are QoS specifications of individual flows, network bandwidth resource management, admission control, QoS verification of individual/aggregated flows, traffic policing, packet forwarding mechanisms, and a QoS routing mechanism that satisfies the given QoS constraints. For more details refer to Section 5.10.

5.7 SIGNALING AND CONTROL PROTOCOLS

Signaling is required to make a connection between two communicating parties. Two protocols used for signaling are H.323 and SIP; these protocols are briefly discussed in this section.

Table 5.3

QoS Requirements for Heterogeneous Traffic Characteristics for Internet Applications

Applications	Traffic Characteristics	QoS Requirements
E-mail (SMTP) File Transfer (FTP) Remote Terminal (Telnet)	Small file size and transferred in batches	Delay: Very tolerant Bandwidth (BW): Low Packet loss: Service type: Best effort
Web Browsing (HTTP)	Series of small file size and bursty transmission in nature	Delay: Moderate Bandwidth (BW): Variable Packet loss: Service type: Best Effort
Client-Server Exchange e-Commerce	Two-way transmission of several small files	Delay: Sensitive Bandwidth (BW): Moderate Packet loss: Highly sensitive Service type: Must be reliable
VoIP Real-time Audio	Constant bit rate or variable bit rate	Delay: Highly Sensitive Bandwidth (BW): Low/moderate Packet loss: Low & predictable Service type: High priority
Streaming Video	Streaming traffic with variable bit rate	Delay: Highly sensitive Jitter: Highly sensitive Bandwidth (BW): High Packet loss: Low & predictable Service type: High priority

5.7.1 H.323

H.323 is a ITU recommendation that describes terminals, equipment, and services for multimedia communication over LAN that do not provide a guaranteed QoS [19]. The terminals and equipment can carry real-time voice, data, and video. Assurance or means of assurance of QoS is out of the scope of this standard. An H.323 terminal can be stand-alone or implemented in a personal computer, it is mandatory that the terminal support voice. Discussing H.323 in depth is impossible in one section. Thus this section will introduce the main points of the specification.

The H.323 standard is an umbrella specification; that is, it refers to other recommendations. These recommendations include signaling protocols: registration admission status (RAS) H.225.0; Q.931 (H.225.0), used for call setup and termination; H.245 control and capabilities exchange; H.261 and H.263 video codecs; G.711, G.722, G.728, G.729, and G.723 audio codecs; and the T.120 series of multimedia communications protocols. These specifications define several network components or entities; these are terminals for real-time bi-directional communication, multipoint control unit (MCU) for support of conference between 3 or more terminals; gateway, which provides connectivity between an H.323 and a non-H.323 network and gatekeeper, which provides services such as AAA for terminals and gateways, bandwidth management, addressing, call routing, billing, and charging. A network architecture is given in Figure 5.6. A protocol architecture is given in Figure 5.7.

The signaling is transported over the Transmission Control Protocol (TCP). The H.323 media is transported using RTP and RTCP using the User Datagram Protocol (UDP).

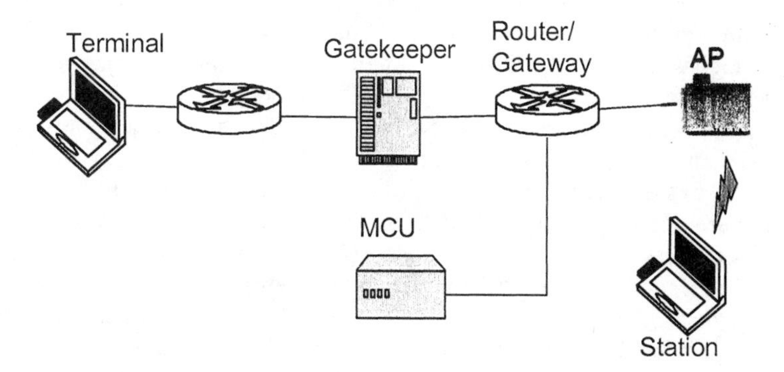

Figure 5.6 WLAN and H.323 network architecture.

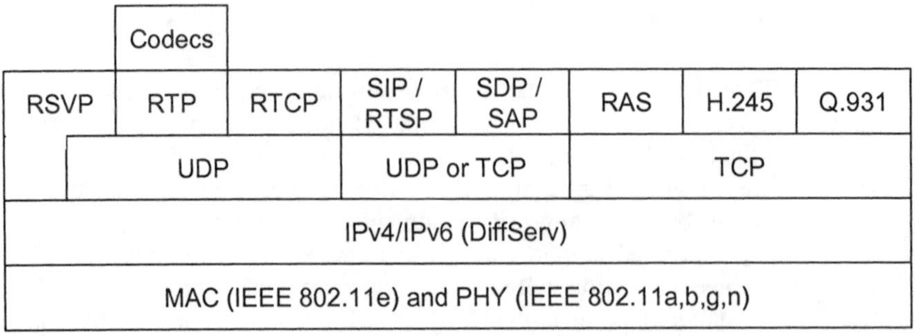

Codecs

RSVP | RTP | RTCP | SIP / RTSP | SDP / SAP | RAS | H.245 | Q.931

UDP | UDP or TCP | TCP

IPv4/IPv6 (DiffServ)

MAC (IEEE 802.11e) and PHY (IEEE 802.11a,b,g,n)

Figure 5.7 Protocol architecture for H.323 and SIP.

5.7.2 Session Initiation Protocol (SIP)

The SIP is a signaling protocol for end to end session setup and control to provide Internet conferencing and telephony. SIP was developed within the IETF Multiparty Multimedia Session Control (MMUSIC) working group. In order to provide SIP-based QoS for audio and video services, reservation of resources may be needed. SIP does not supply the management or control functions by itself but relies on other protocols. Admission control determines whether the network has sufficient resources to support the QoS required for a call, and accepts or rejects the call accordingly. In order to do admission control, the protocol must handle bandwidth management, call management, and bandwidth control. SIP does not support resource reservation by itself. It uses external means and techniques in order to provide QoS, such as the RSVP; this also means that first the user application must be aware of the QoS mechanism used in the access network and the signaling protocols such as RSVP and Common Open Policy Service (COPS). Second, the user terminal also needs to implement the RSVP protocol, which increases the complexity. A SIP-based IP telephony architecture with end-to-end QoS support using the Integrated Services (IntServ) model that suffers from complexity and scalability issues is described in [13].

SIP addresses users by an email-like address and re-uses some of the infrastructure of electronic mail delivery such as DNS MX records or using SMTP EXPN for address expansion. SIP addresses (URLs) can also be embedded in Web pages. SIP is addressing-neutral, with addresses expressed as URLs of various types such as SIP, H.323, or telephone (E.164). SIP can also be used for signaling Internet real-time fax delivery. This requires no major changes. Fax might be carried via RTP, TCP (e.g., the protocols discussed in the Internet fax WG), or other mechanisms. SIP is independent of the packet layer and only requires an unreliable datagram service, as it provides its own reliability mechanism. While SIP typically is used over UDP or TCP, it could, without technical changes, be run over IPX, or carrier pigeons, frame relay, ATM AAL5, or X.25.

SIP utilizes other standards to provide signaling and control functions. These standards are Session Description Protocol (SDP) and Session Announcement Protocol (SAP). SDP provides session announcement and session invitation in a multimedia environment. This allows the recipients of the session announcement to participate in the session [20]. SAP is for advertising multicast conferences and sessions. It is intended to announce the existence of long-lived wide-area multicast sessions.

5.7.3 Real-Time Streaming Protocol (RTSP)

The Real Time Streaming Protocol (RTSP) is used to establish and control time-synchronized streams of continuous audio and video. Sources of data can include both real-time and stored clips. During an RTSP session, an RTSP client may open and close many reliable transport connections to the server to issue RTSP requests. Alternatively, it may use a connectionless transport protocol, such as UDP. The streams controlled by RTSP may use RTP, but the operation of RTSP does not depend on this protocol. RTSP is not typically responsible for the delivery of the data streams, although interleaving of the media stream with the control stream is possible.

5.8 MEDIA GATEWAY PROTOCOLS

Media gateways are used for interconnecting two networks, for us the IP and the PSTN networks. The Media Gateway Control Protocol (MGCP), developed by IETF, and Megaco, developed by IETF and ITU-T (recommendation H.248), are two protocols used by the media gateways.

MGCP defines a gateway that focuses on the audio signal translation function, and a call agent that handles the call signaling and call processing functions. As a consequence, the Call Agent implements the signaling layers of the H.323 standard.

In Megaco/H.248 there are two components: terminations and contexts. Terminations represent streams entering or leaving the media gateway (for example, analog telephone lines, RTP streams, or MP3 streams). Terminations have properties, such as the maximum size of a jitter buffer, which can be inspected and modified by the call agent (media gateway controller, sometimes called a softswitch). Terminations may be placed into contexts, which are defined as when two or more termination streams are mixed and connected together.

5.9 TRANSPORT PROTOCOLS

In this section transport protocols for the purpose of real-time traffic transmission are discussed. The protocol is simply known as RTP; RTP is for data transmission while RTCP is used for control. In this section both RTP and RTCP are discussed.

5.9.1 Real-Time Protocol (RTP)

RTP uses UDP to provide real-time service; the application can make use of multiplexing and checksum provided by UDP. The protocol provides synchronization of data through a time-stamping mechanism, so that the receiver can play back the data in the correct order; it also provides payload identification and loss detection by using sequence numbering. The protocol assumes that the underlying network is reliable. RTP is also used for multicast applications.

The RTP packet header provides the following information:

- Payload type: The payload type is a 7-bit field that specifies two categories of data: audio and video.
- Sequence number: The sequence number of 16 bits is used by the receiver to restore the packet order and detect packet loss. Every RTP packet gets a sequence number.
- Time-stamp: The time-stamp field contains a value that represents the time when the data was sampled. The clock resolution depends on the synchronization accuracy required by the application.
- Synchronization source (SSRC): The synchronization source is a 32-bit number that is randomly generated to uniquely identify a source. The synchronization number is used by the receiver to assemble the packets from a particular source for playback.
- Contributing source (CSRC): The contributing source field contains the contributing sources for the data in the RTP packet. This is used when a mixer has been deployed to combine different streams of RTP packets with the same payload type from different senders.

5.9.2 Real-Time Control Protocol (RTCP)

RTCP is used to provide feedback from the real-time traffic conditions in the network. This information is not available by using RTP alone. RTCP is responsible for providing sender and receiver reports that include information such as statistics and packet counts. The information provided by RTCP can be used to diagnose the network condition by a network operator. It uses a separate UDP port, usually one higher than that of the RTP protocol.

RTCP provides five types of control information used to provide network condition and thus service information; these are listed as follows:

- Sender report (SR): The sender report is sent by the source of an RTP stream to inform the receiver what it should have received.
- Receiver report (RR): The receiver report has the same function as that of the sender report, with information such as the cumulative number of packets lost and the highest sequence number received.

- Source description report: The source description report is used from the RTP sender to provide more information about itself, such as the e-mail address of the user, phone number, and location.
- BYE: This is used when a source leaves the conference.
- APP: The APP message is used for new applications that might appear in future; this can be used for testing the application.

5.10 NETWORK-LEVEL QOS

In this section network-level QoS in the intra-domain and inter-domain is discussed. The two methods are IntServ and DiffServ.

5.10.1 Integrated Services (IntServ)

The IntServ architecture uses an explicit mechanism to signal per-flow QoS requirements to network entities such as hosts and routers. Network entities depending on the available resources implement one of the defined IntServ services guaranteed services or controlled load services based on which QoS will be delivered in the data transmission path. The RSVP signaling protocol was designed as a dynamic mechanism for explicit reservation of resources in IntServ (although IntServ may use other mechanisms as well).

In the context of IntServ a flow defines a stream of packets with the same source and destination address and port numbers. IntServ requires a protocol that would first check for available resources (admission control) and then reserve some amount of flow-specific state in the network. RSVP (the signaling protocol) is used here to communicate the available information about the amount of network resources requested by an application. Some of the RSVP features are summarized below. Figure 5.8 shows the components that are used in an IntServ model.

- RSVP is not a transport, but a network control protocol. As such, it does not carry data, but works in parallel with TCP or UDP data flows. It has no idea what is inside the objects that it transports.
- RSVP supports both unicast and multicast.
- It maintains soft-state operation (reservations must be periodically refreshed, or they will be timed-out, "connectionless").
- Reservations are receiver-based, in order to efficiently accommodate large heterogeneous (multicast) receiver groups.
- Although RSVP traffic can traverse non-RSVP routers, this creates a "weak-link" in the QoS chain where the service falls back to "best effort" (i.e., there is no resource allocation across these links).

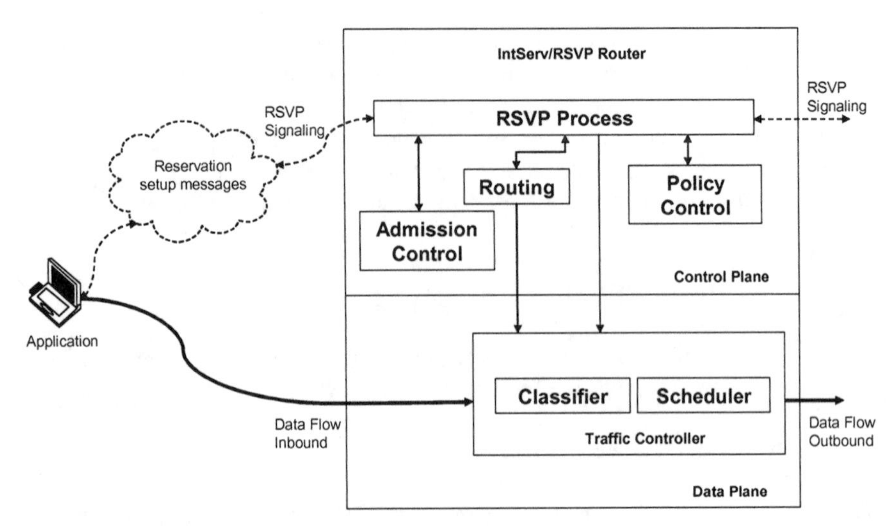

Figure 5.8 Basic components of IntServ/RSVP architecture.

The IntServ/RSVP router supports jointly the RSVP process and the routing process. The RSVP process also needs to interact with the traffic controller block (which consists of traffic classifier and scheduler mechanisms), admission controller, and policy controller. These factors are discussed as follows.

- RSVP process: Is responsible for RSVP state maintenance (path and reservation states) and for processing the corresponding tear down and error messages. A PATH message will trigger the formation of PATH state block (PSB), while RESV message will cause the creation of the RESV state block (RSB). The PSB will maintain the information corresponding to the PATH message, the related outgoing interface, and whenever there will be new PATH messages coming in, the PSB will be checked in order to determine whether these incoming messages are refresh messages, updates, or entirely new. Like the PSB, the RSB will maintain all the relevant information contained in the RESV message and all incoming RESV messages will be compared with the RSB in order to determine whether they are new, updated, or refreshed. For new and updated RESV messages in normal functioning there should always be an applicable PSB, since path state is installed before the RESV is issued. Besides maintaining the PSBs and RSBs, the RSVP process will interact with traffic control mechanisms to establish adequate handling of flows. Furthermore it will perform admission control on an outgoing interface. Finally by installing adequate flow information in the traffic control components, the RSVP process takes care of proper traffic provisioning.
- Routing: Interacts with the RSVP process in the IntServ/RSVP router. Data packets belonging to a flow and all its RSVP messages are

forwarded on the same interface and this is determined by the RSVP process. The routing process will interact with the RSVP process to determine the local previous hop (PHOP) interface.

- Policy control: In the IntServ/RSVP router responsible for enforcing the network administrator's policies on the network resources. This block determines who is eligible to make QoS reservations. Therefore, it interoperates with the RSVP process.
- Admission control: Applied to each node and performs a local accept/reject decision for the reservation request based on a comparison of the resources requested and the available network resources in the specific node (router), along with any additional configured policies relevant to resource allocation and administrative reservation permission (policy control). Admission control is sometimes confused with policing, which is a packet-by-packet function at the "edge" of the network to ensure that a host does not violate its promised traffic characteristics. Policing is considered to be one of the functions of the packet scheduler. To perform admission control, tracking of resource availability is necessary. This can be done using real-time measurements of the network traffic with keeping record of the reservations.
- Traffic Control: This provides specific link layer classification, scheduling, and policing capabilities. These components provide the QoS functionality:
    - Packet classifier: It performs multiple field classification to determine the packet's class and therefore the proper service level. All packets in the same class get the same treatment from the packet scheduler. Choice of a class may be based upon the contents of the existing packet header(s) and/or some additional classification number added to each packet. In addition, the classifier will be updated with the information contained in the filterspec.
    - Packet scheduler: Applies one or more traffic management mechanisms (a set of queues or timers) to ensure the packet is transmitted into the network in time to satisfy the bandwidth and delay constraints of the flow. The scheduler will be updated with the information contained in the flowspec.
    - Flowspec: Information about the QoS that is required by the flow, application's QoS requirements as a list of parameters, that tells each router along a data path which resources it should reserve for that particular flow.
    - Filterspec: This is used by senders in a PATH message to identify themselves, and by recipients in RESV messages to select which senders they are interested in receiving messages from.

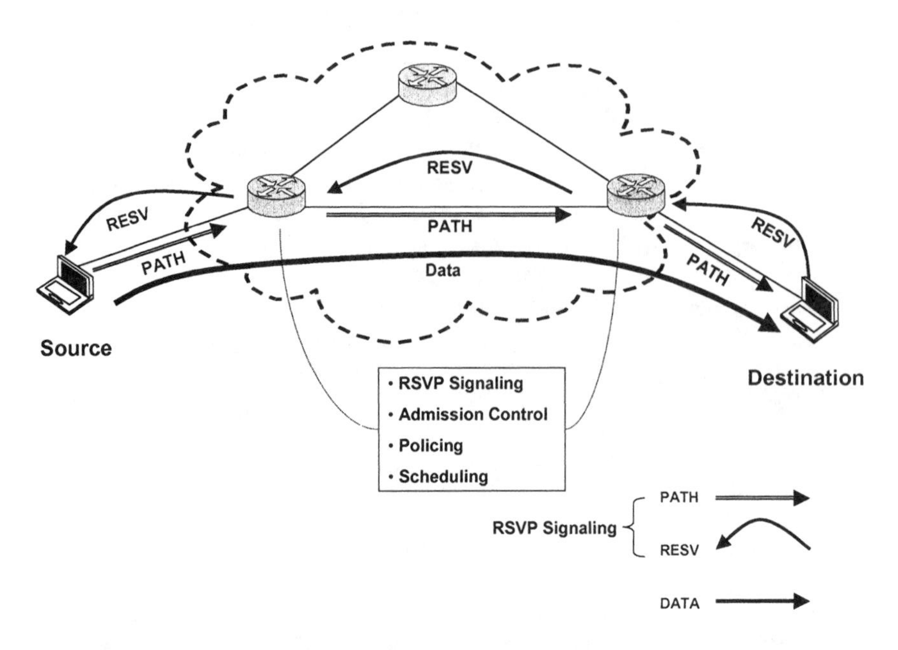

Figure 5.9 Basic message exchange for resource reservation in RSVP.

Currently there are two traffic classes in the IntServ model, guaranteed service (GS) and controlled load (CL). The latter is a more qualitative service and requires a congestion-free network.

5.10.1.1 Resource Reservation Protocol Signaling

The signaling process for the RSVP is illustrated in Figure 5.9. The sender sends a PATH message to the receiver specifying the characteristics of the traffic. The traffic description includes: the identity of the sender, the sending application, the sent traffic profile (bandwidth and burst characteristics), and the classification criteria by which the traffic can be recognized (IP addresses and ports). Every intermediate along the path forwards the PATH message to the next hop determined by the routing protocol. The PATH state is installed at each network device along the route. Upon receiving a PATH message the receiver responds with a RESV message to request resources for the flow. If the request is rejected the router will send an error message to the receiver and the signaling process will be terminated. In the opposite case if the request is accepted, link bandwidth and buffer space are allocated for the flow, and the related flow state information will be installed in the router.

RSVP must generate periodic PATH and RESV messages to refresh the reservation state in the network, or it will time out. If a new PATH is computed (e.g., by dynamic IP routing) due to change in topology, the RSVP message exchange must occur over that new PATH. RSVP supports different reservation styles. One style specifies whether a reservation should be dedicated for a particular sender or shared among different senders in the same session. For example a fixed-filter (FF) reservation style indicates that a single reservation will be allocated for a single sender, which might be useful for videoconference applications requiring minimum bandwidth and tight bounds on delay. On the other hand the shared-explicit (SE) style enables multiple senders in the same session to share a reservation (the person currently speaking, out of the group, is only able to use the reserved bandwidth at that moment).

Data flows for an RSVP session are characterized by senders in the traffic specification (TSpec) contained in PATH messages, and mirrored in the reservation specification (RSpec) sent by receivers in RESV messages. The token-bucket parameters, bucket rate, bucket depth, and peak rate are part of the TSpec and RSpec. Currently there are two traffic classes in the IntServ model, GS and CL. The latter is more qualitative service and it requires a congestion-free network [9, 10].

5.10.1.2 Drawbacks of RSVP/IntServ

There are several practical disadvantages to the IntServ approach [11]; these are described as follows:

- The requirements on routers are high: Every device along a packet's path, including the end systems like server and PCs, needs to be fully aware of RSVP and capable of signaling the required QoS (admission control, multiple field classification, and packet scheduling mechanisms).
- Reservations in each device along the path are "soft," which means they need to be refreshed periodically, thereby adding traffic on the network and increasing the chance that the reservation may time out if refreshed packets are lost.
- The amount of state information, which increases proportionally with the number of flows, places a huge storage and processing overhead on the routers. Therefore using this architecture will not scale well in the Internet core.
- Maintaining soft-states in each router, combined with admission control at each hop, adds complexity of each network node along the path.
- Since state information for each reservation needs to be maintained at every router along the path, scalability with hundreds of thousands of flows through a network core becomes an issue.

5.10.2 Differentiated Services (DiffServ)

The DiffServ (DS) architecture requires that the network may be divided into manageable domains called DS domains as shown in Figure 5.10. A DS domain is a contiguous set of DS nodes that operates with a common service provisioning policy and set of per hop behavior (PHB) groups implemented on each node. A DS domain normally consists of one or more networks under the same administration; for example, an organization's intranet or an ISP. A "service" in this context is defined as some significant characteristics of packet transmission in one direction across a set of one or more paths within a network.

Each data packet that enters a Diffserv network is marked with a Diffserv code point (DSCP) in a newly defined IP header field to indicate which PHB should be applied to the packet. Packets marked with the same code point are considered behavior aggregate, and all receive the same PHB treatment, regardless of the micro-flow to which they belong. The Diffserv framework achieves its scalability by moving complexity out of the core of the network into the edge routers. Per-application flow or per-customer forwarding state is only maintained at domain boundary nodes; it need not be maintained within the core of the network. More specifically, the complex classification and conditioning functions are only installed at the network boundary at ingress and egress nodes, and by applying PHBs to aggregates of traffic in the core routers.

PHBs are defined to permit a rough means of allocating buffer and bandwidth resources at each node among competing traffic streams. Currently two PHBs are within the Diffserv working group: Expedited Forwarding (EF) and the Assured Forwarding (AF). EF PHB is the key ingredient in DiffServ for providing a low-loss, low-latency, low-jitter, and assured bandwidth service. Applications such as VoIP, video, and on-line trading programs belong to this category. EF requires that the departure rate of the aggregate's packets from any DiffServ node be equal or exceed a configurable rate and that it should receive this rate independent of the intensity of any other traffic attempting to transit the node. The rough equivalent of the IntServ controlled load service is the AF PHB. The AF is a means for offering different levels of forwarding assurances based on the following:

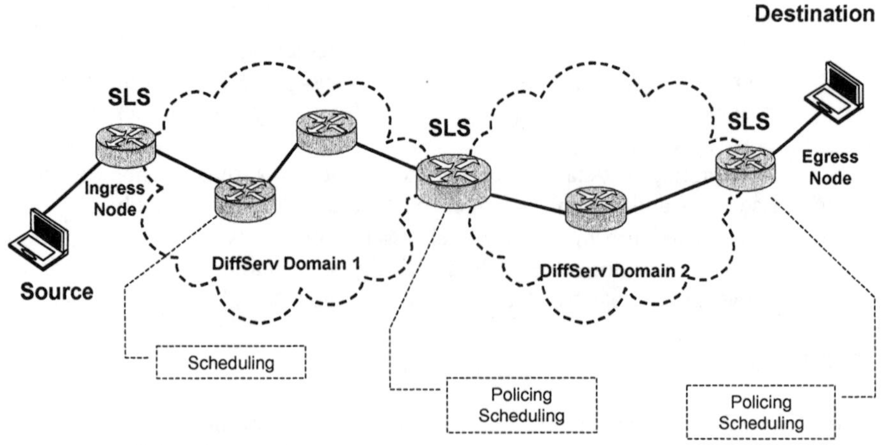

Figure 5.10 DiffServ framework.

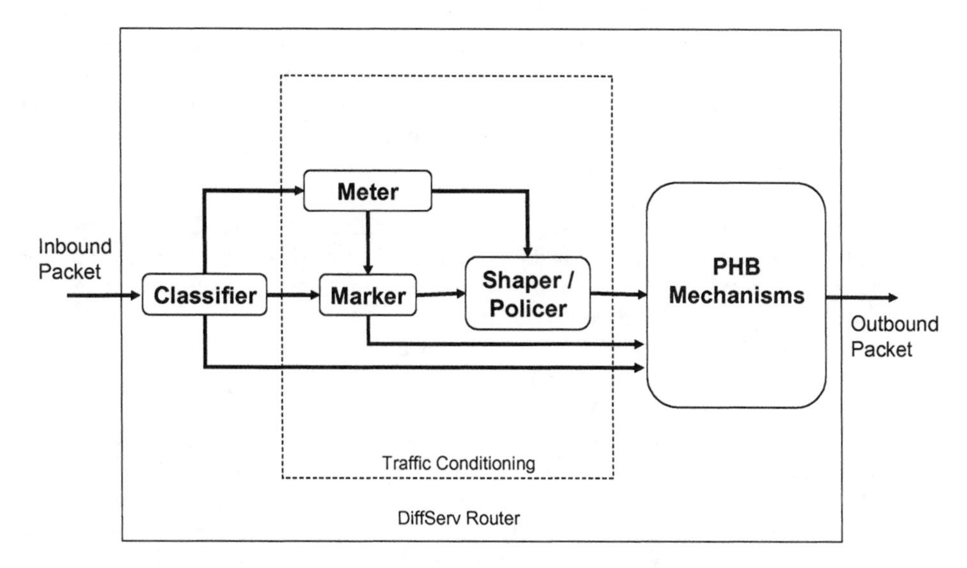

Figure 5.11 DiffServ router architecture with a logical view of packet classifier and traffic conditioner.

1. How much forwarding resources have been allocated to the AF class that the packet belongs to.
2. Current load of the AF class.
3. The drop precedence of the packet. Individual network elements implement these PHBs through a variety of mechanisms and queuing disciplines.

Two types of classifiers are defined. The behavior aggregate (BA) classifier classifies packets based on the DSCP only. The multi-field (MF) classifier selects packets based on the value of a combination of one or more header fields, such as source address, destination address, and DS field. PHBs may be specified in terms of their resource (e.g., buffer or bandwidth) priority relative to other PHBs, or in terms of their relative observable traffic characteristics (e.g., delay or loss). These PHBs may be used as building blocks to allocate resources and should be specified as a group (PHB group) for consistency.

5.10.2.1 Differentiated Services Architecture

The DiffServ architecture is based on a simple model where traffic entering a network is classified and possibly conditioned at the boundaries of the network and assigned to different behavior aggregates. The traffic conditioning components are able to classify, mark, shape, and drop packets as they enter and leave a Diffserv domain. An ingress router performs per-flow policing and marking, which is shown in Figure 5.11. As marked packets flow downstream, they are combined with similarly marked Diffserv packets into behavior aggregates. In order to ensure that the burst characteristics of aggregated data conform to the traffic specification for the required service, each Diffserv domain must have the ability to perform traffic shaping on each aggregate traffic as it exits the domain.

In order for a customer to receive differentiated services from its ISP, it must have a SLA with its ISP. An SLA basically specifies the service classes supported and the amount of traffic allowed in each class. An SLA can be either static or dynamic and exists only at the boundaries between domains. In general SLAs are complex business-related contracts that cover a wide range of issues, including network availability guarantees, payment models, and other necessities. The SLA may specify the PHB to be applied. Those actions can involve packet classification, policers (re-)marking, and shapers and their parameters.

For static SLA, routers can be manually configured with the classification, policing, and shaping rules. Other customers can share unused resources. Dynamic SLAs allow customers to request services on demand without subscribing to them. RSVP is used to request resources from ISPs. Admission control is needed for dynamic SLAs. In both cases, the ISPs' core routers must be shielded from the requests to avoid the scalability problem. SLAs may also include certain non technical guarantees and issues that do not stand directly on packet handling. An SLA is changed by the (human) parties involved in the agreement.

To facilitate QoS specification within the contract, an SLA will contain a service-level specification (SLS), also known as traffic conditioning agreement (TCA) which characterizes traffic profiles (often based on token bucket parameters) and the PHB to be applied to each aggregate. It specifies how transit traffic from one DS domain to another is conditioned at the boundary between the two DS domains. A DS ingress node is responsible for ensuring that the traffic entering the DS domain conforms to any SLS between it and the other domain to which the ingress node is connected. A DS egress node may perform traffic conditioning functions on traffic forwarded to a directly connected peering domain, depending on the details of the SLS between the two domains. The SLS between the domains is derived (explicitly or implicitly) from this SLA. The content of an SLS [12] includes the essential QoS-related parameters, including the following:

- Scope and flow identification (expected throughput, drop probability);
- Traffic conformance parameters (marking, shaping, mapping services);
- Service guarantees.

Metering, marking, shaping, and policing are discussed below:

- Meter: The meter's functionality is based on the temporal profile and the conformance levels, related to certain metering outputs. The temporal profile (i.e., average rate, and burst size) is chosen by the Diffserv network providers based on SLS, which is specified in a Traffic Conditioning Agreement (TCA), within which the customers receive a service for their traffic. Based on this profile the meter determines whether the customer's traffic is conforming or not (i.e., whether it is in-profile or out-of-profile). Depending on the policy decisions of the Diffserv network providers the conformance levels may be used to trigger actions in the other traffic conditioning blocks, i.e., marker or shaper/dropper.

- Marker: Packet markers set the DS field of a packet to a particular code point, adding the marked packet to a particular DS behavior aggregate. The marker may be configured to mark all packets, which are steered to it to a single code point, or may be configured to mark a packet to one of a set of code points used to select a PHB in a PHB group, the amount of traffic sent by the customer and also traffic load on the network according to the state of a meter. The marker has only a single parameter (i.e., 6 bit DSCP that needs to be marked).
- Shaper: The shaper delays some or all of the packets in a traffic stream in order to bring the stream into compliance with a traffic profile. A shaper usually has a finite-size buffer, and packets may be discarded if there is not sufficient buffer space to hold the delayed packets.
- Policer: The policer discards some or all of the packets in a traffic stream in order to bring the stream into compliance with a traffic profile. Only policing can be applied to inbound traffic on an interface. Note that a policer/dropper can be implemented as a special case of a shaper by setting the shaper buffer size to zero (or a few) packets. In the case where no traffic profile is in effect, packets may only pass through a classifier and a marker.

5.10.2.2 Inter Domain Resource Management for DiffServ Networks

The DiffServ domain is often outfitted with a bandwidth broker (BB), for resource managing purposes as a scalable QoS provisioning over DiffServ architecture. The BB architecture automates the provisioning of a DiffServ service between network domains (it negotiates SLAs between different ASs). A BB can be a host, a router, or a software process on an exit router. The BB is responsible for ensuring that resources within the DiffServ domain and on links connecting adjacent domains are properly provisioned and not oversubscribed.

A BB maintains information relating to the SLSs that are defined between a DiffServ domain and its customers. Customers include local users as well as the adjacent networks that provide connectivity to other parts of the Internet. BB automates inter domain SLA negotiations and maintains bilateral SLA agreements with its neighboring BBs to allocate resources to traffic crossing domain borders. The BB uses this SLS information to configure the routers in the local DiffServ domain and to make admission control decisions. The BB allows separately administrated DiffServ domains to manage their network resources independently, yet still cooperate with other domains to provide dynamically allocated end-to-end QoS.

5.10.3 Drawbacks of DiffServ Mechanism

Even though DiffServ is a good framework for providing scalable and coarse-grained QoS throughout the network, it has some drawbacks:

- Unlike RSVP/IntServ, DiffServ needs to be provisioned. Setting up the various classes throughout the network requires knowledge of the applications and traffic statistics for the aggregates of traffic on the network. This process of application discovery and profiling can be time-consuming.
- Management is still a big issue (billing and monitoring are still difficult).
- Even though QoS assurances are being made at the class level, it may be necessary to drill down to the flow level to provide the requisite QoS. For example although all HTTP traffic may have been classified as EF, and a bandwidth of 100 Mbps assigned to it, there is no inherent mechanism to ensure that a single flow does not use up that allocated bandwidth.

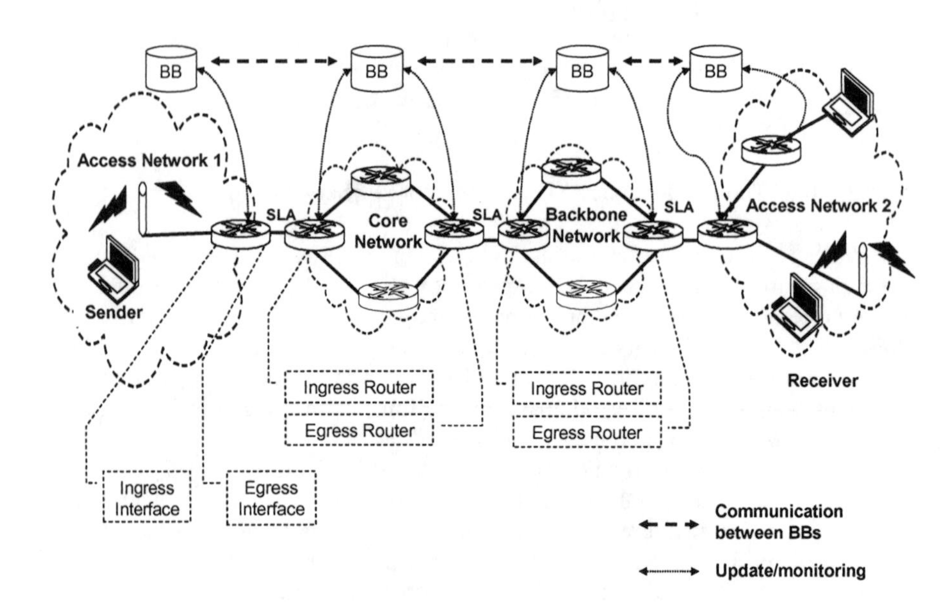

Figure 5.12 Inter domain resource management for DiffServ networks.

5.10.4 IntServ over DiffServ

IntServ and DiffServ architecture is designed to deploy QoS by means of different mechanisms for differentiating services over the Internet and each has its own advantages and disadvantages as mentioned in Sections 5.10.1 and 5.10.2. The benefit of IntServ over DiffServ architecture is obtained by combining the advantages of both mechanisms. IntServ provides means for flow-based end-to-end network-based QoS over different heterogeneous networks (only when all intermediate nodes must support IntServ mechanism) and DiffServ aggregates traffic and guarantees end-to-end services that are not supported by RSVP/IntServ due to the scalability issue. Combining IntServ and DiffServ will provide a scalable end-to-end network QoS. In fact DiffServ may use RSVP mechanism to properly provision QoS services across the networks as described below:

- In a DiffServ-enabled network a static provisioning policing mechanism is used for admission control at network nodes but using RSVP will enable DiffServ to apply resource-based admission control and/or policy-based admission control and thereby optimize the resource utilization in the network.
- In DiffServ a DCLASS object is used to carry DSCP information between a DS network and upstream nodes that may wish to mark packets with DSCP values. A DCLASS object when used with RSVP signaling will affect the static nature of DiffServ admission control by applying QoS policy with a DS network [18].
- IntServ network nodes perform flow-based traffic admission control. Therefore in the IntServ over DiffServ network only traffic that is admitted by IntServ nodes gets to enter the DS network. This enhances the DS nodes ability to provide guaranteed services in the ensuing aggregated traffic provision.

5.10.5 Policy Management and Billing

The allowance of one user to get a better service than another requires QoS policy enforcement that will require a policy management infrastructure. Policy management, Authentication Authorization Accounting (AAA), and billing are essential to the success of providing different services (QoS) [21, 22]. All of them present technical challenges and another very important scaling consideration that applies to all three of these services is managing peering arrangements between various ISPs. Before one can enable end-to-end QoS, bilateral agreements must be in place so ISPs sharing QoS responsibilities for common flows can share the necessary policy, authentication, and accounting/billing information.

5.10.5.1 Common Open Policy Service (COPS)

The COPS protocol [14] is a simple client/server query and response protocol that allows policy servers [policy decision points (PDP)] to communicate policy decisions over QoS signaling protocols to network devices [policy enforcement points (PEP)]. The model does not make any assumptions about the methods of the policy server, but is based on the server returning decisions to policy requests. The main characteristics of the COPS protocol are as follows:

- A client/server model where the PEP sends requests, updates to a remote PDP, and the PDP returns decisions back to the PEP.
- The protocol uses TCP as its transport protocol for the reliable exchange of messages between policy clients and server. The PEP is responsible for initiating a persistent TCP connection to a PDP.

Two main models are supported by the COPS protocol: the outsourcing model and the provisioning model:

- COPS outsourcing model: The outsourcing model is sometimes referred to as "Pull" mode, or "reactive" mode, since the PEP pulls policy decisions from the PDP, and the PDP reacts to PEP events. The outsourcing model is used when there are "trigger events" in the PEP that require a policy decision (e.g., dynamic request to admit a new flow). The PEP delegates this to an external policy server (PDP). It sends a query message to the PDP, typically waiting for the response decision before admitting the new flow. The outsourcing model is for instance used for the RSVP client type defined in [15].
- COPS provisioning model: The provisioning model is almost the reverse of the outsourcing model [16]. The PDP typically predicts future configuration needs, and foresees that the PDP proactively configures the PEP so that it knows how to run its QoS mechanisms. Rather than responding to PEP events, the PDP prepares and "pushes" configuration information to the PEP, as a result of an external non-PEP event, such as change of applicable policy, time of day, expiration of account quota, or a result of third party (non PEP) signaling. In the provisioning model, either there are no "trigger events" at the PEP (i.e., only packet classification, marking, or scheduling) or these events must be handled using local information (i.e., mapped in the available resources provisioned by the PDP).

5.10.5.2 Open Settlement Protocol (OSP)

The Open Settlement Protocol (OSP) [17] is a client-server protocol (that consists of a set of protocols and associated profiles) defined by the ETSI TIPHON to establish authenticated connections between gateways/servers, and allows gateways and servers to transfer accounting and routing (optionally) information securely. OSP allows service providers to deliver multimedia services without establishing direct peering agreements with other service providers. It has XML based technical specifications and relies on HTTP and Transport Layered Security (TLS) for transport and is further independent of the call signaling protocols. With OSP a service provider is able to contact a clearinghouse, bandwidth brokers, and resource management servers and receive multiple routes that match the service provider's criteria. The actual route costing and route selection is determined by a higher layer application. OSP is only used for transporting such information, not determining it. Billing or settlement information is sent in a UsageIndication message. This contains certain standard data elements, but also contains a sequence of UsageFields that can be extended as needed to carry additional data elements. There is a mechanism for a server to indicate specific UsageFields that it prefers or that it requires. UsageIndication can be sent multiple times during a call to send partial information or update previously sent data. Servers can specify call events for which a client should send a UsageIndication. When calls are authorized, the server returns a token that proves this authorization. The token is carried in subsequent call signaling for that call and can be verified by the destination administrative domain. OSP is highly flexible in deployment: The different protocol features may be used independently of one another — for example, OSP could be used for call detail reporting, while call authorization is being done by a separate mechanism. The protocol provides access to the clearinghouse's database of available service providers, but does not require any constraints on the policies implemented by the clearinghouse.

5.10.5.3 Clearinghouse (Billing Settlement Agent)

A clearinghouse is basically a translator of different operations support systems (OSSs). OSS is used to support such jobs as service ordering, provisioning, trouble administration, and billing transactions. If different network providers have to communicate with each, and each of them uses a different OSS system then communication is impossible. A solution for the network providers is to communicate with all existing network providers by having one-to-one contracts; such a solution obviously is not scalable. A clearinghouse can solve this by translating messages from and to different OSSs. One of the clear benefits of a clearinghouse is the customer who gets one bill even though he or she has used different networks.

5.11 QOS SUPPORT ACROSS HETEROGENEOUS ACCESS NETWORKS

A heterogeneous network is shown in Figure 5.13. The provision of QoS in such a network is discussed in this section.

5.11.1 Top-to-Bottom System QoS Model

In this section a method to provide top-to-bottom QoS is discussed [12]. Top-to-bottom QoS means each layer in the protocol stack has QoS provision, understands QoS requirements, and has either a similar understanding of QoS or understands the QoS parameters of the layers higher and lower. End-to-end QoS means provision of QoS from one terminal to another or more specifically from the senders' application layer to the receivers' application layer. Without provision of the top-to-bottom QoS, the end-to-end QoS is not possible. In Figure 5.14 different protocol layers for the top-to-bottom QoS are given.

An explanation of communication of top-to-bottom is given below:

1. The application should be capable of providing QoS services like video, audio, or voice and should be capable of calling on a QoS-based application program interface (API). The API should be built such that it can call on the QoS sockets.

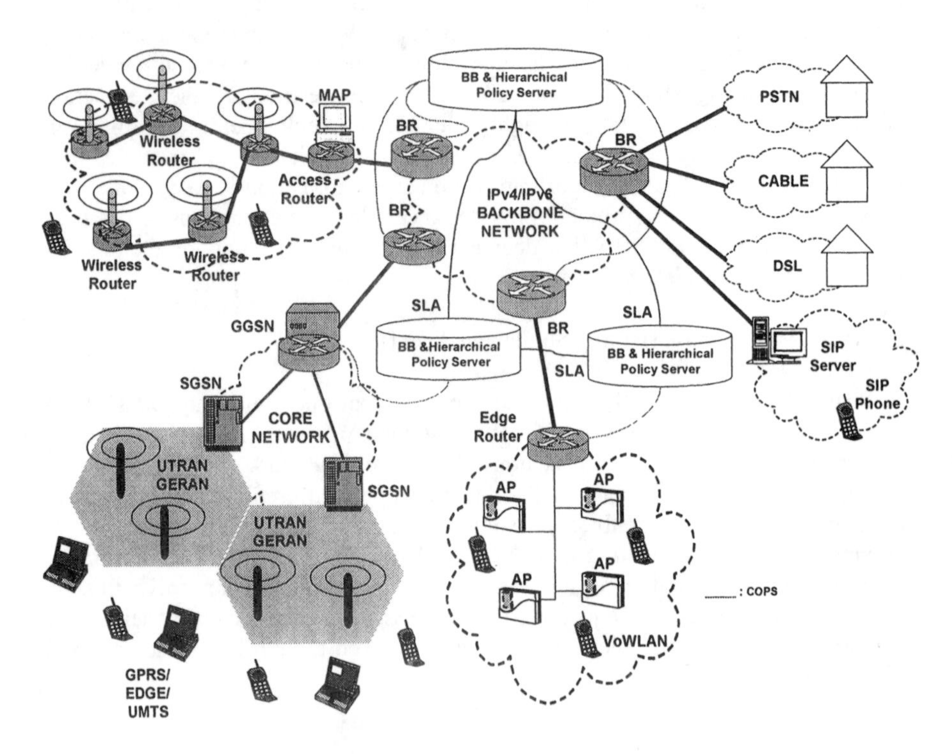

Figure 5.13 QoS support for IP based heterogeneous access networks using hierarchical bandwidth broker and resource management.

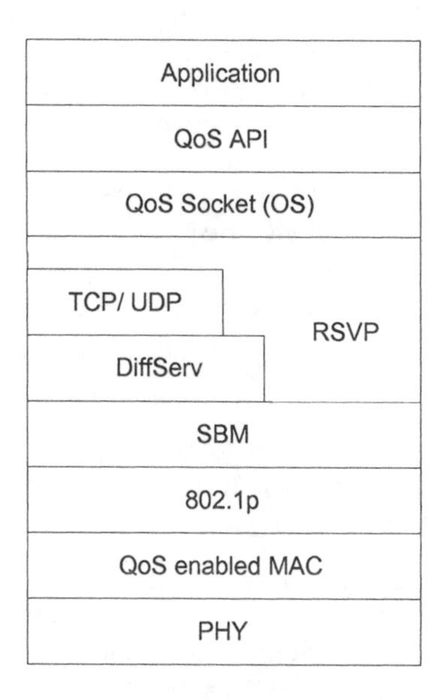

Figure 5.14 QoS protocol stack in wireless stations (end-to-end system QoS).

2. The QoS information can thus by using the API and the socket be mapped on the RSVP protocol, [8], or another protocol.
3. The information is then mapped to the Subnet Bandwidth Manager (SBM), [9], for bandwidth allocation and admission control.
4. The SBM maps the quality needed on the IEEE 802.1p, which in turn maps the required QoS on the QoS MAC.
5. The QoS MAC then performs medium access so as to provide the required QoS.

Figure 5.14 combined with Figure 5.7 gives a complete picture of the protocols discussed in this chapter.

5.11.2 Intra Domain and Inter Domain End-to-End QoS for Heterogeneous Access Networks

The guarantee of end-to-end QoS requires an efficient resource management mechanism that can reserve and control resources like bandwidth, delay, and jitter according to a policy for immediate and future resource utilization. This requires strict admission control and delay calculation is necessary; also the bandwidth usage has to be strictly policed within intra domain and inter domain. The various mechanisms and protocols needed to provide QoS in IP networks, and managing and coordinating them across several networks can be a difficult task. It is not possible to manually configure every network device with the right queuing and traffic processing mechanisms to provide consistent, priority-based service everywhere necessary in a large-scale network. In addition, QoS applications must continue to work properly even if the network is dynamic and network topologically changes frequently. The "traditional" network management applications cannot meet those requirements. To fully automate the decision-making process, using a third-party hierarchical policy-based networking offers a new way of controlling the QoS capabilities in the network. All this requires QoS routing. The main purpose of QoS aware routing solutions is twofold:

1. Satisfy the QoS requirements for every admitted connection;
2. Achieve global efficiency in resource utilization.

For QoS routing the key issue is to select network routes with sufficient resources for the requested QoS parameters. Before this can be realized, the following challenges must be considered:

- Distributed applications such as Internet phone and distributed games may have very different QoS constraints on delay, jitter, packet loss, bandwidth, and so on. Multiple constraints make the routing problem often intractable.
- Any future integrated service network is likely to carry both QoS and best-effort traffic, which makes the issue of performance optimization complicated.
- The network state (topology) changes dynamically due to transient load fluctuation.

A possible approach is based on the hierarchical policy server architecture for managing inter domain and the local policy server for intra domain QoS services. Hierarchical policy server architecture attempts to provide higher QoS assurance levels and higher network utilization, as offered by stateful networks (e.g., IntServ), while maintaining the scalability and robustness found in stateless network architecture (e.g., DiffServ).

In order to provide truly end-to-end intra domain QoS across wireline and wireless IP-based heterogeneous networks belonging to the same administrative domain first, the QoS mechanisms at the link layer must be in place for the wireless systems. Sections 5.3 and 5.4 addressed the link layer QoS support for the 802.11 system, namely MAC enhancement for QoS support. Second, the network-QoS as discussed in Section 5.6–5.10 needs to be provisioned so that the network is QoS-aware so that the intermediate nodes are able to differentiate priorities of individual flows and provision them accordingly. Finally the End_System_QoS which deals with top-to-bottom system/terminal QoS provisioning at each protocol layer from the application layer of the sender to the application layer of the receiver must be considered; this has been discussed in detail in Section 5.11.1.

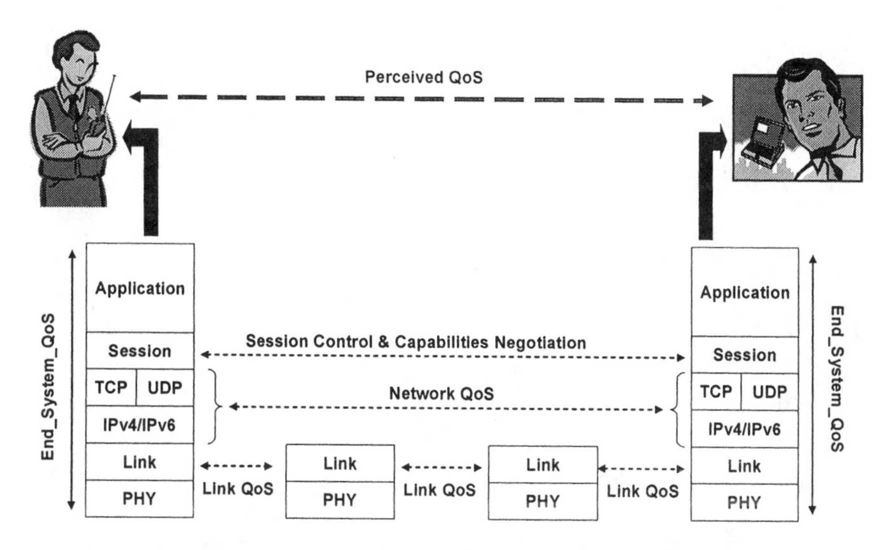

Figure 5.15 Perceived QoS for wireline and wireless heterogeneous networks.

In the case of inter domain where QoS guarantees are required across multiple administrative domains belonging to different stakeholders, hierarchical policy server architecture allows efficiently coordinating distributed resource management decisions to adapt the reservation of inter domain trunks. This in turn leads to flexible management of resources allocated across multiple domains and provides scalable and more predictable end-to-end QoS.

Perceived QoS by the end users (sender and receiver) at both ends of the communication depends on the link layer QoS mechanism, network level QoS techniques/architecture, and the end system top-to-bottom QoS interworking. End user perceived QoS is no stronger than the weakest link QoS (see Figure 5.15). Perceived QoS by the user = min {End_System_QoS, Network QoS, min (Link QoS)}. In order to deliver truly end-to-end QoS at the user it is crucial that the above mentioned QoS categories together with the hierarchical policy server are considered as part of the scalable QoS architecture.

Although this chapter discusses various protocols, methods, and network architectures to provide end-to-end QoS across heterogeneous access networks, there are also issues of security and mobility, which are discussed in Chapter 4 and Chapter 6, respectively.

5.12 VOICE OVER WLAN PRODUCTS

The WLAN network may significantly enhance the value proposition of VoIP. Wired VoIP reduces or eliminates the need for a separate circuit-switched voice network. Wireless VoIP may extend phone coverage throughout the WLAN coverage area and enable numerous applications. The 802.11e standard, which remains in development, will allow WLAN networks to achieve the QoS levels necessary for carrying voice traffic. At present there are products on the market that are being deployed primarily in the enterprise market segment offering a proprietary QoS solution [23–27].

References

[1] Prasad, N.R., and A.R. Prasad, editors, *WLAN Systems and Wireless IP for Next Generation Communication*, Norwood, MA: Artech House, 2002.

[2] IEEE Std 802.16-2003/DO, Part 16: Air Interface for Fixed Broadband Wireless Access Systems, 2003-08-01.

[3] IEEE P802.11e, Draft Supplement to IEEE Std 802.11, 1999 Edition, Draft Supplement to STANDARD FOR Telecommunications and Information Exchange Between Systems - LAN/MAN Specific Requirements - Part 11: Wireless Medium Access Control (MAC) and physical layer (PHY) specifications: Specification for Medium Access Control (MAC) Enhancements for Quality of Service (QoS).

[4] IEEE, "802.11, Wireless LAN Medium Access Control (MAC) and Physical Layer (PHY) specifications," November 1999.

[5] Gu, D., and J. Zhang, QoS Enhancement in IEEE 802.11 Wireless Local Area Networks, *IEEE Communications Magazine*, June 2003.

[6] Braden, R., D. Clark and S. Schenker, "Integrated Services in the Internet Architecture: an Overview," RFC 1633, June 1994.

[7] Schenker, S., C. Partridge, and R. Guerin, "Specification of guaranteed Quality of Service," RFC 2212, Sept. 1997.

[8] Wroclawski, J., "Specification of the controlled-load Network Element Service," RFC 2211, Sept. 1997.

[9] Wroclawski, J., "The use of RSVP with IETF Integrated Services," RFC 2210, Sep. 1997.

[10] Braden, B., et al., "Resource ReSerVation Protocol (RSVP) -- version 1," RFC 2205, Oct. 1997.

[11] Xia, X., and L.M. Ni, "Internet QoS: A Big Picture," *IEEE Network Communication Magazine*, Apr. 1999.

[12] Goderis, D., "Service Level Specification Semantics and Parameters," Internet Draft, Nov. 2000.

[13] Marshall, W., et al., "Integration of Resource Management and SIP," IETF RFC 3312, Oct. 2002.

[14] Durham, D., et al., "The COPS (Common Open Policy Service) Protocol," RFC 2748, Jan. 2000.

[15] Durham, D., et al., "COPS Usage for RSVP," RFC 2749, Jan 2000.

[16] Reichmeyer, F., "COPS usage for Policy Provisioning," work in progress, Oct. 2000.

[17] ETSI, "Telecommunications and Internet Protocol Harmonization Over Networks (TIPHON); OSP for Inter-domain pricing, authorization, and usage exchange," Technical Specification 101 321 version 2.1.1, Aug. 2000.

[18] Bernet, Y., "Format of the RSVP DCLASS object," RFC 2996, Nov. 2000.

[19] Open H.323 Project: http://www.openh323.org/, 29 June 2004.

[20] Handley, M., and V. Jacobson, "SDP: Session Description Protocol," RFC 2327, April 1998.

[21] Metin, E., M. Alam, and N. R. Prasad, "A Novel Model for Inter-Domain QoS Management for Real-Time Applications," 5th *International Symposium on Wireless Personal Multimedia Communications (WPMC'02)*, October 27 - 30, 2002, Honolulu, Hawaii.

[22] Prasad, N. R., M. Alam and M. Ruggieri, "A Framework for QoS and Security for Wireless Computing using AAA Architecture," *WPMC 2004*, September 12-15, 2004, Abano Therme, Italy.

[23] Colubris WLAN, Colubris Networks, http://www.colubris.com/ NewsItem. aspx?id=74.

[24] Cisco Wireless IP Phone 7920, Cisco Systems, Inc. http://www.cisco.com/ en/US/products/hw/phones/ps379/products_data_sheet09186a00801739bb.h tml.

[25] Cisco Aironet 1200 Series, Cisco Systems, Inc. http://www.cisco.com/en/ US/products/hw/wireless/ps430/products_data_sheet09186a00800937a6.htm l.

[26] NetLink Wireless Telephone e340 and i640, SpectraLink http://www. spectralink.com/products/nl-wts.html.

[27] PDT 8146, Symbol Technologies based on Intel XScale processor http://www.symbol.com/products/whitepapers/whitepapers_vot.html.

Chapter 6

Roaming, Handover, and Mobility

This chapter talks about mobility and handover; in essence they mean the same thing with a slight difference: Mobility usually is used for wired systems, in particular for Internet Protocol (IP) networks, while handover is used for wireless systems. Handover is sometimes also known as handoff. Achieving inter domain and intra domain handover is a must for the future growth of WLAN technology. Another term is roaming which is used for handover in WLAN [1–40]. In this chapter handover in WLANs and mobility methods in IP are discussed. The issues of QoS and security are also discussed in the chapter.

6.1 HANDOVER AND MOBILITY MANAGEMENT

In this section the basics of mobility management and handover are detailed. Handover is a component of mobility management and happens at the physical and data link layers.

6.1.1 Mobility Management

Mobility management supports roaming users to enjoy their services on progress through *different* access technologies either simultaneously or one at a time. Mobility management enables communication networks to do the following:

- Static scenario: Locate mobile node (MN) in order to deliver data packets;
- Dynamic scenario: Maintain connections with MN moving into new areas.

It contains two distinct managements:

- Location management: How to locate a MN, track its movement, and update the location information;

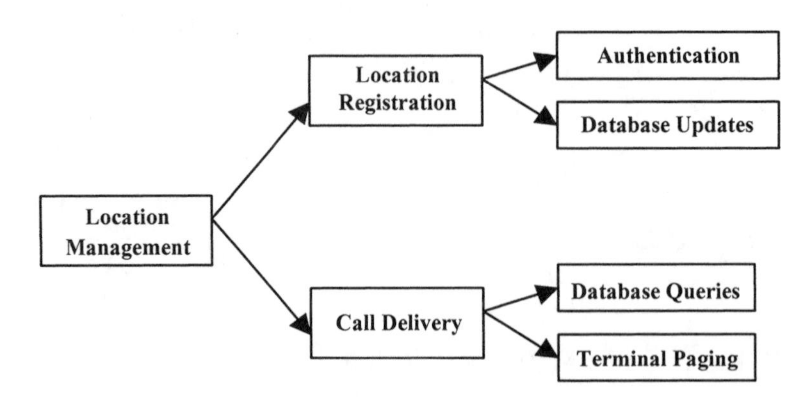

Figure 6.1 Location management operation [1].

- Handoff management: focuses mostly on the control of the change of a MN's access point (AP) during active data transmission.

The operations of the two components are shown in Figure 6.1.

Mobility affects the whole protocol stack, from the physical, data link, and network layers up to the transport and application layers. An example includes radio resource reuse at the physical layer, encryption and compression at the data link layer, congestion control at the transport layer, and service discovery at the application layer. The network layer offers routing from one network to another through an independent link. Since mobility, modeled as changing node's point of attachment, network layer supports mobility by changing the routing of packets destined to the mobile node to arrive at the new point of attachment. Implementing mobility management at the network layer may also shield the upper layer protocols from the nature of the physical medium and make mobility transparent to applications and higher level protocols such as TCP.

Besides the basic functions that implement the goal of mobility management, there are many other requirements on performance and scalability that should be carefully taken into account when trying to design or select a mobility management scheme, including the following:

1. Fast handoff: The handoff operations should be quick enough in order to ensure that the MN can receive IP packets at its new location within a reasonable time interval and so reduce the packet delay as much as possible.
2. Seamless handoff: The handoff algorithm should minimize the packet loss rate to zero or near zero, which, together with fast handoff, is something referred to as smooth handoff.
3. Signaling traffic overhead: The control data load (e.g., the number of signaling packets or the number of accesses to the related database), should be lowered to within an acceptable range.

4. Routing efficiency: The routing paths between the communication nodes to the MN should be optimized to exclude redundant transfer or bypass paths (e.g., triangle routing).
5. QoS: The mobility management scheme should support the establishment of a new QoS reservation in order to deliver a variety of traffic, while minimizing the disruptive effect during the establishment.
6. Fast security: The mobility scheme should support different levels of security requirements such as data encryption and user authentication, while limiting the traffic and time of security process (e.g., key exchange).
7. Special support required: It is better for a new mobility mechanism to require minimal special changes on the components (e.g., MN, router, communication media, networks, and other communication nodes).

There are many distinct but complementary techniques especially for mobility management to achieve its performance and scalability requirements listed above, including the following:

- Buffering and forwarding: To cache packets by the old attachment point during the MN in handoff procedure, and then forward to the new attachment point after the processing of MN's handoff.
- Movement detection and prediction: To detect and predict the movement of the mobile host between different access points so that the future visited network is able to prepare in advance and packets can be delivered there during handoff.
- Handoff control: To adopt different mechanisms for the handoff control, (e.g., layer two, or layer three-triggered handoff, hard, or soft handoff, mobile-controlled or network-controlled handoff).
- Paging area: To support continuously reachable with low overhead on location update registration through location registration limited to the paging area.
- Domain-based mobility management: To divide the mobility into micro-mobility and macro-mobility according to whether the mobile host movement is intra domain or inter domain.

6.1.2 Handover

Handover (HO) is a basic mobile network capability for the dynamic support of terminal migration. HO management is the process of initiating and ensuring a seamless and lossless HO of a mobile, from the region covered by one base station to another base station [2, 3]. On the other hand in mobile communications the term "roaming" is used for access and the use of the network of an operator different from the one the accessing user has a contract with (home network). Foreign and home network operators typically do have roaming agreements based on which the subscribers of the home operator are allowed to access and use the network of the foreign operator. For IEEE 802.11 WLANs roaming is the synonym for handover. In this chapter roaming and handover will be used interchangeably.

In any wireless communications system HO plays an essential part of radio resource management (RRM), radio resource being the bread and butter of the wireless communications industry.

There are a variety of issues related to HO. These issues are divided into two categories: architectural issues and HO decision time algorithms [2, 3]. Architectural issues are those related to the methodology, control, and software/hardware elements involved in rerouting the connection. Issues related to the decision time algorithms are the types of algorithms, metrics used by the algorithms, and performance evaluation methodologies. In the following sections a few of the architectural and algorithm issues are discussed.

The main requirements that HO must fulfill are latency, scalability, minimum drop off and fast recovery, QoS maintained or renegotiated, minimal additional signaling, goodput (i.e., actual data throughput), and maintenance of security. The performance of HO is based on call blocking probability, handover blocking probability, call dropping probability, rate of unnecessary handover, rate of handover, duration of interruption and delay.

6.1.3 Handover Metrics and Initiation Algorithms

Several algorithms are being employed or investigated to make the correct handover decision. Traditional HO algorithms are based on received signal strength (RSS) or received power P. Others are based on carrier-to-interference ratio (CIR), bit error rate (BER), and block error rate (BLER). There are of course more intelligent techniques being discussed in the literature like prediction techniques, pattern recognition based on neural networks or fuzzy logic, and enhanced techniques that are based on system parameters like data rate, and service type.

6.1.4 Handover Protocols (Control)

Handover procedures involve a set of protocols to notify all the related entities of a particular connection that an HO has been executed and that the connection has to be redefined. When a mobile moves from one point to other, it executes handover from one point of attachment to another. This usually means informing the old attachment point of change, known as disassociation, and reassociation at a new point of attachment. Network elements involved in providing the session should also be informed so that the service can be provided seamlessly. It is possible that a new connection is made after the old one is broken, in hard handover or two connections are simultaneously maintained until one of them improves above a threshold, in soft handover. The decision to perform handover can be at the network, in network-controlled handover (NCHO), at the mobile, in mobile-controlled handover (MCHO) or information can be sent by the mobile and used by the network entity in making the handover decision. This is called mobile-assisted handover (MAHO). In any case, the entity that decides on the HO uses some metrics, algorithms, and performance measures in making the decision as discussed in earlier sections.

6.1.5 Handover Methodology

There are vertical and horizontal handovers. Vertical handover is handover between two different systems (e.g., GPRS and WLAN), while horizontal handover is handover within same type of system. We can further classify handover as inter cell, inter sector, etc. All these handovers are based on two basic methods: hard or soft handover.

6.2 IEEE 802.11 HANDOVER SCENARIOS

IEEE 802.11 defines three different handover scenarios [5] (see Figure 6.2):

1. No-transition: In this type, two subclasses that are usually indistinguishable are identified:
   a. Static—no motion;
   b. Local movement—movement within the Basic Service Set (BSS) (i.e., coverage of the AP).

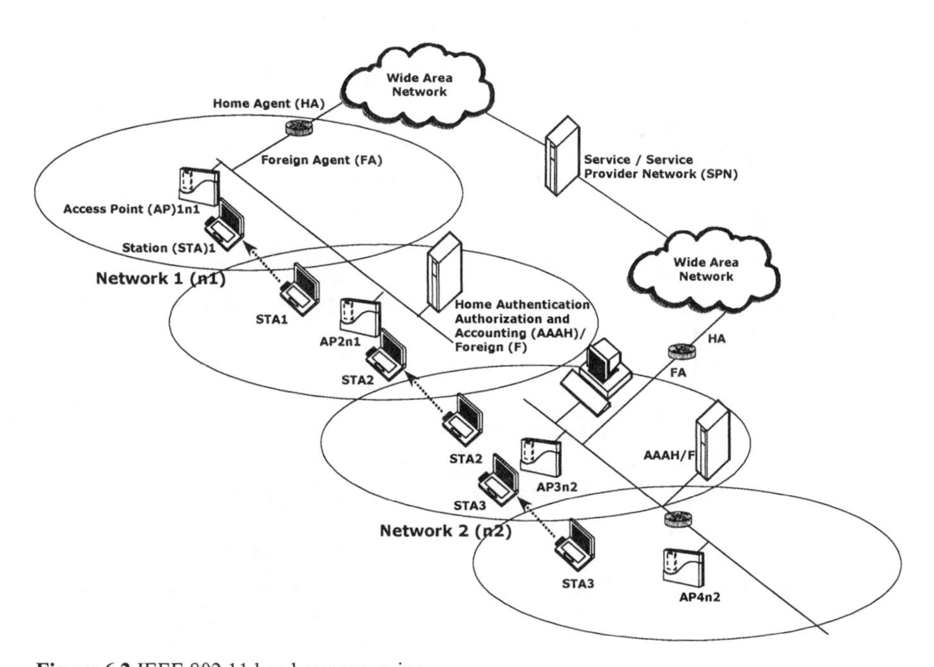

Figure 6.2 IEEE 802.11 handover scenarios.

2. AP transition: This type is defined as a station movement from one AP to another within the same extended service set (ESS).
3. ESS-transition: STA movement from a BSS in one ESS to a BSS in a different ESS. Usually a WLAN network is within one ESS and in an IP subnet. This case could mean the following:
   a. Inter subnet handover.
   b. Inter domain handover (e.g., between two different networks).

Most of the solutions available in the market provide handover for the first two scenarios; these can be done using layer-2, which is the medium access control (MAC) layer. The last scenario requires the involvement of higher layers. Solutions for all the scenarios will be discussed in this chapter.

6.3 IEEE 802.11 ROAMING

The standard also allows roaming between APs either in the same channel or a different channel. The standard does not define an exact procedure for this purpose [3–6].

6.3.1 Synchronization

Synchronization in IEEE 802.11 is done by the timing synchronization function (TSF) of the beacon. It is used for [3–6]:

- Power management:
  - Beacons sent at well known intervals;
  - All station timers in BSS are synchronized.
- Acquisition:
  - Stations scan for beacons to find networks.
- Superframe timing:
  - TSF timer used to predict start of contention free burst.
- Hop timing for FH PHY.

6.3.2 IEEE 802.11 Roaming Mechanism

IEEE 802.11 provides the following mechanisms for the support of roaming and initial AP selection:

- The MAC TSF offers a beaconing mechanism. For the infrastructure environment, it involves the sending of beacon frames on a regular

schedule by the APs. These beacon frames can be used by stations to home in on a particular AP, and to base decisions for handover.
- The MAC passive scanning function involves listening for beacon frames with the purpose of keeping stations in contact with their current AP. It can also be used to listen for beacons from other APs to find a better quality connection.
- The MAC active scanning function involves the sending of probe frames by stations and the responding of APs to these probes, for the purpose of finding a better quality connection to a new AP. This function can be used to scan across multiple channels.
- The BSS-ID and ESS-ID form the basis for the cellular structure used for roaming; the BSS-ID is the identification of a cell and the ESS-ID defines the boundary of the roaming territory.

6.3.3 General Roaming-Related Functions

In the following sections the various functions for roaming are discussed.

6.3.3.1 Communication Quality (CQ) Analysis

WLANs should have the capability to do an instant analysis of the communication quality (CQ) of each receipt of a message. The following attributes could potentially be made available as inputs to the analysis:

- Receive signal strength indicator (RSSI): A measure of the RF energy received, provided by the PHY;
- Number of retransmissions by this station to its AP (due to ACK time outs): This measure is provided by the MAC;
- Number of double received messages from the AP (due to missed ACKs): This measure is provided by the MAC.

The algorithm for the CQ determination is to keep a weighted running average over *n* measurements of the beacon SNR. Beacon SNR is defined as the difference between the RSSI and the silence level at the AP or the station (the worse of the two) for the reception of a beacon frame.

6.3.3.2 Cell Search Thresholds

A threshold mechanism is deployed for handover decision support. The mechanism includes a hysteresis to avoid rapid successive handovers while operating a station in the borderline between two cells. The following thresholds should be defined:

- Start cell search;

- Fast cell search;
- Stop cell search.

The threshold values are considered configuration parameters for a station.

6.3.3.3 Buffering Messages While Scanning

The station should use the buffering facility of the power management functions to ensure that no messages are lost during the periods it is searching in different channels other than the one used for its current AP.

6.3.3.4 Scan Channel

A set of channels can be used in an ESS. Stations should be aware of that information to avoid having to scan all possible channels (in the regulatory domain). Only the ones used for this ESS need to be scanned.

6.3.3.5 Passive Scanning

Passive scanning is the function of listening to a frequency channel for a predetermined period of time to detect one or more beacon messages from one or more APs, and repeating that process for a number of different frequency channels. When operating on a single channel, it is not disrupting normal data transmission and reception; when operating with multiple channels, the stations involved in the scan are temporarily not capable of normal transmits or receives.

6.3.3.6 Active Scanning

In active scanning, the station sends out a probe message on a specific frequency channel to find out the communications quality (and other information) with all APs operating in that channel. The APs receiving the probe will send a probe response back. This process is repeated across all channels to be scanned. The successive probing of all intended channels and waiting for the reply is called a scan sweep.

When operating in a single channel system, normal communication continues during the active scan. When operating in a multichannel system, the station cannot be used for normal communication during the scan of other channels other than the one used by the current AP.

6.3.3.7 Cell Switch

The result of a scanning action can be that the station has found a better AP. In that case, the following happens:

- The station will make a best attempt to get buffered messages from its current (soon to be "old") AP forwarded, using the normal procedures;
- The station will issue a "reassociate" message to the new AP and wait for the response;
- After a positive reassociate response, the station will start applying the new BSSs attributes for further roaming related processes.

6.3.3.8 Handover

Handover covers the process and associated protocols for a new AP to communicate to another AP when a station has done a reassociate. The handover will include a certain amount of administration and clean-up in the old AP (e.g., a "disassociate").

In some cases, the APs will have stored messages during the cell search period; these will not be forwarded via the new AP.

6.3.3.9 Forwarding Buffered Messages After Handover

When at the moment of the handover messages are still buffered at the "old AP," it is possible to forward these messages via the distribution system (e.g., the backbone LAN) to the "new AP," which can insert them in the communication flow to the station involved. This will require measures to ensure the proper sequence of messages in case later messages were already received by the "new AP."

6.3.4 Initial AP Association

When a station enters an ESS for the first time (or is powered up), it needs to establish an association with the nearest (best) AP. It does so by executing a number of passive scan sweeps complying with regulatory requirements and selecting the best AP. When the station is aware of its environment it can select appropriate channels to scan; otherwise it must go through all possible channels.

To satisfy regulatory requirements, it may be required that an initial scan is always done passively.

6.3.5 Single and Multichannel Roaming

A WLAN system provides roaming within the coverage boundaries of a set of APs that are interconnected via a (wired) distribution system. The APs send beacon messages at regular intervals. Stations can keep track of the conditions at which the beacons are received per individual AP. The running average of these receive conditions is determined by a CQ indicator (Figure 6.3). The different zones within the full-range CQ scale refer to various states of activities at which a station tracks or tries to find an AP. When the CQ is poor, then the station has to expend more effort to quickly find another AP that gives a better CQ.

The APs, which are interconnected by a wired distribution system, can use channel frequencies from a basic set within all supported channel frequencies within 802.11b. A station can search for an AP giving a better CQ by looking at all channel frequencies selected for the interconnected APs. The searching station can initiate an active mode by sending a so-called probe request message referencing the target set of interconnected APs. Each AP will respond to a probe request with a probe response message. This will serve as a solicited beacon.

When a station's CQ with respect to its associated AP decreases, this station starts searching more actively. After the station has found a second AP that gives a sufficiently good CQ, the station arrives in a handover state and will reassociate to this second AP. The APs deploy an IAPP to inform each other about station handovers and to correct any (intermediate) MAC bridge filter tables.

6.3.5.1 Single-Channel Roaming

In a WLAN single-channel system, roaming goes as follows:

1. All APs of an ESS use the same frequency channel and are configured to transmit beacons.

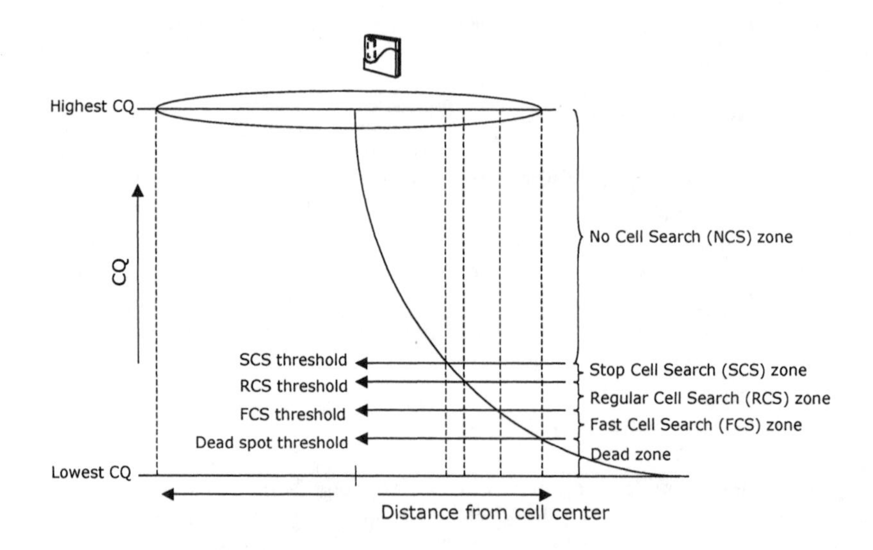

Figure 6.3 CQ scale and cell search zones.

2. A station in a BSS receives beacons from its AP and analyzes the CQ associated with them.
3. When CQ drops below the "start cell search" threshold, a passive scan is initiated for all APs within the ESS.
4. During the passive scan, communication with the current AP in both directions continues as normal. When an AP is found with a CQ better than the "stop cell search" threshold, the station initiates a cell switch procedure, and leaves the passive scanning mode.
5. When CQ drops below the "fast cell search" threshold, a single active scan sweep is carried out in the (single) channel.
6. During the active scan, communication with the current AP continues (to the extent still possible). When an AP is found with a CQ better than "fast cell search," the station initiates a cell switch procedure, and returns to passive scanning mode.
7. The cell switch to a new AP involves a reassociation between the station and the new AP. The new AP will allocate resources and will use a handover IAPP to communicate with the old AP. The old AP will disassociate the station. The handover must also update the data path in the distribution system for messages destined for the station.
8. Any messages that were still in process in the old AP will be dropped.

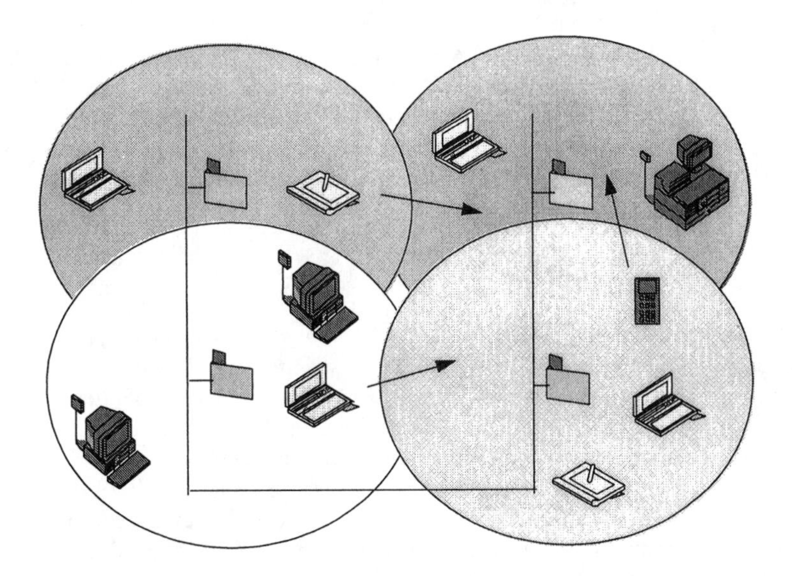

Figure 6.4 Wired infrastructure between access points and multichannel operation with three channel frequencies.

6.3.5.2 Multichannel Roaming

In a WLAN multichannel system, roaming goes as follows (see Figure 6.4):

1. All APs of an ESS are configured to transmit beacons; all stations are aware of the set of channels used in this ESS.
2. A station in a BSS receives beacons from its AP and analyzes the CQ associated with them.
3. When CQ drops below the "start cell search" threshold, a single active scan sweep is done across all channels of this ESS; before this active scan, the current AP is informed that the station cannot receive messages via a power save (PS) mode state change for the station.
4. When an AP is found with a CQ better than "stop cell search," the station initiates a cell switch procedure.
5. When no AP is found with a CQ better than "stop cell search," the station returns to the current AP's channel and polls for buffered messages. The scan sweep (step 3) is repeated on a regular basis.
6. When CQ drops below the "fast cell search" threshold, a single active scan sweep is done on all channels of the ESS; before this active scan, the current AP is informed that the station cannot receive messages via a power save (PS) mode state change for the station. In this mode, any other AP with a CQ above "fast cell search" is accepted and handover is initiated.
7. The cell switch to a new AP involves a re-association between the station and the new AP. The new AP will allocate resources and will use a handover IAPP to communicate with the old AP. The old AP will disassociate the station. The handover must also update the data path in the distribution system for messages destined for the station.

6.3.6 IEEE 802.11 Handover Delays

In the previous subsections the steps for handover are discussed. The handover process has several delays; these are discussed in the following for handover within a network (same IP subnet) where IEEE 802.11i is not used; the delays are given for single and multichannel deployments. In case of multichannel deployment a good implementation will take care that the STA scans only the non-overlapping channels. As given in Chapter 3, there are only three non overlapping channels for the 2.4GHz band while a maximum of five channels can be used [4, 8]. The study given below is for three channels in the 2.4GHz band; in the 5GHz band there are no overlapping channels so the delay will be much more.

The steps that cause delay for handover within a network are listed as follows:

1. Detection of the need for handover;
2. Active or passive scan ;

3. Re-authentication;
4. Re-association.

In the case of single-channel there is no detection and (passive) scan delay. The STA will perform handover as soon as it has determined a new AP with better signal and within the handover zone. The other thing, besides the signal level, that causes delay is the number of unacknowledged frames before the system determines the need for handover. The unacknowledged frames are not considered here because it is implementation-dependent.

Steps 3–4 take a minimal time when we consider the use of IEEE 802.11 using WEP as discussed in Chapter 3. Association is two messages and depending on type of authentication used, re-authentication is at a maximum four messages. In case of IEEE 802.11i there can be a re-authentication delay when complete authentication is needed, see Section 6.5.

Now let us consider multichannel deployment. Here the main delay will be due to active scanning. Depending on how heavy the traffic is in the channel there will be a delay in accessing the channel by the STA to send the probe request plus the probe delay before which the STA is not allowed to transmit the request. Then there is the delay in sending, from the AP to the STA, the probe response, which is also dependent on the traffic in the channel. The next step is for the STA to re-authenticate and re-associate; these two steps will take a minimal time with a maximum of six frames depending on the authentication type used. The re-authentication and re-association will also depend on the traffic at the AP. A measurement done in [9] shows that the ORiNOCO card has a delay of about 87ms for step 2 active scan and 1 ms for the steps 3–4. Thus a total of 88 ms. Detection for the need for handover is implementation dependent. Here it is considered that the implementation is such that the STA starts searching for an AP as soon as the signal strength is below the desired level. Note that this desired level or threshold is also dependent on the data rates that are to be used for a given deployment.

6.4 INTER-ACCESS POINT PROTOCOL: IEEE 802.11F

In July 2003, IEEE 802.11f, a recommended practice for multivendor interoperability by the inter-access point protocol (IAPP), got standardized. The recommended practice describes a service access point (SAP), service primitives, a set of functions, and a protocol for APs to interoperate on a common distribution system (DS), using Transmission Control Protocol (TCP) or User Datagram Protocol (UDP) to carry IAPP packets between APs as well as describing the use of the Remote Authentication Dial-in User Service (RADIUS) Protocol, so APs may obtain information about one another. A proactive caching method is also described for faster roaming. This recommended practice also affects layer-2 devices like switches and bridges. An AP architecture is shown in Figure 6.5.

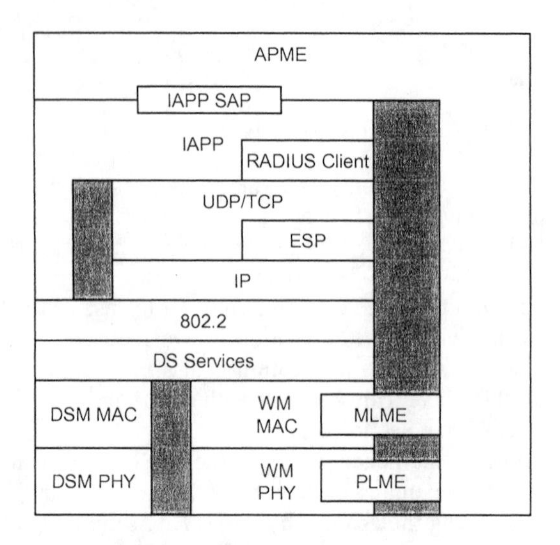

Figure 6.5 AP architecture with IAPP (reprinted with permission from IEEE 802.11f-2003).

In the figure the AP management entity (APME) refers to functions external to IAPP but implemented in an AP. The gray areas indicate areas where there is an absence of connection between blocks. The IAPP services are accessed by the APME through the IAPP SAP. Some IAPP service primitives require RADIUS protocol; in particular the AP should be able to find and use a RADIUS server to look up the IP addresses of other APs in the ESS when given the basic service set Identifier (BSSIDs) of those other APs (if a local capability to perform such a translation is not present), and to obtain security information to protect the content of certain IAPP packets.

There are three main steps in IAPP, these are:

1. IAPP ADD procedure: This is when a new STA associates with the AP and is added in the list;
2. IAPP Move procedure: This procedure comes in action as soon as the STA re-associates with a new AP;
3. IAPP Cache: This procedure happens when context is used when proactive caching is used. Proactive caching means sending and caching of the context information of the STAs at the APs in the neighborhood graph. The neighborhood graph contains the APs in the near vicinity of the AP with which the STA is currently associated and the AP has a secure connection with.

In this section the above three steps of IAPP are explained. Figure 6.6 shows the complete IAPP procedure.

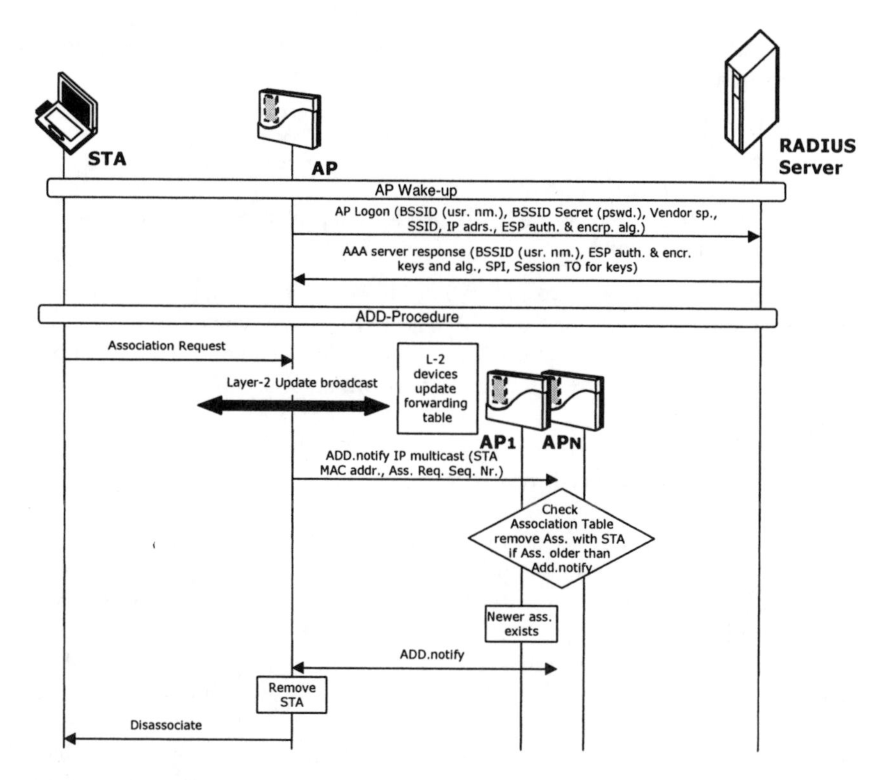

Figure 6.6 IAPP wakeup and ADD procedures.

6.4.1 AP Wakeup, ESS Formation, and RADIUS

For the formation of ESS the IAPP support three levels; these are:

1. No administrative or security support;
2. Support of dynamic mapping of BSSID to IP address;
3. Support of encryption and authentication of IAPP messages.

The first level is achieved with each AP being configured with BSSID to IP address mapping of all other APs in the ESS. This works fine for small ESSs. The other two levels require RADIUS support.

When using RADIUS support each client (AP) and the RADIUS server are configured with a shared secret and each other's IP address. Note that the shared secret is different for each AP. Since the roaming STA sends an 802.11 re-association request frame to the new AP containing the BSSID it is roaming from, each RADIUS server must also be configured with the following information for each BSSID. From an IAPP point of view, this set of BSSID entries defines the members of an ESS:

1. BSSID;
2. RADIUS BSSID secret at least 160 bits in length;
3. IP address or DNS name;
4. Cipher suites supported by the AP for the protection of IAPP communications.

The IAPP entity is invoked by the APME to initiate STA context transfer between the old AP and the new AP. The IAPP may invoke RADIUS to obtain mapping of the old BSSID to the distribution service medium (DSM) IP address of the old AP and the security information with which to secure the communications with the peer IAPP entity.

6.4.2 IAPP-ADD Procedure

When the STA sends an association request an IAPP-ADD.request is generated which in turn leads to the sending of, by the AP, an IAPP-ADD.notify packet and a layer-2 update frame. The IAPP-ADD.notify packet is an IP packet with a destination-IP-address of the IAPP IP multicast address, the source IP address, and MAC address of the AP. The message body contains the MAC address of the STA and the sequence number from the association request sent by the STA. On receiving this message an AP should check its association table and remove an association with the STA if it exists and is determined to be older than the association indicated by the ADD.notify packet. The purpose of the IAPP-ADD.notify packet is to remove stale associations, not to modify the forwarding table. The forwarding table update is done by the layer-2 update frame. This frame has the source MAC address of the associating STA. This frame is used by receiving APs and other layer-2 devices to update their forwarding tables with the correct port to reach the new location of the STA.

6.4.3 IAPP-Move Procedure

Once the STA roams to a new AP it sends a re-associate request (see Figure 6.7). The re-associate request triggers, at the AP, an IAPP-MOVE.Request; as a result the AP sends an IAPP-MOVE.Notify packet to the old AP. The old AP then responds with a MOVE-Response. The IAPP MOVE-Response carries the context block for the STA's association from the old AP to the new AP. Context could be information like the security related data for fast re-authentication at the new AP.

The IAPP-MOVE.Notify and MOVE.Response are IP packets carried in a TCP session between APs. The IP address of the old AP is found by mapping the BSSID from the re-associate message to its IP address using RADIUS exchange or locally configured information.

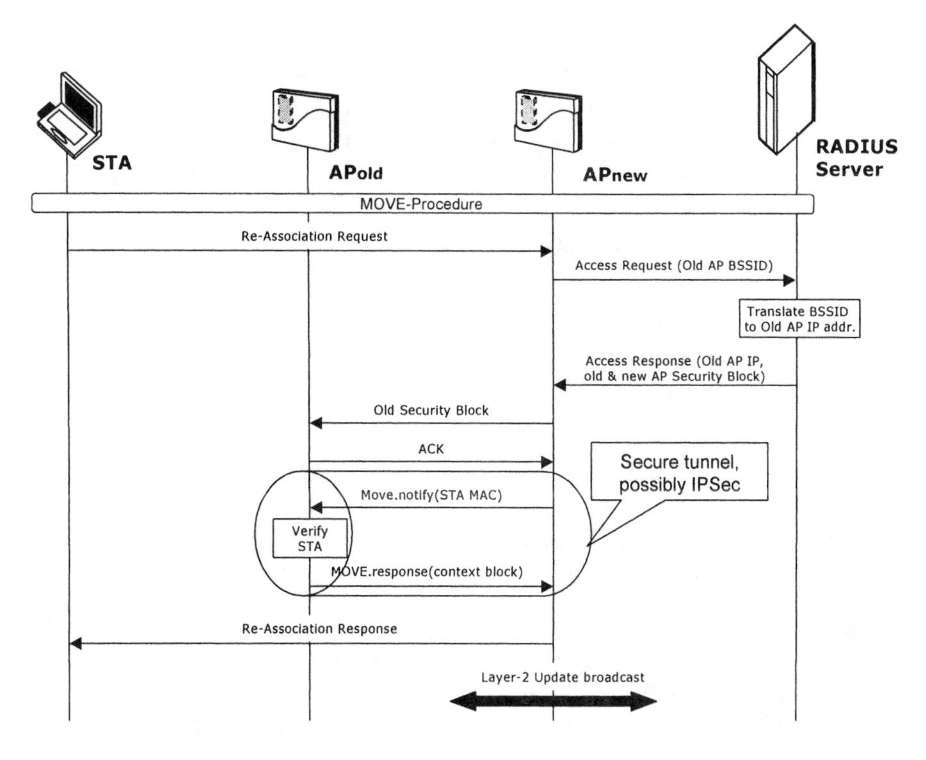

Figure 6.7 IAPP-MOVE procedure.

If it is desired to encrypt the IAPP-MOVE.Response packet, the RADIUS reply to the new AP will include, in addition to the IP address of the old AP, reply items with security blocks for both the new and old APs. The security blocks each contain information for securing the AP-AP connection. This information is dynamically generated by the RADIUS server as the security blocks are constructed. The security blocks are encrypted using the APs' BSSID user password in the RADIUS registry. The RADIUS server would have to have an add-on to create the security block. The new AP sends the security block for the old AP, which it received from the RADIUS server, as a send security block packet. This is the first message in the IAPP TCP exchange between the APs. The old AP returns the ACK-Security-Block packet. At this point both APs have the information to encrypt all further packets for this exchange between the APs.

6.4.4 IAPP-Cache

IAPP-CACHE-NOTIFY.request from the APME in an AP means that the given context should be sent to each of the APs in the neighbor graph (see Figure 6.8).

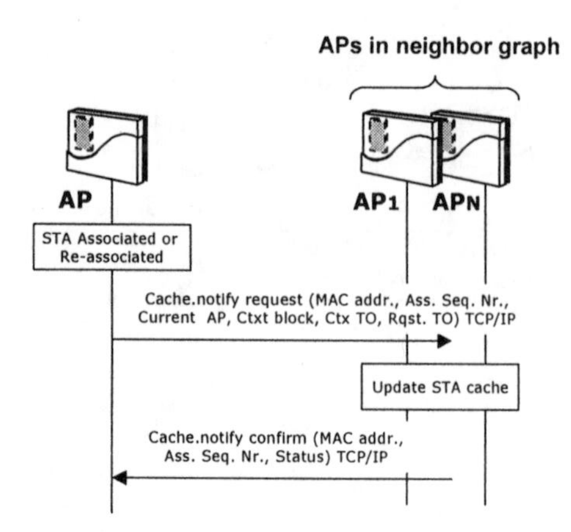

Figure 6.8 IAPP-Cache procedure.

On reception of an IAPP-CACHE-NOTIFY.request the receiving APs update the STA context cache by adding or updating the STA context entry corresponding to the IAPP-CACHE-NOTIFY.indication. After the neighboring APs have updated their cache, a CACHE-response packet goes back to the original AP. When the original AP IAPP entity receives the CACHE-response packet, it issues an IAPP-CACHE-NOTIFY.confirm to the original APME.

If CACHE-response packets are not received from each of the neighboring APs before the IAPP-CACHE-NOTIFY.request{RequestTimeout} expires, the IAPP entity should delete the neighboring APs that did not respond before the expiration of the RequestTimeout from the neighbor graph. Only when all neighboring APs fail to respond before the expiration of the RequestTimeout will the original IAPP entity issue an IAPPCACHE-NOTIFY.confirm {Status= TIMEOUT} to the original APME. If any CACHE-response packets are received with a status of STALE_CACHE, the IAPP should issue an IAPP-CACHE-NOTIFY.confirm {Status=STALE_CACHE} and the APME should delete the corresponding STAs' context entry from the local cache.

Security for this message should be provided as for the IAPP-MOVE.

6.4.5 Neighbor Graph

A neighbor graph is the set of neighbors relative to a given AP. This set is kept by an AP so that its neighbors can be identified quickly. Rather than incur the management overhead of manually listing the neighbors for a given AP, the AP can learn its neighbors dynamically through the course of operation from information in REASSOCIATION-REQUEST frames, and IAPP-MOVE.Request frames. The AP can prevent the addition of bogus neighbors by adding only those APs where a RADIUS Access-Accept message is returned by the RADIUS server.

The exact form of the implementation of neighbor graphs is vendor-dependent, but it is suggested that a least recently used (LRU) cache be used since some neighbors will be misidentified due to STA moves without radio operation (e.g., when a laptop is closed). In these cases, the STA will fail to disassociate and then will re-associate at another AP, which may or may not be a valid neighbor. Since these events will occur less frequently than handovers to valid neighbors, a LRU cache will, over time, push the invalid entries from the neighbor graph. The added benefit is that the neighbor graph size can be fixed permitting easier memory management.

6.5 IEEE 802.11 HANDOVER DELAYS

The handover process has several delays as discussed in Section 6.3.6. One of the major causes of delay is channel access, which is dependent on the traffic in the WLAN network. So we propose that the handover delay study should be done without consideration of channel access delay. One can use the number of frames transmitted to get an idea of the overall delay. This way one gets, from a simple method, a realistic result. In this section handover delay is given for a system using IEEE 802.11f and IEEE 802.11i.

The handover steps, which are listed as follows, each cause delay [4].

1. Active or passive scan;
2. Re-authentication;
3. Re-association;
4. IEEE 802.11i authentication;
5. Depending on handover scenario, layer-3 and higher layer handoff.

As discussed in Section 6.3.6, Steps 2–3 take minimal time when we consider the use of IEEE 802.11 using WEP. In the case of IEEE 802.11i there will be considerable re-authentication delay if complete authentication is performed. Extensible Authentication Protocol (EAP)-Transport Layer Security (TLS) delay is shown in Figure 6.9.

Figure 6.9 shows multichannel deployment using IEEE 802.11i and IEEE 802.11f with the delays. Active scanning will cause delay in multichannel deployment due to probe request and probe response in each channel. A good design will take care that the STA scans only the non-overlapping channels; maximum 3 in the 2.4GHz band. In the 5GHz band there are no overlapping channels so the delay will be much more [4, 8]. The next step is for the STA to re-authenticate and re-associate; these two steps will take a maximum of six frames depending on the authentication type used. In the case of IEEE 802.11i there will be considerable re-authentication delay.

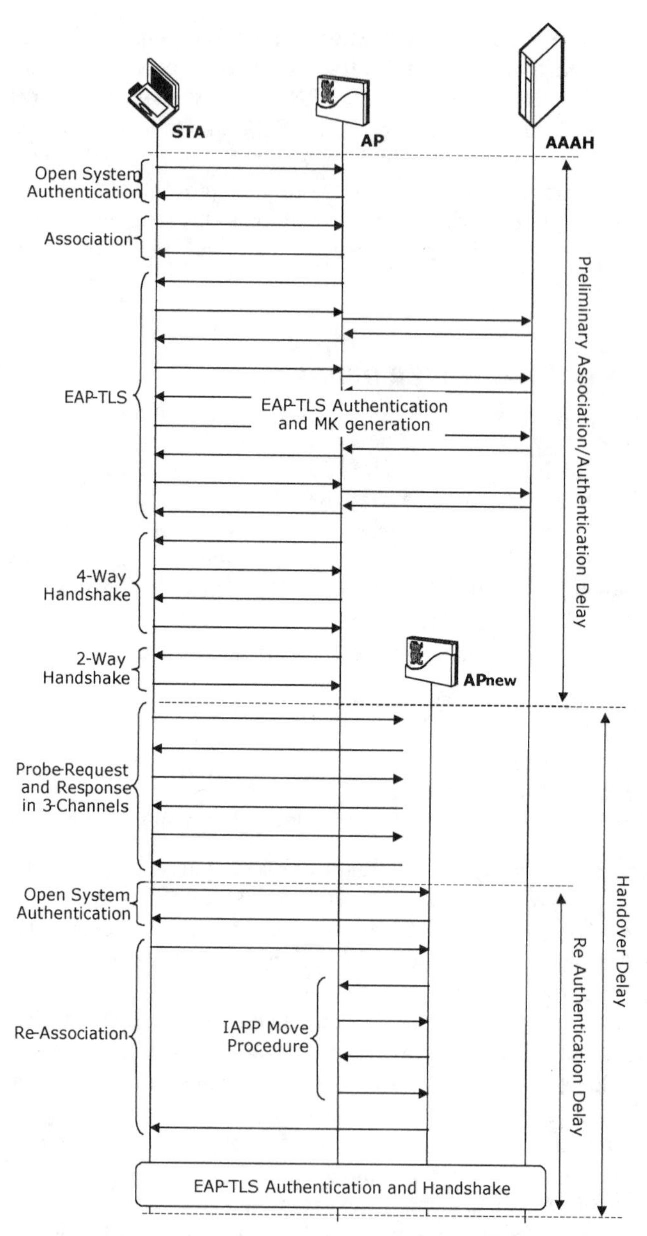

Figure 6.9 IEEE 802.11 authentication, re-authentication, and handover delay with IEEE 802.11i with EAP-TLS and IEEE 802.11f. Note: In Japan there are 4 non overlapping channels; therefore the STA will send probe request in 3 channels for handover; in other countries there are 3 non overlapping channels; thus the STA will send probe request in other 2 channels.

Associated to the IEEE 802.11 delay, which is in layer-2, is also the layer-3 delay, which occurs due to Mobile-IP (MIP) when performing inter subnet or inter domain handover. A study done in [10] shows the delays for each MIP step: (1) advertisement—500 ms, (b) registration request—36 ms, and (c) registration response—2 ms.

Further to this there is the delay related to Authentication, Authorization, and Accounting (AAA) communication when the STA moves from one domain to other. The delay caused by re-/authentication will be by far the most prominent one. This delay will be because authentication can involve several network elements and communication between different domains for inter domain handover.

6.6 IP MOBILITY

There are two kinds of IP mobility, which can be defined as follows and will be discussed in this section:

- Macro mobility: When a MN, STA in IEEE 802.11, moves from one IP domain to another it is a significant network event — usually a new IP address is required —; other processes including re-authentication are needed, and the air interface technology may change. This is termed macro mobility. Mobile IP (MIP) [13, 14] is an IP protocol that supports macro mobility. In Macro-mobility, users expect delays and disruption to service and it does not provide anything like the functionality required for a mobile network supporting real-time services and requiring an efficient use of the spectrum available.
- Micro mobility: Protocols providing micro mobility are used to provide support for real-time handovers in IP networks. Such protocols usually also integrate with the layer-2. Knowledge, and the consequences, of the MN's mobility are confined to the access network domain — to all communicating parties the MN seems to be static — with a static IP address and a QoS guaranteed packet delivery service. Hierarchical Mobile IP (HMIP) [22, 23] is one of the protocols developed for IP micro mobility management although the original purpose of HMIP was to solve the problem of location privacy in MIP.

6.6.1 Macro Mobility: Mobile IP

MIP has been specified for wide area macro mobility management [13, 14]. It enables a node to move freely from one point of connection on the Internet to another, without disrupting end-to-end connectivity. MNs are required to securely register a Care-of-Address (CoA) with their Home Agent (HA) while roaming in a foreign domain. If however security mechanisms are not employed, the network can be compromised through remote redirection attacks by malicious nodes. In addition, mechanisms are needed that allow Foreign Agents (FAs) in the visited domain to verify the identity of MNs and authorize connectivity based on local policies or the ability to pay for network usage.

The basic requirements of Mobile IP protocol are defined as follows:

- A MN must be able to communicate with other nodes after changing its link-layer point of attachment to the Internet, yet without changing its IP address.
- A MN must be able to communicate with other nodes that do not implement these mobility functions. No protocol enhancements are required in hosts or routers that are not acting as any of the new architectural entities.
- All messages used to update another node as to the location of a MN must be authenticated in order to protect against remote redirection attacks.

The elements and terminology that make up a MIP enabled network are summarized in Figure 6.10. The MN has an IP address assigned to it, which belongs to a subnet on the MN's HA. The HA advertises the subnet that the MN belongs to in periodic routing updates. When a MN roams, the HA is responsible for intercepting IP packets sent to the MN and tunneling them to the new location of the MN. A Correspondent Node (CN) is another IP endpoint (user or application) that wishes to communicate with a MN. CN sends IP packets to the MN's IP address and they are routed to the HA router (via normal IP routing) to delivery to the MN (over a MIP tunnel if the MN has roamed). When a MN roams

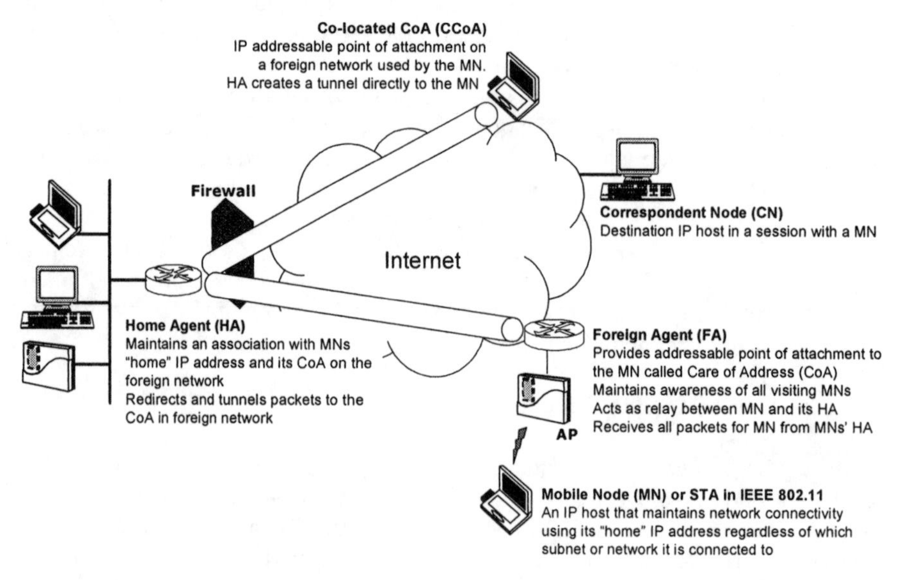

Figure 6.10 Mobile IP components.

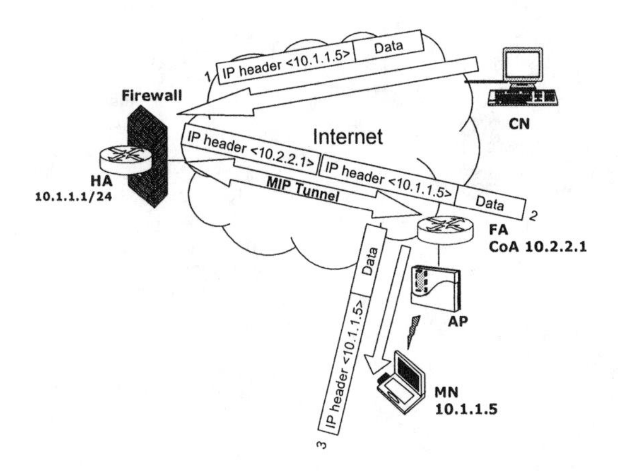

Figure 6.11 Mobile IP tunneling FA CoA example.

to a foreign subnet it obtains a CoA. A CoA is the IP address the HA uses when it forwards IP packets to a MN on a foreign subnet. The process of encapsulating the original IP packet with an additional IP header is called tunneling. An example of the HA tunneling IP packets to a MN via a CoA is shown Figure 6.11 and Figure 6.12.

There are two types of CoA:

- Foreign agent CoA: The edge router has MIP FA functionality enabled, and provides an IP endpoint for the MIP tunnel. The FA strips the additional IP header from the tunneled packet and forwards the original packet on the local subnet.
- Co-located Care of Address (CCoA): The mobile node acquires a topologically correct IP address (in addition to its mobile IP address) for use as a tunnel endpoint. The topologically correct IP address is obtained via DHCP or can be statically configured.

6.6.1.1 Discovering the CoA

The CoA discovery procedure used in MIP is based on the ICMP router advertisement standard protocol, specified in RFC 1256. In MIPv4, the router advertisements are extended to also contain the required CoA. These extended router advertisements are known as agent advertisements. Agent advertisements are typically broadcasted at regular intervals (e.g., once a second, or once every few seconds) and in a random fashion, by HAs and FAs. However, if a mobile needs to get a CoA instantaneously, the MN can broadcast or multicast a solicitation that will be answered by any FA or HA that receives it.

The functions performed by an agent advertisement are the following:

- Allows the detection of HAs and FAs;

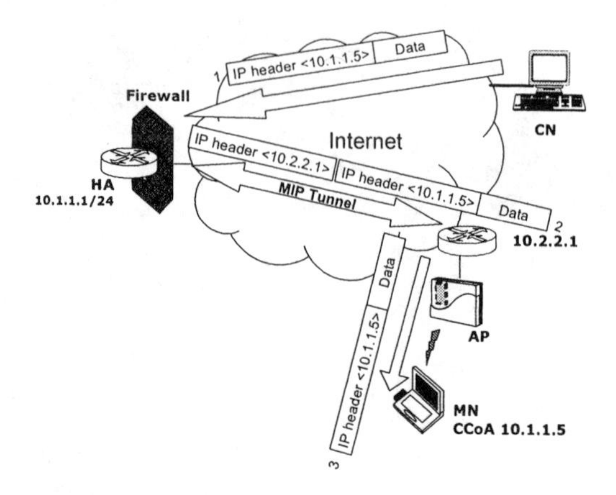

Figure 6.12 Mobile IP tunneling CCoA example.

- Lists one (or more) available CoAs;
- Informs the MN about special features provided by FAs (e.g., alternative encapsulation techniques;
- Permits MNs to determine the network number and congestion status of their link to the Internet;
- Lets the MN know, whether it is in its home network or in a foreign network by identifying whether the agent is a HA, a FA, or both.

ICMP router solicitations (agent solicitation) procedures are defined in RFC1256. If the MN does not receive agent solicitation advertisements from a FA anymore, it will presume that this FA is not anymore within the range of its network interface. The process of FA discovery and MIP registration is shown in Figure 6.12.

6.6.1.2 Registering the CoA

After the MN gets the CoA it will have to inform the HA about it. The MN sends a registration request (using UDP) with the CoA information. This information is received by the HA and normally, if the request is approved it adds the necessary information to its routing table and sends a registration reply back to the MN through the foreign agent.

The flags and parameters required to characterize the tunnel, through which the HA will deliver packets to the CoA, are contained in the registration request message. After accepting a registration request, the HA begins to associate the home address of the MN with the CoA for a pre specified time duration, called registration lifetime. The group that contains the home address, CoA, and registration lifetime is called a binding for the MN. This binding is updated by the MN at regular intervals, sending a registration request to the HA.

During the registration procedure, there is a need to authenticate the registration information. The reason is that a malicious node could cause the HA to alter its routing table with erroneous CoA information, and then the MN would be unreachable. Therefore, each MN and HA must share a security association. The process of MIP registration is shown in Figure 6.12.

6.6.1.3 Tunneling to the CoA

The tunneling to the CoA is accomplished by using encapsulation mechanisms. All mobility agents, (i.e., HAs and FAs), using MIPv4 must be able to use a default encapsulation mechanism included in the IP within IP protocol [RFC2003]. By using this protocol, the source of the tunnel, (i.e., HA), inserts an IP tunnel header in front of the header of any original IP packet addressed to the MN's home address. The destination of this tunnel is the MN's CoA. In IP within IP [RFC2003] there is a way to indicate that the next protocol header is again an IP header. This is accomplished by indicating in the tunnel header that the higher level protocol number is "4." The entire original IP header is preserved as the first part of the payload of the packet. By eliminating the tunnel header the original packet can be recovered.

6.6.1.4 Proxy and Gratuitous Address Resolution Protocol (ARP)

The IP nodes located in the home network of a MN are able to communicate with the MN while it is at home, by using the Address Resolution Protocol (ARP), RFC 826, cache entries for this MN. When the MN moves to another subnetwork, the HA will have to inform all IP nodes in the home network that the MN moved away. This is accomplished, by sending gratuitous ARP messages. These messages will update all ARP caches of each node in the home network. After that moment the packets sent by these IP nodes to the MN will be intercepted by the HA by using proxy ARP. The intercepted packets are then tunneled to the CoA.

6.6.1.5 Route Optimization in Mobile IP

The operation of the base MIP protocol is extended to allow for more efficient routing procedures, such that IP packets can be routed from a CN to a MN without going to the HA first. These extensions are referred to as route optimization, wherein new methods for IP nodes (e.g., CNs) are provided. The CN receives a binding (the association of the home address of a MN with a CoA) update message from the mobile's node HA that contains the MN's CoA. This binding is then stored by the CN and is used to tunnel its own IP packets directly to the CoA indicated in that binding, bypassing the MN's HA. In this way, the triangular routing situation, explained previously is eliminated. However, in the initiation phase, the IP packets sent by the CN still use the triangle routing until the moment that the binding update message sent by the MN's HA, is received by the CN.

Extensions are also provided to allow IP packets sent by a CN with an out-of-date stored binding, or in transit, to be forwarded directly to the MN's new CoA. All operation of route optimization that changes the routing of IP packets to the MN is authenticated using the same type of mechanisms also used in the base MIP protocol. This authentication generally relies on a mobility security association established in advance between the sender and receiver of such messages.

The route optimization protocol operates in four steps:

- A *binding warning* control message is usually sent by a node (e.g., MN or CN) to the HA (i.e., recipient) along with the remaining lifetime of that association. CN (i.e., target) seems unaware of the MN's new CoA;
- A *binding request* message is sent by a CN to the HA at the moment it determines that its binding should be refreshed;
- Typically an authenticated *binding update* message is sent by the HA to all the CNs that need them, containing the MN's current CoA;
- A *binding acknowledgment* message can be requested by a MN from a CN that has received the *binding update* message.

6.6.2 Mobile IPv6

MIPv6 was developed by the IETF to manage IPv6 nodes' mobility. It contains a number of advantages over the MIPv4 protocol. Most of these advantages are due to the inherent features of IPv6 and the ability to integrate the MIPv6 protocol into the IPv6 stack from day one. Mobility support is mandated in every IPv6 stack, which is essential for producing an effective and interoperable protocol that can be widely deployed within the Internet.

MIPv6 provides an end-to-end mechanism for route optimization between a MN and a CN. Such a feature is important for optimizing routing within the Internet and minimizing delays in packet delivery.

Due to the very high availability of IPv6 addresses, MIPv6 allows MNs to obtain a topologically correct address when moving to a new subnet. This eliminates the need for additional mobility management functions in the network when compared to FAs in MIPv4. Furthermore, this allows network administrators to use mechanisms like ingress filtering to prevent users from using topologically incorrect addresses within an operator's administrative domain.

Since IPv6 mandates the support of IPSec in every stack, IPSec is used to secure MIPv6 signaling. This allows for an interoperable solution that can be integrated within the IPv6 stack.

MIP is currently the most accepted solution for IP mobility. MIPv4 is currently supported within the 3GPP standards by the inclusion of the FA functionality in the Gateway GPRS Support Node (GGSN). Since no FAs are required in IPv6, MIPv6 does not add any additional requirements on the 3GPP architecture. GPRS Tunnel Protocol (GTP) is used in 3GPP networks for intra-domain mobility. MIPv6 is an attractive candidate to complement this for inter-domain and inter-access technology mobility.

Future mobile terminals are expected to implement a number of interfaces for use with different access networks. For example, apart from the obvious support for the cellular interfaces, terminals may also support other radio technologies, like Bluetooth, Infrared, etc. Assuming that those access technologies are connected to different access routers, a terminal's IP address may change when moving between those different media. Hence, to ensure seamless mobility and maintenance of ongoing connections, MIPv6 can be utilized. MIPv6 can also be used when roaming between different 3GPP networks, thus allowing a device to be reachable in a route optimized manner.

6.6.3 Mobile IP and AAA

Information on how MIP should work with a AAA server is given in [15, 16]. The network elements involved in a handover using MIP and AAA is given in Figure 6.13. Note that this kind of handover will be necessary when the user moves from one administrative domain to another.

There are three AAA elements: the Home AAA (AAAH), the Foreign AAA (AAAF), and the Broker AAA (AAAB). It is possible that AAAH and AAAF have some prior relation (e.g. roaming contract between two ISPs); in that case AAAB is not required. When there is no trust relation between AAAH and AAAF then it is possible that they have a trust relation with an AAAB. In that case AAAB can create a trust relation between AAAF and AAAH. The basic point about trust is that there is a security association (SA) between two network elements.

In Figure 6.13 there is SA1 between AAAF and AAAB, SA2 between AAAH and AAAB, and SA3 between AAAH and AAAF. SA3 can be created through AAAB as explained earlier or it can exist. There is also SA4 between FA and AAAF, SA5 between HA and AAAH, and SA6 between MN and HA. Now when the MN requests registration at the FA, the FA will send the request to the AAAF that will start the AAA protocol. Of course this means that the user will be involved and thus there is no possibility of service continuity.

The easiest way of course would be that the MN on authentication request sends its network address identifier (NAI). Using the NAI the AAAF can find the AAAH. Before any negotiations can take place between AAAF and AAAH the SA3 should be created. The AAAH verifies the authentication response and then passes the registration request to the HA. The HA then sends the registration response to the AAAH, which in turn proxies it together with an authentication response to the AAAF. The AAAF then authenticates the user and sends the registration response through the FA. Now the communication between HA and FA can take place. During this exchange a SA (SA7) is also created between FA and HA.

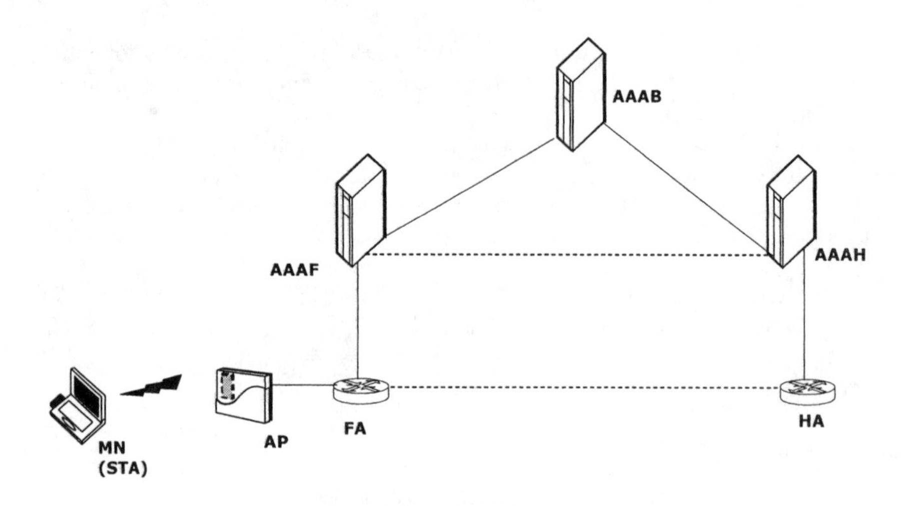

Figure 6.13 Mobile IP and AAA interaction.

6.6.4 Mobile IP Security Issues

Security issues of Mobile IP have been discussed in [1, 39, 40]; some interesting parts are extracted here.

- *Ingress filtering*: In an ISP any border router may discard packets that contain a source IP address that is not being configured for one of the ISP's internal network. This issue is called ingress filtering. In MIPv4 the MNs that are away from home, (i.e., in a foreign ISP) use their home address as the source IP address, which is different than the IP addresses configured in the ISP's internal network.
- *Minimise the number of required trusted entities:* Security may be enhanced, if the number of the required trusted entities, (i.e., HA or FA), in a MIP scenario is decreased.
- *Authentication:* The recipient of a message should be able to determine who the actual (real) originator of the message is. Therefore authentication procedures between mobile agents and MNs should be provided. The MIPv4 authentication techniques between MNs and FAs are not reliable enough.
- *Authorization:* An organization that owns or operates a network would need to decide who may attach to this network and what network resources may be used by the attaching node.
- *Non-repudiation:* In the future wireless Internet, a recipient of a message should have the opportunity to prove that a message has been originated

by a sender. In other words, the sender of a message should not be able to falsely deny that it originated a message at a later time.

- *Encryption key distribution:* The authentication, integrity, and non-repudiation can only be accurately provided (enforced) by using some form of cryptography that requires the distribution/exchange of encryption key information amongst message senders and receivers. In other words key management procedures should be supported by MIP. Two methods can be used for this purpose. One method for distributing the key information is to manually load it into each node. For a small number of nodes this is possible but it runs into administrative problems. Another method to distribute the key information is dynamical, using basic IETF security protocols.
- *Location privacy:* A sender of a message should be able to control which, if any, receivers know the location of the sender's current physical attachment to the network. Location privacy is concerned with hiding the location of a MN from CNs.
- *Use one single subscription for all service types:* In RFC 2468 [18] the NAI is defined; it is used to identify ISP subscribers during roaming operations. Regarding second and third generation cellular networks, an interesting approach for cellular service providers would be to evolve their home cellular networks to provide second and third generation cellular services, IP packet data services, and so on with a single subscription using NAIs. MIPv4 does not provide solutions to this issue.
- *Firewall support in MIP:* If a MN has to enter a private Internet network (intranet) that is securely protected by a firewall, then MIP-aware support at this firewall is required. In MIP this support is not provided.
- *ARP issues:* ARP is not authenticated, and can potentially be used to steal another host's traffic. The use of "gratuitous ARP" brings with it all of the risks associated with the use of ARP.

6.6.5 Mobile IP QoS Issues

As shown in [10] each step of MIP causes lots of delay. There are publications available that show methods to improve MIP for QoS provision. In this section we will look at request for comments (RFCs) or other standards instead of research papers because standards have a more global effect and lead to products.

The problem simply stated is that the path traversed by the packets in the network will change as the MN moves. Depending on the mobility of the MN the impact could be large or small. As the MN changes its CoA it is possible that some functions in the path do not recognize the handover [17]. Further as mentioned in scenarios section, Section 6.2, the handover can be between two administrative domains.

Within an administrative domain the possible solution is to use the solutions presented by the IETF Seamless Mobility (Seamoby) group [24]. This is discussed in Section 6.6.10. Besides Seamoby solutions it is possible to optimize handover by different techniques; some of these are discussed in [10]. There are two ideas, one of them [10] is to trigger pre-registration, (i.e., layer-3 handover happens before layer-2 handover and thus the CoA of the new FA is sent to the MN before the actual layer-2 handover happens). This means that the old FA keeps in track the advertisement from different neighboring FAs and caches them. Another solution presented in [10] is to use post registration where the MN does not register with the new FA but instead uses the old FA as anchor FA and continues communication. The MN performs registration at a later time. There is also the possibility of combining pre and post-registration.

6.6.6 Mobile IP and IPSec

This section describes the problems in deploying MIPv4 when working with IPSec-based VPN. There are several ways in which a VPN gateway (GW) and MIP HA can be deployed. Scenarios where IPSec is encapsulated by MIP do not face problems; the issue in such a case is multivendor support. Scenarios where MIP is encapsulated by IPSec have serious problems; this issue will be discussed here.

6.6.6.1 Scenarios

To start with we will briefly look at different scenarios of deploying MIP and VPN and after that we will discuss in depth the scenario of concern.

Possible ways of deploying MIP and IPSec are listed as follows [19]:

1. MIPv4 HA(s) inside the intranet behind an IPSec-based VPN gateway: Requires MIP inside IPSec; this means that traffic between the MN and the VPN server is encrypted. Thus if a FA is being used it cannot inspect and relay the packet. A CCoA might work but it means that the VPN tunnel should be renegotiated every time the MN changes its point of attachment.
2. VPN Gateway and MIPv4 HA(s) in parallel at the network border (i.e., VPN and HA are separate): This scenario can work with MIP in IPSec or IPSec in MIP. MIP in IPSec will have the same problem as in 1. IPSec inside MIP will have no problem though there will be routing logic modification needed at the VPN gateway or the HA.
3. Combined VPN gateway and MIPv4 HA: This way IPSec in MIP can be easily used but it does not support multi-vendor interoperability.
4. MIPv4 HA(s) outside the VPN domain: Same as 3 except that the HA is separate and placed away from the VPN gateway outside the home network.
5. Combined VPN gateway and MIPv4 HA(s) on the local link, (i.e., using NAT at the firewall and VPN/HA inside the intranet). It can be possible to

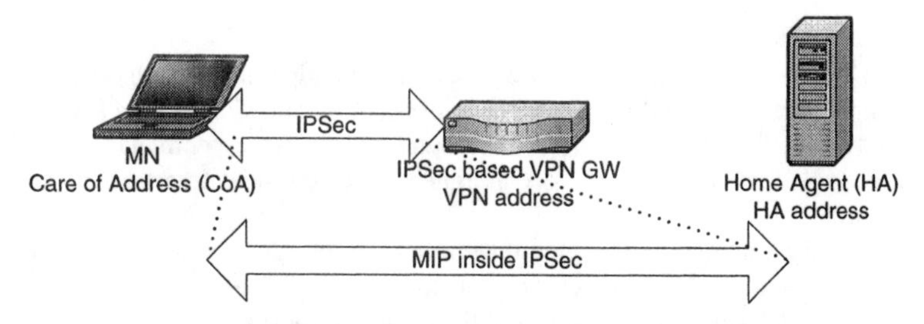

Figure 6.14 MIP with collocated address.

give the user IPSec connectivity using solutions; now this scenario is similar to 3. In the case of MIP inside IPSec, the problem is the same as in 1.

As scenario 1 is the one supposed to be most practical [19], its issues are further discussed below.

6.6.6.2 MN Registers with Its MIPv4 HA Using CCoA

Figure 6.14 shows the MIPv4 and the IPSec tunnel endpoints in co-located mode. MN's CoA (most likely obtained through DHCP) is used as both the IPSec and MIP tunnel outer addresses at the MN end.

The MN obtains a CoA at its point of attachment (via DHCP or some other means), and then first sets up an IPSec tunnel to the VPN gateway, after which it can successfully register with its HA through the IPSec tunnel. The problem is that in an end-to-end security model, an IPSec tunnel that terminates at the VPN gateway must protect the IP traffic originating at the MN. If the IPSec tunnel outer address is associated with the CoA, the tunnel SA must be refreshed after each IP subnet handoff, which could have noticeable performance implications on real-time applications. As MIPv6 uses CCoA, the issues discussed above are also valid for IPSec usage with MIPv6.

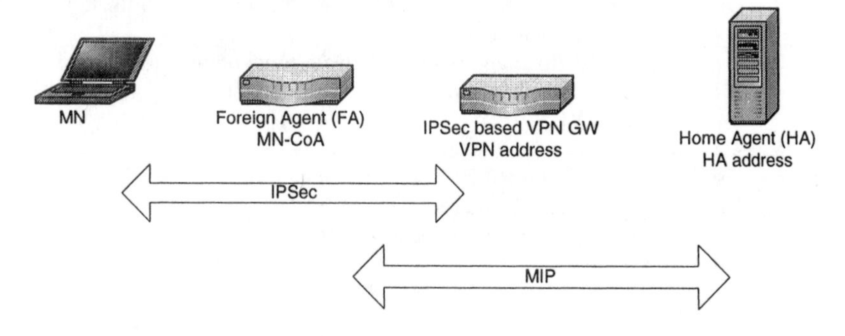

Figure 6.15 MIP with FA.

6.6.6.3 MN registers with Its HA Through a FA

Figure 6.15 shows the MIPv4 and the IPSec tunnel endpoints in a hypothetical (but impossible) noncollocated mode. MN's home address and CoA (i.e., a FA address) are used as the IPSec and the MIP tunnel outer addresses, respectively. Please note that the MN does not have a CoA assigned to its physical interface in non-co-located mode.

There are a number of problems with this. Simply put, you could say that the FA needs to see the MIP tunnel outermost, while the VPN-GW needs to see the IPSec tunnel outermost. A more detailed explanation follows.

First, the MN must have an IPSec tunnel established with the VPN-GW in order to reach the HA, which places the IPSec tunnel outside the MIP traffic between MN and HA. The FA (which is likely in a different administrative domain) cannot decrypt MIPv4 packets between the MN and the VPN gateway, and will consequently not be able to relay the MIPv4 packets. This is because the MIPv4 headers (which the FA should be able to interpret) will be encrypted and protected by IPSec.

Second, when the MN is communicating with the VPN-GW, an explicit bypass policy for MIP packets is required, so that the MN can hear FA advertisements and send and receive MIP registration packets. Although not a problem in principle, there may be practical problems when VPN and MIP clients from different vendors are used.

6.6.6.4 Solutions

Reference [20] discusses pros and cons of the solutions available in the open literature. Details will not be given in this document. Solutions discussed in [4] are listed as follows:

1. Dual HA: This solution says that two HAs should be used, for internal and external, respectively. This leads to three layers of tunnels: external HA, IPSec and internal HA.
2. Optimized dual HA: The motivation of this solution is to eliminate use of double MIP encapsulation discussed in 1.
3. Use of Mobile IP signaling to VPN gateway (route optimization).
4. MIP proxy: This solution aims at introducing a MIP proxy for seamless traversal across VPN.
5. Making VPN GW accept outer IP changes.
6. Use IPSec instead of GRE/IP-IP for MIP tunneling.
7. Host routing and end-to-end security.
8. Explicit signaling to update IPSec endpoint.
9. Use FA to route ESP.

6.6.7 MIP and NAT Issues

As NAT is often used, MIP's basic assumption fails: MN and FA are uniquely addressable, and need global IP addresses.

MIP relies on sending traffic from the home network to the MN or FA through IP-in-IP tunneling. IP nodes, which communicate from behind a NAT are reachable only through the NAT's public address(es). IP-in-IP tunneling does not generally contain enough information to permit unique translation from the common public address(es) to the particular CoA of a mobile node or foreign agent that resides behind the NAT; in particular there are no TCP/UDP port numbers available for a NAT to work with. For this reason, IP-in-IP tunnels cannot in general pass through a NAT, and Mobile IP will not work across a NAT.

MIP's registration request and reply will on the other hand be able to pass through NATs and NAPTs on the MN or FA side, as they are UDP datagrams originated from the inside of the NAT or NAPT. When passing out, they make the NAT set up an address/port mapping through which the registration reply will be able to pass in to the correct recipient. The current Mobile IP protocol does not however permit a registration where the mobile node's IP source address is not either the CoA, the Home Address, or 0.0.0.0.

What is needed is an alternative data tunneling mechanism for MIP that will provide the means needed for NAT devices to do unique mappings so that address translation will work, and a registration mechanism that will permit such an alternative tunneling mechanism to be set up when appropriate. A NAT traversal-based solution is discussed in [21].

6.6.8 Hierarchical Mobile IP

Micro-mobility protocol, HMIP, has been designed to support fast, seamless, and local handoffs between adjacent cells without contacting the MN's HA. In this protocol mobile hosts send mobile IP registration messages (with appropriate extensions) to update their respective location information. Registration messages establish tunnels between neighboring foreign agents along the path from the mobile host to a gateway foreign agent. Packets addressed to mobile hosts travel in this network of tunnels, which can be viewed as a separate routing network overlay on top of IP. The use of tunnels makes it possible to employ the protocol in an IP network that carries non-mobile traffic as well. Typically one level of hierarchy is considered where all foreign agents are connected to the gateway foreign agent. In this case, direct tunnels connect the gateway foreign agent to foreign agents that are located at access points.

Paging extensions for HMIP allow idle mobile nodes to operate in a power saving mode while located within a paging area. The location of mobile hosts is known to home agents and is represented by paging areas. After receiving a packet addressed to a mobile host located in a foreign network, the home agent tunnels that packet to the paging foreign agent, which then pages the mobile host to reestablish a path toward the current point of attachment. Paging a mobile node can take place using a specific communication time-slot in the paging area similar to the paging channel in second generation cellular systems. Paging schemes increase the amount of time a mobile host can remain in a power-saving mode. In this case, the mobile host only needs to wakeup at predefined time intervals to check for incoming paging requests [36–38]. HMIP has the following design attributes: fast handoff, paging, fast security, hierarchical mobility, and hierarchical tunneling.

6.6.8.1 Architecture

HMIP provides micromobility support by installing a hierarchy of mobility agents and assigning a "virtual" COA on each hierarchy level. When an MN moves locally, only those mobility agents directly affected by the move must be notified. The HA (and the CNs in case of route optimization) has to be notified only if the MN moves between the domains of different top-level mobility agents [12]. HMIP architecture (see Figure 6.16) is hierarchical in two points. First, it separates the local mobility management from the global one. Local handoffs are managed locally and transparently to mobile's correspondent hosts. Second, it clearly separates the protocols managing local mobility from the protocols managing global mobility.

The protocols designed to provide the functionalities of mobility management and authentication, authorization, and accounting, include the registration protocol, tick payment protocol, and accounting protocol.

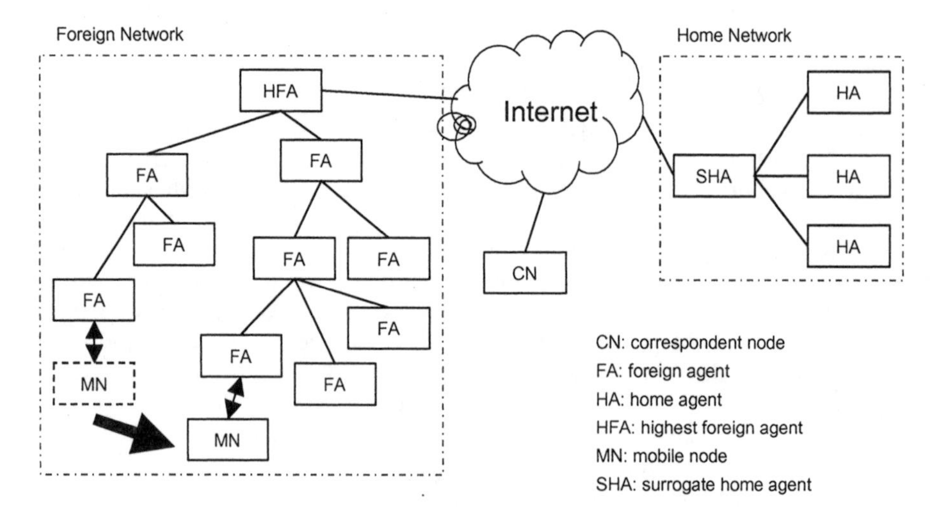

Figure 6.16 Hierarchical mobile IPv4 architecture.

This hierarchy is furthermore motivated by the significant geographic locality in user mobility patterns, since most of a user's mobility is local. According to a study, 69% of a user's mobility is within its home site (within its building and campus) and 70% of all professionals can be classified as mobile. It is therefore important to design a hierarchical mobility management architecture that optimizes local mobility.

6.6.8.2 Security

HMIP provides micromobility support by installing a hierarchy of mobility agents and assigning a "virtual" COA on each hierarchy level. When an MN moves locally, only those mobility agents directly affected by the move must be notified. The HA (and the CNs in case of route optimization) has to be notified only if the MN moves between the domains of different top-level mobility agents. Main advantages and disadvantages of this approach are the following:

- Security (advantage): MNs can have (limited) location privacy because lower level CoAs are hidden.
- Transparency (disadvantage): Additional infrastructure and changes to the MN protocol stack are required.
- Efficiency (disadvantage): In case of (inter-domain) handover, all hierarchy levels have to be reconfigured by the MN, which may be connected to the network by a slow wireless link.
- Security (disadvantage): Routing tables are changed based on messages sent by mobile nodes, and there are no provisions for control message authentication.

6.6.9 Next Generation All-IP Mobility Management Requirements

The most important requirements regarding mobility management of next generation all-IP networks are the following [35]:

1. Support of different types of mobility:

   - *Terminal mobility*: Refers to an end user's ability to use her/his own terminal regardless of location and the ability of the network to maintain the user's ongoing communication as she/he roams across radio cells within the same subnet and subnets either in the same administrative or in different administrative domains.
   - *Service mobility*: Refers to the end user's ability to maintain ongoing sessions and obtain services in a transparent manner. The service mobility includes the ability of the home service provider to either maintain control of the services it provides to the user in the visited network, or transfer their control to the visited network.

- *Personal mobility*: Refers to the ability of end users to originate and receive calls and access subscribed network services on any terminal in any location in a transparent manner, and the ability of the network to identify end users as they move across administrative domains.

2. Hierarchical topology: This means separation of global from local mobility. The term global mobility refers to the movement of mobile hosts across different networks/domains (inter-domain) whereas the term local mobility is used to describe the movement within a specific subnet. For local mobility, it is important to differentiate active from idle systems to improve performance and scalability.
3. Use of multiple protocols: Currently the IETF standardizes the Mobile IP protocol to support dynamic mobility across Internet domains for mobile hosts. There are two variations of MIP, for IPv4 and IPv6 networks, respectively. However, Mobile IP is struggling with the problem of triangular routing, which adds delay to the traffic towards a mobile host, but not from the mobile host. For delay sensitive traffic (e.g., voice or multimedia over IP) this is not acceptable, due to the high latency in the network. Mobile IP route optimization solves the problem, but on the other hand requires modifications in the IP stack of the end hosts. To alleviate the problems involved with Mobile IP, application layer mobility is proposed, by using the Session Initiation Protocol (SIP). SIP [27–30] is an emerging protocol, designed to provide basic call control and application-layer signaling for voice and multimedia sessions in a packet-switched network. Supporting both real-time and non-real-time multimedia applications in a mobile environment may require mobility awareness on a higher layer above IP, in order to utilize knowledge about the traffic type. Several wireless technical forums (e.g., 3GPP, and 3GPP2) have agreed upon SIP utilization to provide session management and a means of personal, as well as service mobility.
4. Adaptive service-aware security: A mobile user should be able to achieve the required security level depending on the services, device capability, battery lifetime, and access technologies from the security manager located with the AAA server at each stakeholder without the need to contact the home network.

6.6.10 Seamless Mobility (Seamoby)

Within IETF, a seamless mobility or Seamoby group aims at facilitating the realization of real-time services over the IP wireless infrastructure [24]. These activities are in three technical areas, namely context transfer, handover candidate discovery, and IP paging. Seamoby work does not affect MIP; they can work together. Note that the Seamoby solutions are only for intra-domain handover.

As in IEEE 802.11f, Seamoby is also looking at the benefit of transferring context in the IP layer. The Seamoby working group has analyzed the requirements and trade-offs for the goal of transferring context information from a MN's old to new access router (AR). Currently the Seamoby working group has prepared a draft [25] of the so-called Context Transfer Protocol (CTP), to transfer the context information across ARs.

Further Seamoby WG is developing a Candidate Access Router Discovery (CARD) protocol [26]. This protocol can be used to facilitate fast handover.

Seamoby also plans to develop a protocol for Dormant Mode Host Alert (DMHA), which is particularly of interest for wireless devices that can go in dormant mode to save power. The device should thus be tracked and packets delivered when they become available.

6.7 HIGHER LAYER MOBILITY

In this section we discuss mobility protocols in layers higher than IP layer. This includes the session and transport layers. Protocols discussed are TLS, Stream Control Transmission Layer Protocol (SCTP), and SIP. First we start with issues in Mobile IP and service continuity.

6.7.1 Mobile IP Issues

One of the basic issues of MIP is that the IP address of the MN can change. This means that after connection is established between two attachment points, the local application end-point will only accept packets addressed from the remote attachment point. Here we see the relation between application end-point and IP address. So when the user moves to a new place, a new attachment point, and tries to make connection the old connection with the end-point cannot be used anymore. So, to continue communications between the same two end-points, a new connection must be established between the new attachment points. This means that packets will be lost while the new connection is made. Creating a new connection, we have already seen, means that there should be discovery phase to find the new attachment point and then binding phase, informing the remote end-point(s) of the change.

The above problems arise because the IP routing model does not allow machines to dynamically change their address and open sockets do not allow a change in address. In addition, IP does not allow open connections to change physical location. If a host were to keep its address and move to a different location, IP routing would fail and the connection would be lost.

Another issue relates to Domain Name Service (DNS), which statically maps machine names to addresses (dynamic DNS also exists). Anyway, when a host moves it cannot be found by DNS and as DNS does not use username, it cannot locate the user. SIP overcomes this limitation of the DNS. SIP provides a directory lookup of user location; however, it does not provide methods for finding devices when they change addresses.

Further the triangular routing in MIP creates unnecessary and unwanted traffic in the home network. For applications having strict QoS requirements there is a need to ensure resources such as bandwidth and delay boundaries. Today, two models are commonly considered for making reservations, IntServ and DiffServ. In both cases, the tunneling breaches the end-to-end semantic and makes reservation cumbersome, if not impossible.

MIP also has a weakness which relates to its lack of support for multiple network interfaces at the MN. In future networking environments, it is very likely that a MN will have access to more than one network simultaneously; thus there is a need for multi-homing. It is also very likely that network availability will be dynamic so that during a session another network type might become available. All different networks will have different characteristics. This in turn will lead to users wanting to dynamically select different network connections for different traffic flows. In Mobile IP all traffic to a MN's network interface is aggregated and tunneled to the same address, so Mobile IP makes no distinction between different flows. Thus if a user wants one of the open flows to use another interface, all flows must be redirected to use that interface.

6.7.2 Stream Control Transmission Layer

SCTP, from 3GPP, has been accepted by IETF as a general-purpose transport layer protocol. The protocol has the TCP features of reliable service delivery over an IP network. Besides the features of TCP, it has other features that support mobility such as partial reliability and multi-homing. Partial reliability mechanism allows SCTP to configure the reliability level. This helps, for example, traffic that has high QoS requirements and has only a limited time to live.

Using multihoming SCTP allows a MN to give a set of IP addresses as primary IP addresses and another set as secondary IP addresses, with a port for primary and secondary set, to the CN and/or HA. This way one of the addresses can be used as secondary address. The protocol also allows the deletion and addition of new addresses on-the-fly, (i.e., autoconfiguration). This way when a handover is detected the secondary address can be automatically used.

6.7.3 Transport Layer Security

Session mobility implements VPN and mobility functions at the session layer without making any attempt to keep the transport level connections alive during network roaming [33, 36–38]. Instead the solution relies on recovery mechanisms at the session layer for fast transport connection re-establishment.

TLS can bring in strong end-to-end security on an application level, discussed in Chapter 4. The end could also be the edge of the network as for IP VPN. With TLS or wireless TLS (WTLS) one can set application-based policy; thus a strong security in a device will allow a highly sensitive e-mail download while another will not. Secure connection establishment is a process that requires heavy computations and many messages need to be sent between the two communicating peers. If the communication device is a PDA with limited processing power, a session setup may take several minutes to perform. Therefore it is of paramount importance that the session is able to survive periods during which network services are unavailable. TLS addresses this issue by supporting re-establishment of lost connections by using the session resume functionality. If a connection is lost, a new connection can be set up without any complex computations, at a fraction of the original setup time. Since the reconnection is performed in the background, no user interaction is required, (i.e., the user does not have to re-logon after a session resumes). On the other hand IPSec lacks the session resume functionality; a full-scale handshake including key exchange mechanisms and capability negotiations must be performed each time the client reconnects. As connection instabilities are quite common in wireless communication, the user experience is dramatically improved with session resume functionality. By using the session resume mechanism in TLS/WTLS, a new TCP session is opened and secured very quickly, totally transparent to the application layer [9].

6.7.4 Session Initiation Protocol

Maintaining multimedia sessions by means of SIP [27–29, 33, 39] (see Chapter 5), signaling has often been termed as application layer mobility management scheme. There have been numerous proposals to support different types of mobility using SIP signaling, which is generally used to set up and tear down the multimedia stream. Mobility in a heterogeneous network plays a very important role since the user could move between multiple types of access networks involving many service providers during a multimedia session. These access networks could be 802.11b-, GSM-, GPRS-, or UMTS-based networks supporting DHCP or PPP servers in the networks. The movement of the mobile host can be between the access networks where each access network may belong to the same subnet and same domain but different subnets or a different domain altogether and under a different stakeholder. In most cases the end-client would have access to both the networks at the same time, but connectivity to the network would be determined by any local policy defined in the client itself such as signal strength or any other measurement based on the QoS parameter of the traffic. SIP supports personal mobility, [i.e., a user can be found independent of location and network device (e.g., PC, laptop, and IP phone)]. The step from personal mobility to IP mobility support is basically the roaming frequency, and that a user can change location (IP address) during a traffic flow. SIP has the ability to move while a session is active. It is assumed that the mobile host belongs to a home network, on which there is a SIP server (in this example, a SIP redirect server), which receives registrations from the mobile host each time it changes location. This is similar to home agent registration in Mobile IP. Note that the mobile host does not need to have a statically allocated IP address on the home network. When the correspondent host sends an INVITE to the mobile host, the redirect server has current information about the mobile host's location and redirects the INVITE there. If the mobile host moves during a session, it must send a new INVITE to the correspondent host using the same call identifier as in the original call setup. It should put the new IP address in the contact field of the SIP message, which tells the correspondent host where it wants to receive future SIP messages. To redirect the data traffic flow, it indicates the new address in the SIP Description Protocol (SDP) [31] field, where it specifies the transport address. IPSec ESP as specified in [6] can provide confidentiality protection of SIP signaling at the IP level between the user's device and the home SIP server or SIP redirect server. ESP confidentiality is applied in the transport mode between user's device and corresponding SIP server. The level of security provided by SIP over IPSec is similar to the IPSec with MIP since the same level of security is applied to the IP layer.

6.8 ROAMING IN THE PUBLIC WLAN

Roaming between public WLAN networks is a goal to achieve a bigger footprint for small wireless ISPs (WISPs). The Wi-Fi Alliance has developed a recommendation for roaming between WISPs known as WISP roaming or WISPr [27]. This recommendation proposes the use of universal access method (UAM). In this section basic methods for roaming and the WISPr recommendation are briefly explained.

6.8.1 Inter-WISP Roaming Methods

To roam between WISPs there has to be some business relationship. There are three basic ways this can be done:

In an inter WISP relationship each WISP has a roaming contract with every other WISP. Thus a user of one WISP can roam to another and still receive one bill. This of course means too many separate contracts that can lead to overloading of the network as the number of roaming users increases.

In the roaming consortium approach, a consortium is built of which different WISPs become members. Such a consortium can set the contract relationship between different WISPs and can also act as a clearinghouse.

The broker method is maybe the most flexible approach. Here different WISPs have a contract with a broker that allows roaming from one WISP to another depending on such factors as the service level agreement (SLA). The broker can be in the position to authorize a user or if needed pass the user credentials to the appropriate WISPs' AAA. This is the most flexible method for roaming.

6.8.2 Universal Access Method and WISPr

UAM is the browser-based user authentication and authorization method used widely in many public hotspots. With this method, any IP-based device with a Web browser that supports SSL can login and be authenticated to the hot spot network. The network basically consists of a STA, which communicates through an AP to a DHCP server, if an IP address is needed, to a public access controller [PAC (AAA client)], a Web server, and an AAA server.

After the STA is associated it is given an IP address. Now the user starts the Web browser that leads to a Hypertext Transport Protocol (HTTP) request. The HTTP request is captured by the PAC and sent to the Web browser which displays a logon page to the user. This also starts a SSL connection. The user then types the username and password, which are passed to the AAA server. On authentication the AC is informed. IEEE 802.11i using IEEE 802.1x and EAP methods can also be used to give a higher level of security.

So as to provide roaming with UAM, WISPr recommends a roaming intermediary. The roaming intermediary is like the AAAB discussed in Section 6.6.3.

6.9 FAST HANDOVER IN WLAN

IEEE 802.11r [7] is looking into methods for fast handover within an ESS. The idea is to make VoIP available to users. As seen in Section 6.3.6, the major cause of delay is re-authentication. There are several solutions being proposed in IEEE 802.11r to overcome this issue while maintaining the security level. As handover within an ESS usually means a single subnet, it does not involve MIP delays.

Solutions dealing with seamless inter-domain handover are currently not much studied. One of the reasons is the issue of fast AAA communication without user intervention.

With the available technology and some fine-tuning it should be possible to perform fast and most probably seamless handover within one administrative domain, even between different subnets. Achieving handover between different administrative domains, even fast handover, will take some time as this involves not only technical challenges but also business interests. Then again the business arena is unpredictable; no one would have guessed that Vodafone would be a global operator within a decade, and alliances such as FreeMove [32] appearing to meet the challenge.

References

[1] Prasad, N. R., *Adaptive Security for Heterogeneous Networks*, Ph.D. Thesis, University of Rome "Tor Vergata," Rome, Italy, 2004.

[2] Prasad, R., *Universal Wireless Personal Communications*, Norwood, MA: Artech House, 1998.

[3] Pahalavan, K. et al., "Handoff in hybrid mobile data networks," *IEEE Personal Communications*, April 2000, pp. 34-47.

[4] Prasad, N. R., and A. R. Prasad (eds.), *WLAN Systems and Wireless IP for Next Generation Communication*, Norwood, MA: Artech House, 2002.

[5] ISO/IEC 8802-11, ANSI/IEEE Std 802.11, First Edition 1999-00-00, Information Technology – Telecommunications and information exchange between systems – Local and metropolitan area networks – Specific requirements - Part 11: Wireless LAN Medium Access Control (MAC) and Physical Layer (PHY) specifications.

[6] IEEE Std 802.11f/D1, March 2001, Draft Supplement to IEE Std 802.11, 1999 Edition, Draft Recommended Practice for Multi-Vendor Access Point Interoperability via an Inter-Access Point Protocol Across Distribution Systems Supporting IEEE 802.11 Operation.

[7] IEEE 802.11r Web site: http://grouper.ieee.org/groups/802/11/Reports/tgr_update.htm

[8] Prasad, A. R., *Wireless LANs: Protocols, Security and Deployment*, Ph.D. Thesis, Delft University of Technology, Delft University Press, 2003.

[9] Velayos, H., and G. Karlsson, *Techniques to Reduce IEEE 802.11b MAC Layer Handover Time*, KTH Technical Report, ISSN 1651-7717, Stockholm, Sweden. April 2003.

[10] Sharma, S., N. Zhu, and T. Chiueh, "Low-Latency Mobile IP Handoff for Infrastructure-Mode," to appear in *IEEE Journal on Selected Areas in Communication*, Special issue on All IP Wireless Networks, Vol. 22, No. 4, May 2004. http://www.ecsl.cs.sunysb.edu/tr/latency.pdf.

[11] Itani, S., "Use of IPSec in Mobile IP," May 2001. http://ntrg.cs.tcd.ie/htewari/papers/ipsec_itani.pdf.

[12] http://www.cs.pdx.edu/research/SMN/.

[13] Perkins, C., "IP Mobility support for IPv4," August 2000, RFC 3344.

[14] Mobile IP WG: http://www.ietf.cnri.reston.va.us/html.charters/mobileip-charter.html.

[15] Glass, S., T. Hiller, S. Jacobs, and C. Perkins, *Mobile IP Authentication, Authorization, and Accounting Requirements*, October 2000, RFC 2977.

[16] Dommety, G., et al., "AAA requirements from Mobile IP," http://www.ietf.org/proceedings/99jul/slides/mobileip-aaa-99jul.pdf.

[17] Chaskar, H., editor, "Requirements of a Quality of Service (QoS) Solution for Mobile IP," September 2003, RFC 3583.

[18] I REMEMBER IANA. V. Cerf. October 1998, RFC 2468.

[19] Problem Statement: Mobile IPv4 Traversal of VPN Gateways <draft-ietf-mobileip-vpn-problem-statement-req-01>.

[20] Mobile IPv4 Traversal Across IPSec-based VPN Gateways <draft-ietf-mobileip-vpn-problem-solution-00>.

[21] Mobile IP NAT/NAPT Traversal using UDP Tunnelling <draft-ietf-mobileip-nat-traversal-07.txt>.

[22] Soliman, H., et al., Hierarchical Mobile IPv6 mobility management (HMIPv6), June 2004, <draft-ietf-mipshop-hmipv6-02.txt>.

[23] Bandai, M., and I. Sasase, "A Load Balancing Mobility Management for Multilevel Hierarchical Mobile IPv6 Networks," PIMRC 2003.

[24] Seamoby WG: http://www.ietf.org/html.charters/seamoby-charter.html.

[25] Loughney, J. (ed.), et al., "Context Transfer Protocol," Internet draft <draft-ietf-seamoby-ctp-08.txt>, expired: July 21, 2004.

[26] Liebsch, M., et al, "Candidate Access Router Discovery," Internet draft <draft-ietf-seamoby-card-protocol-06.txt>, expires: June 2004.

[27] Anton, B., B. Bullock and J. Short, "Best Current Practices for Wireless Internet Service Provider (WISP) Roaming," Wi-Fi Alliance, February 2003, v1.0.

[28] Rosenberg, J., H. Schulzrinne, G. Camarillo, A. Johnston, J. Peterson, R. Sparks, M. Handley, and E.Schooler, SIP: Session Initiation Protocol, June 2002, RFC 3261.

[29] Rosenberg, J., and H. Schulzrinne, "Reliability of Provisional Responses in Session Initiation Protocol (SIP)," June 2002, RFC 3262.

[30] Rosenberg, J., and H. Schulzrinne, "Session Initiation Protocol (SIP): Locating SIP Servers," June 2002, RFC 3263.

[31] Rosenberg, J., and H. Schulzrinne, "An Offer/Answer Model with Session Description Protocol (SDP)," June 2002, RFC 3264.

[32] http://www.freemovealliance.net/.

[33] 3GPP, "UTRAN Iub interface: Signalling transport," 3GPP TS 25.432, version 6.0.0, Dec. 2003.

[34] Schulzrinne, H., and E. Wedlund, "Application-layer mobility using SIP," *ACM Mobile Computing and Commun. Rev.*, Vol. 4, No. 3, pp. 47–57, July 2000.

[35] Snoeren, A. C., and H. Balakrishnan, "An end-to-end approach to host mobility," *ACM Mobicom'00*, Boston, MA, Aug. 2000.

[36] Stewart, R., et al., "Stream Control Transport Protocol," Oct. 2000, IETF RFC 2960.

[37] Stewart, R., et al., "Stream Control Transmission Protocol (SCTP) dynamic address reconfiguration," draft-ietf-tsvwg-addip-sctp-08.txt, Sept. 2003, work in progress.

[38] Riegel, M., and M. Tuexen, "Mobile SCTP," draft-riegel-tuexen-mobile-sctp-03.txt, issued Aug. 2003, work in progress.

[39] Koh, S. J., et al., "Mobile SCTP for transport layer mobility," draft-sjkoh-sctp-mobility-03.txt, issued Feb 2004, work in progress.

[40] Dutta, A., et al., "Secure Universal Mobility for Wireless Internet," ACM WMASH 2004, Oct. 1, 2004, Philadelphia, PA., http://wmash2004.intel-research.net/program.html.

Chapter 7

WLAN Deployment and Mobile Integration

The deployment of WLANs requires the understanding of both wireless and IP network deployment issues. WLAN deployment issues are not the same as mobile because of the spectrum used (unlicensed versus licensed) and the difference in protocol (distributed versus centralized). Still we are seeing a development towards WLAN and mobile network interworking and integration. This chapter will discuss the issues and solutions for WLAN and IP network deployment by enterprise, public, and mobile operators. A discussion on interworking and the integration of WLAN with 3G networks based on 3GPP technology will be presented [1–32]. Security will be given attention during discussions on deployment in this chapter because it is a major concern for IEEE 802.11-based WLANs.

7.1 DEPLOYMENT ISSUES AND REQUIREMENTS

In this section we present the issues and requirements that should be considered for WLAN deployment. These are presented as general considerations and for the wireless part of the network. This section also discusses briefly some other practical issues of deployment like power supply and gives the requirements from the user and operator points of view.

7.1.1 General Network Deployment Considerations

The general network deployment considerations are those that are valid for most network deployments [1–4]:

- *Utilization*: Deployment of a network, be it WLAN, depends on the kind of users and the kind of application they will run. This issue will not be discussed further although a point related to this issue is discussed in 7.1.3.

- *Mobility:* The users should have the chance to move within one network and have access to services even when in a foreign network. The other step is that the user should have service continuity when performing handover from one network to other. Mobility is discussed in Chapter 6 and will be considered again in this chapter.
- *Security:* One cannot be blind to this issue. Hackers breaking into the networks are common these days. WLAN deployment has to take care of security. This becomes an even more important issue when the network interfaces with the Internet although it is worth noting here that most of the attacks come from inside the network.

  Applying security at a later phase basically means patchwork, solving the problem after damage has happened. Details of security for WLAN are discussed in Chapter 4; in this chapter deployment of various security solutions will be discussed.
- *Scalability:* A well designed network should be scalable, so as to grow and accommodate new network/business needs. Introduction of new hosts, servers, or networks to the network should not require a complete redesign of the network topology.
- *Bandwidth:* Adequate bandwidth should be available for the types of service and users in the network.
- *Quality of service (QoS):* The network should be ready to provide the required QoS by users. QoS here is not only valid for real-time traffic but also for non-real-time traffic like Web access. A delay in Web access will not be appreciated by users.
- *Availability/reliability:* This in a sense also relates to service level agreement (SLA). A network is required to be available and reliable. A stock trading system based on a network that guarantees transaction response times of three seconds is meaningless if the network is down three out of seven days a week!

  The mean time between failures (MTBF) of the components must be considered when designing the network, as must the mean time to repair (MTTR). Designing logical redundancy in the network is as important as physical redundancy. It is too late and costly to consider redundancy and reliability of a network when you are already halfway through the implementation stage.
- *Modularity:* This basically means dividing the network in smaller parts so that they are easy to handle and support scalability. Making modularization also makes the network safe in the sense that only a module fails.
- *Network management and configuration:* Network management should be taken into account from the beginning as is valid for security. The network management should be able to monitor the network situation,

isolate faults, and allow easy configuration of network elements. Within the WLAN world there are two groups showing up, one that believes in removing some of the MAC functionalities to a centralized place or switch and another that believes in having a management machine that controls and configures the normal APs.

- *Performance:* There are two types of performance measures that should be considered for the network. One is the throughput requirement and the other is the response time. Throughput is how much data can be sent in the shortest time possible, while response time is how long a user must wait before a result is returned from the system.

  Both of these factors need to be considered when designing the network. It is not acceptable to design a network only to fail to meet the organization's requirements in the response times for the network. The scalability of the network with respect to the performance requirements must also be considered, as mentioned above.
- *Metering, accounting, charging, and billing*: Appropriate solutions should be placed for collecting statistical data or metering of the traffic and the service being provided by the network. In addition to this is of course accounting.

  Metering is done on the network resource usage. A user should be charged based on flat-rate, per-minute use or the bandwidth used. Charging depends on the SLA. Authentication, authorization, and accounting (AAA) protocols have been used for accounting. The accounting protocol checks if there is inconsistency in the record collected and processes the session record collected. Accounting depends on the charging model applied and if different organizations are used the accounting information is passed to appropriate organizations. The billing function prepares the bill and sends the invoice to the user.

Due to limited space we will not consider all the issues in this chapter.

7.1.2 Wireless Deployment

Planning a network that fulfills the requirements of the user is a major issue. In this section some critical points are discussed for WLAN deployment; details can be found in [1, 2]. Some of the details are given in later sections but everything is not discussed in this chapter.

- *Data rate:* This is a very important issue for WLANs. Factors such as the available bandwidth and the frequency restrict the maximum achievable data rate. The data rate is further restricted by the interference in the system. The optimum choice of the data rate depending on the channel conditions and the communicating devices is extremely important. The IEEE 802.11 WLAN standards can communicate at various data rates. A rate control algorithm and the threshold levels are discussed in Section 7.2. Throughput results can be found in [1].

- *Coverage:* Factors like band restrictions and output power controlled by regulatory bodies affect the coverage. The coverage is also affected by the interference and the type of building. The material used in a building affects the propagation of the radio waves which in turn affect the wireless network. As WLANs are used in a variety of businesses, the propagation characteristics must be taken into account during the network deployment. The coverage is directly related to the propagation environment; some thoughts on coverage are given in Section 7.3.1; details can be found in [1].
- *Interference (coexistence/interoperability)*: In any wireless system one is haunted by the interference from other systems working at the same frequency; this becomes an even more important issue when discussing systems working in the industrial, scientific, and medical (ISM) band. Thus, the study on interference for the coexistence of different devices in the ISM band and the interoperability among WLANs following the same standard is bread and butter for a network deployment engineer. This issue is discussed briefly in Section 7.3.2; details are given in [1].
- *Power management:* Battery life is a major issue for wireless communication devices. The size and the weight of the wireless systems depend on power consumption. For optimum battery usage, power management is used in the WLANs, which means that the station goes to sleep mode when not in use. This can save the battery but in certain situations increase the delay and decrease the system throughput; thus the optimum power management setting should be done depending on the user and the traffic in the system. This topic is not discussed further in the chapter; see [1].
- *Cell planning (cell size/frequency planning/capacity):* This factor is dependent on the density of the users, the base stations, and last but not least the available bandwidth. As discussed in Section 7.2 cell planning is also dependent on the threshold values. A good network deployment will require an adequate cell size planning which in turn will give an optimum capacity. Thus good cell planning must make optimal use of the already scarce wireless spectrum for which several points must be taken care of together with the frequency planning. Good planning is extremely important to increase the coverage and the capacity. The cell planning is also dependent on the part of the world where the WLAN is being deployed because the regulatory bodies control the frequency band allocation. Within this issue also comes frequency reuse, which plays an important role in cellular deployment. For WLAN deployment, frequency reuse becomes an even bigger issue because there is only a limited number of channels, which have to be shared by all users or operators.

Further, being in an unlicensed band, there is far more interference in the channel [1].

7.1.3 Other Deployment Considerations

Although some important technical considerations for network deployment are given in the previous two sections, there are several others that should be taken into account. One of the basic things is the budget; a network that can fulfill all the requirements but needs several times the available budget does not make sense. Besides budget, one should consider the time to finish the work and the resources available to get the job done.

There are other practical issues that should be considered. When placing the access point (AP) the power supply plays an important role although this could be of no concern if power over Ethernet (POE) is used. Another point is the possibility and availability of a good place to hang the APs. In many cases the APs should not destroy the looks of a place.

User and operator needs should also be considered. Here an operator could be the mobile operator, enterprise, or simply a WISP. The user needs and utilization are considered in the following section.

7.1.4 Wireless Network User Needs and Utilization

Before one starts deploying WLANs it is important to study the target environment. Analysis of the environment should lead to the study that will finally give the required number of APs. The maximum range of the WLAN (as measured from point to point) determines the number of APs required. In the following the criteria that determine the maximum range are given [4, 5]:

- The AP density.
- The environment, in which the WLAN equipment will be installed.

In this section these points are discussed.

7.1.4.1 User Need and Access Point Density

In the networking environments where there are either data-intensive users, or a large number of users in a small area, one may wish to balance the throughput performance versus the cost of the investments. Three different AP densities —low, medium, and high— that fulfill different user requirements, are considered in the following [4, 5].

- Low AP density provides a maximum wireless coverage with a minimum number of access points. This option that is typically used for single-cell networks and Point-to-Point link, will also provide an efficient and cost effective solution for most networks that include multiple wireless cells.

- Medium AP density can be used for environments where the stations experience slow network response times even though the quality of the radio communications is rated as excellent. The slow response times might be experienced in areas where the following are true:
  - A high number of the wireless stations are located close to one another, causing other stations within the same cell around the AP to defer the data transmissions.
  - A number of wireless stations engaged in heavy network traffic are causing other stations to defer the data transmissions.
- High AP density can be used when one is designing a wireless infrastructure where the total cost of the hardware investments is less critical than the maximum data throughput per cell. Per definition, a high density network will include the highest concentration of APs.

7.1.4.2 Type of Radio Environment

Subject to the type and nature of the building materials, the WLAN radio signals may either pass obstacles in the radio signal path, or be absorbed or reflected by the RF barriers. According to the number and severity of the RF barriers in the radio signal path, the wireless networking environments can be classified as one (or a combination) of the following types of radio environments [1, 4, 5]:

- Free space: No physical obstruction in the signal path;
- Open office: Similar to free space but obstruction can occur sometimes;
- Semi-open office: Shoulder-height partitions of wood or synthetic material;
- Closed office: Floor-to-ceiling walls of brick and plaster.

7.2 SYSTEM CONSIDERATIONS

For IEEE 802.11-based WLAN deployment one has to consider certain system parameters during both the design and deployment phases [1, 2]. Some of these are related to roaming/handover as discussed in Chapter 6; other considerations are power management, discussed in Chapter 3, security, and QoS, discussed in Chapter 4 and 5, respectively. Besides these considerations there are also issues related to various thresholds one can set in IEEE 802.11; these are discussed in this section. Note that carrier sense multiple access with collision avoidance (CSMA/CA) is used for all variants of IEEE 802.11 physical layers.

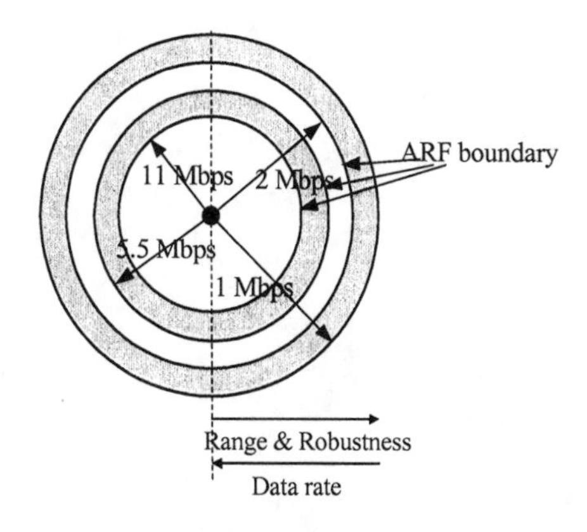

Figure 7.1 Relation between data rate and cell regions.

7.2.1 Automatic Data Rate Control Algorithm

The different modulation techniques used for the different data rates of the IEEE 802.11 can be characterized by more robust communication at the lower rate. In this section the discussion is presented from the point of view of IEEE 802.11b but the same is valid for other IEEE 802.11 physical layers. The robustness at different data rates translates into different reliable communication ranges for the different rates, 1 Mbps giving the largest range. Figure 7.1 shows the four concentric cell regions associated with the four data rates of IEEE 802.11b. Stations moving around in such a large cell will be capable of higher speed operation in the inner regions of the cell. To ensure usage of the highest practicable data rate at each moment an automatic rate fallback (ARF) algorithm is proposed in this section. Although the name ARF means decreasing the data rate, this algorithm also increases the data rate automatically as the situation changes. The IEEE 802.11 standard does not define any data rate control algorithm [1].

The ARF algorithm causes a fallback to the lower data rate when a station wanders to the outer regions or has a high level of interference and should increase the data rate when it moves back into the inner region or into better channel conditions [2, 5]. The fallback algorithm prevents a ping-pong effect. The ARF functions come into play when the ARF boundary is crossed in either direction. These boundaries are the result of the fallback scheme with a fallback in data rate after a few successive retries in transmission of a frame and change to higher data rate after a number of successful frame transfers. In this way the data rate shows a carrier-to-interference (C/I) dependent behavior and leads to a difference in coverage range. The upgrade in the data rate will be done step-wise, i.e., if in 1 Mbps the WLAN will go to 2 Mbps but not to 5.5 or 11 Mbps. Besides resulting in a bigger range, the lower rates will also be more robust against other interfering conditions like high path-loss, high background noise, and extreme multipath effects.

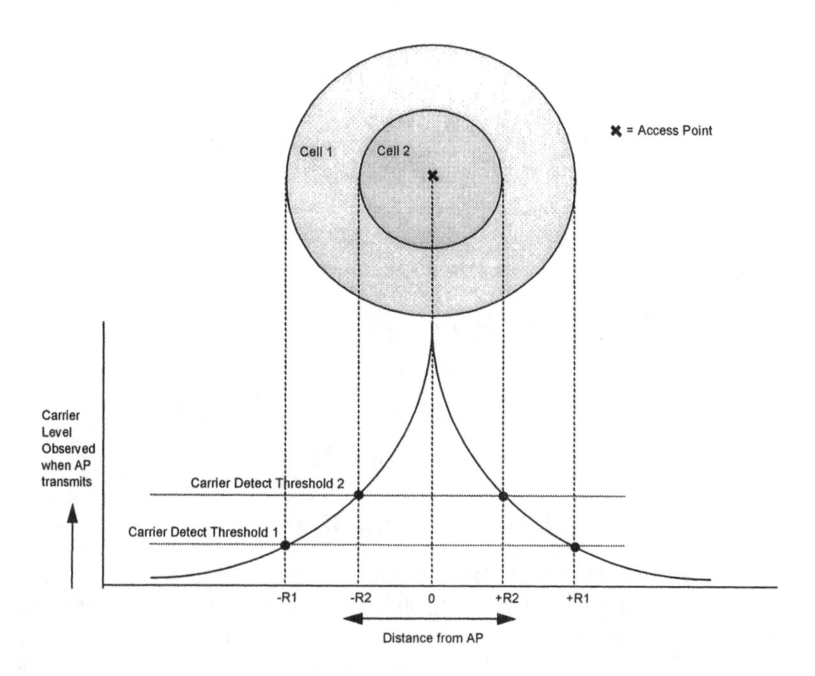

Figure 7.2 Carrier detect threshold (CDT) impact on cell size.

Introduction of different data rates means the CQ indicator mechanism used for roaming must be modified and levels for each data rate must be introduced. Although the data rate control algorithm, ARF, is based on MAC frames, in certain deployments it will not make sense to fall back to lower data rates; for example an office scenario where an 11 Mbps data rate is required and thus dense deployment of AP is done. A "cell search zone" in this case should start when the communication quality (CQ), see Chapter 6, is barely good enough to maintain 11 Mbps. While in the case of a warehouse where 1 Mbps is sufficient the "cell search zone" should start at the level where CQ is coming to a level below which 1 Mbps can barely be maintained.

7.2.2 Thresholds and System Scalability

A WLAN system can be deployable in a broad variety of situations, posing different and often conflicting requirements on the system behavior. In particular, the situation of a stand-alone single-cell network versus an infrastructure network consisting of multiple overlapping cells will have different requirements on both transmit and receive behavior. To accommodate these varying operational situations, some WLAN products have a number of built-in provisions to create scalable systems, optimized for environment and network usage needs.

When an AP transmits, the observed levels of the carrier signal by a station will decrease with the distance. Figure 7.2 illustrates the typical curve for the signal level in two opposite directions. In this section threshold values that should be understood for the system design and deployment are discussed. These values can either be fixed by the vendor or left for the user to control.

7.2.2.1 Carrier Detect Threshold

The CDT is defined as: "the carrier signal level, below which the WLAN receiver will not receive." Figure 7.2 shows that the CDT at level 1 crosses the curves at distances −R1 and +R1. This implies an associated cell size for this CDT value with radius R1. This is shown as the dark ring area above the curves. The other example at level 2, which is higher (less sensitive) than level 1, shows a smaller cell (radius R2). The range for meaningful CDT levels has a lower boundary determined by the sensitivity of the WLAN receiver circuitry. Setting the CDT to a lower value will result in a number of meaningless receive attempts, which will have a high failure rate. The importance of configurable carrier detection is that it allows the WLAN cards to be configured at smaller cell sizes than the receiver is capable of handling. Small cell sizes play an important role when considering the possibilities for reuse of the same channel in a relatively small area.

7.2.2.2 Defer Threshold

The 802.11 medium access rules (CSMA/CA) are based on defer and random backoff behavior of all the stations within range of each other as discussed in Chapter 3. The defer decision is based on a configuration entity called the defer threshold (DT). When a carrier signal level is observed above the DT level, the WLAN card will hold up a pending transmission request.

Taking the example of the CDT level 2 (Cell 2) from Figure 7.2, the ideal value of the DT is such that it produces the double radius. This means that a station on one edge of the cell defers for a station at the other edge. This is shown by plotting the curve for one edge station and ensuring that the DT level crosses this curve at the other cell edge. Choosing this relation between the CDT and the DT level gives a cell in which all stations defer to each other, and where each station can communicate with the AP.

The range for the DT level has a lower boundary determined by the sensitivity of the WLAN carrier detect circuitry. Below a certain level, the signal will not be detected and no defer will be done; however, this can be a little lower than the receiver sensitivity that marks the level for reliable data reception. The ideal relation shown in Figure 7.3 cannot be achieved in the case where the CDT is set to the lowest (and most sensitive) level. In that case the lowest meaningful DT will not guarantee the wanted deferral between the two "edge stations." If a low CDT value is chosen, a large cell size with radius R is created; this is shown with the large circle in Figure 7.4. The lowest meaningful DT level has a smaller size; this

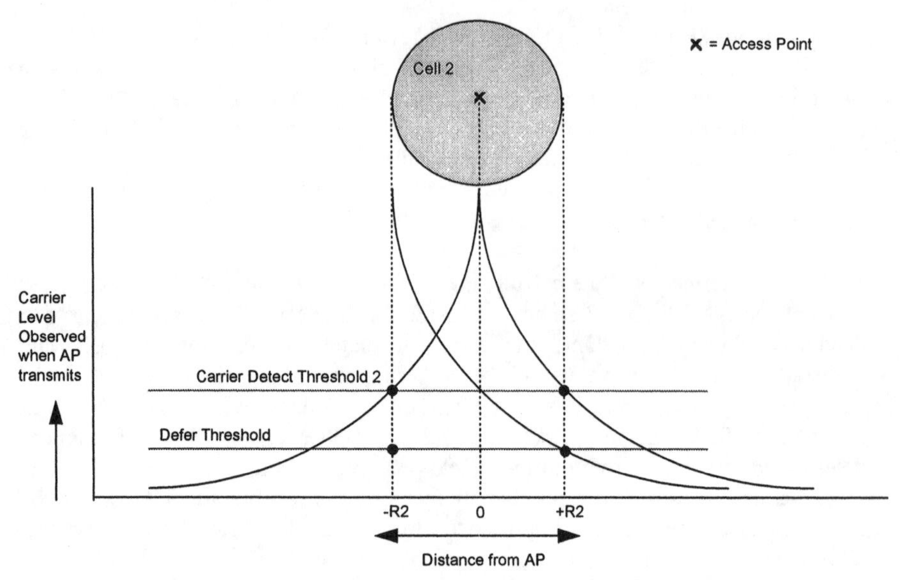

Figure 7.3 Ideal relation between DT and CDT.

is shown with the smaller and light gray circle with radius R'. The outer area of the cell (dark ring) will not have guaranteed deferral.

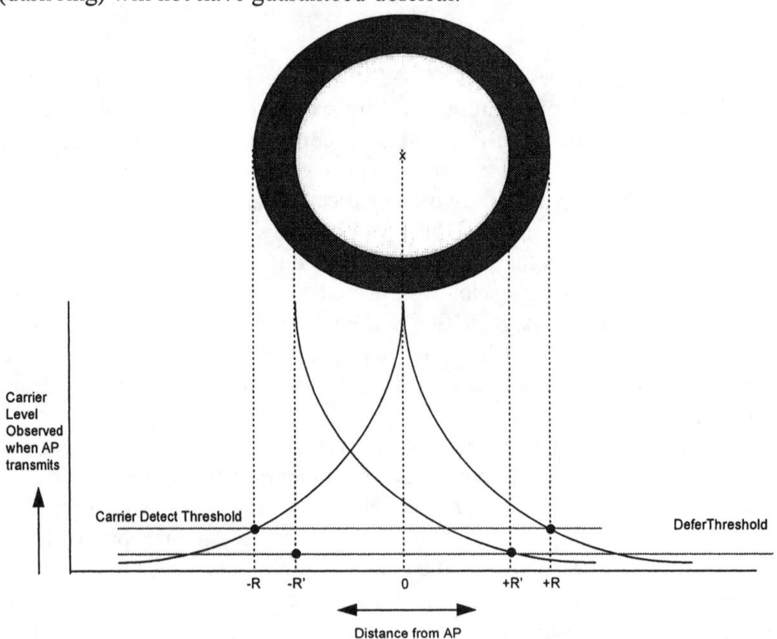

Figure 7.4 Large cell characteristics.

Similar to the CDT, the DT plays an important role when considering the possibilities to reuse the same channel. If a station is at the edge of a cell it will differ from the signal received from a station or an AP that is working on the same frequency and is a cell away. Cell planning is discussed in Section 7.3.

7.2.2.3 Roaming Thresholds

When creating a cellular infrastructure system with the above defined thresholds for the low-level receiver and transmitter control, there must be a proper balance with the earlier discussed roaming thresholds. While the CDT and the DT determine the transmit/receive behavior of the stations and the APs that belong to the same cell, the roaming thresholds determine the moments for deciding to start or stop participation in a cell. A station should base its handover decisions on the currently configured capabilities of the receiver. In particular when small cell sizes are required, the roaming thresholds must be set such that stations will start searching for a new (better) AP in advance of the moment that the receiver becomes physically incapable of receiving messages from the current access point.

7.3 WLAN MAC AND PHY LAYER DEPLOYMENT

In this section issues related to coverage, interference, cell planning, and frequency planning are discussed [1, 2, 5]. Note that although the results are often given for IEEE 802.11b, similar calculations can be made for the 5GHz band. The major difference that one can consider is that the interference in the 5GHz band is either nonexistent or minimal.

7.3.1 Coverage

In Figure 7.5, signal strength curves are shown for different path-loss models [1, 2, 5]. These results can be used to find the range covered for a given receiver sensitivity. With a transmit power of 15 dBm (30 mW) the receive level at 1m is –25 dBm. The receiver sensitivity for different data rates, as given in Table 7.1, can give us the achievable distance for different data rates, e.g., to find the range for 11 Mbps the –84 dBm while reserving room for 10dB fading margin gives –74 dBm. This –74 dBm gives with lower solid line (coefficient 4.5) a cutting point at the distance 29m. Experience shows that the Ericsson multibreakpoint model gives a good fit for most indoor environments where the fading margin is already included. For most realistic scenarios the achievable range can vary between the points found through the upper and lower bound of the Ericsson multi-breakpoint model.

The reliable coverage analysis is based on path-loss modeling for environments like open plan building, semi-open office, and closed office with respectively, path-loss coefficients of 2.2, 3.3, and 4.5 above the 5 meter breakpoint.

On top of this modeling with path-loss dependent on the transmit-receive (TX-RX) distance there will be a margin of 10 dB required in relation to variation due to fading. With two antennas and a Rayleigh fading channel the 10dB margin reflects a reliability of 99%.

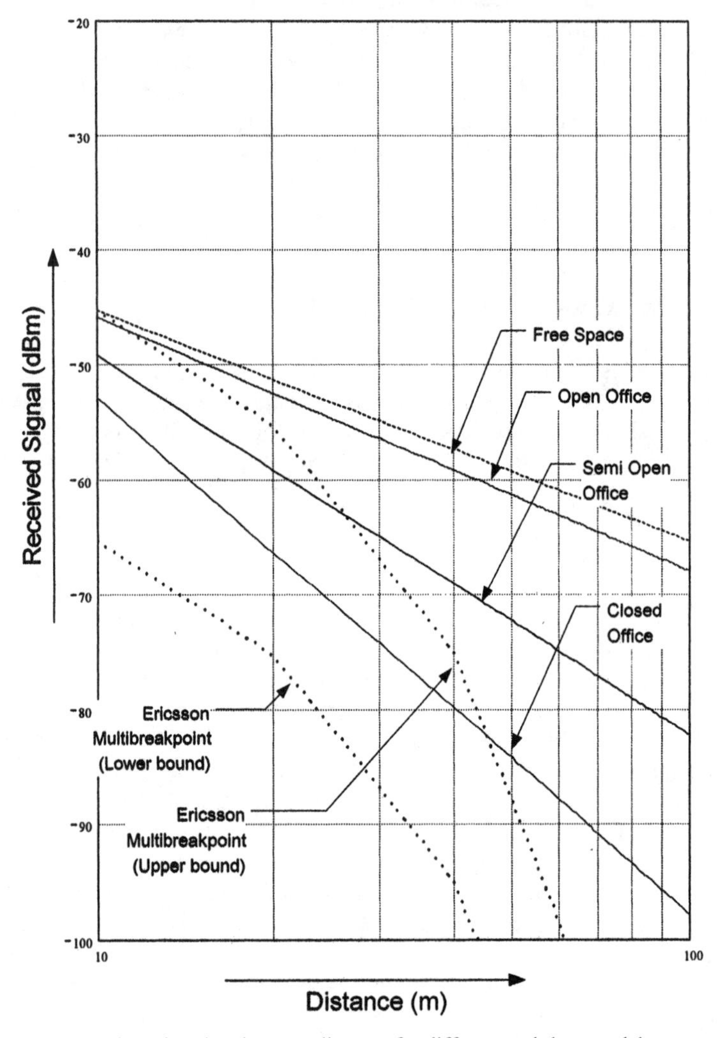

Figure 7.5 Typical receive signal versus distance for different path-loss models.

Table 7.1

Reliable Ranges According to Path Loss Models

Data rate	1 Mbps	2 Mbps	5.5 Mbps	11 Mbps
Receiver sensitivity for BER 10^{-5}	-93 dBm	-90 dBm	-87 dBm	-84 dBm
Range covered 99% point TX power 15 dBm				
Open plan building (range factor per dB: 1.110)	485m	354m	259m	189m
Semi open office (range factor per dB: 1.072)	105m	85m	69m	56m
Closed office (range factor per dB: 1.053)	46m	40m	34m	29m

7.3.2 Interference

There are various sources of interference of concern for WLANs. The basic ones are of course adjacent channel and cochannel interference. Then again there are also microwave oven interference and interference due to other devices working in the same frequency band.

7.3.2.1 Adjacent Channel and Cochannel

Studies show that adjacent cells will not interfere with each other when the channel spacing uses channel center frequencies that are 15 MHz apart. With fully overlapping cells, the separation has to be 25 MHz to avoid interference and medium sharing. Channel rejection is the combined effect of the transmitter spectrum output shaping, filtering, and detection at the receive side. In particular the IF filter (mostly surface acoustic wave, SAW, filter) at the receive side is one of the key components. The required capture ratio (6 dB at 2 Mbps, 12 dB at 11 Mbps) is fundamental in terms of how robust the scheme is with respect to co-channel interference from a neighbor cell that wants to use the same channel; the defer threshold gives the point from where to allow channel reuse.

The defer threshold level and the required capture ratio give the basis of medium reuse planning. The focus could be among other smaller cells with a more dense reuse that need more APs, or larger cells to limit the number of APs. At 2 Mbps the channel frequency can be reused when there is one other cell in between that is not using that channel frequency.

7.3.2.2 Microwave Oven and FHSS

Microwave ovens also work in the ISM band and create a lot of noise. In the case of professional microwave ovens a nearby WLAN device will not be able to work. For home microwave ovens the idle period of the duty cycle can be used for transmission. This means that a higher data rate should be used as the packets at a higher data rate traverse the medium much faster and possibly within the idle period of the oven. The ARF on the other hand decreases the data rate depending on the number of consecutive erroneous packets and signal level; thus the microwave oven type interference should be considered during system design. Of course the distance from the oven is also a consideration; the farther you are the less the interference.

Within the ISM band there are several other sources of interference. Two of them, Bluetooth and frequency hopping spread spectrum (FHSS) based WLAN, bring in the need for coexistence. Direct sequence spread spectrum (DSSS) is more robust against in-channel interference because of its despreading (correlation) process. FHSS rejects many of the DSSS signals because of its narrower filtering; however, its low-modulation GFSK scheme is much more sensitive to in-channel interference. With single-cell DSSS and FHSS systems, the channel overlap risk is limited because FHSS hops through the whole 2.4GHz band. Roughly the tolerable interference for both systems in case of channel overlap is 10 dB.

7.3.3 Cell Overlap

The network could be deployed in a single floor, that is, when one talks about horizontal overlap. In reality there are WLANs deployed in different floors and in neighboring rooms thus leading to vertical and horizontal overlap. In the following sections the two overlap types are discussed.

7.3.3.1 Horizontal Overlap

With the results in the previous sections it is possible to find the point-to-point distance. If the location includes multiple types of construction materials, for example a combination of shoulder-height partitions and floor-to-ceiling brick walls, one must identify the range associated with the type of RF obstacle(s). For the stations (STAs) to roam between various locations, they should be able to detect other access points prior to leaving the coverage area. To avoid out of range situations, the network should be designed in such a way that the wireless coverage areas of individual APs overlap one another slightly. Now the question arises: How much overlap is needed?

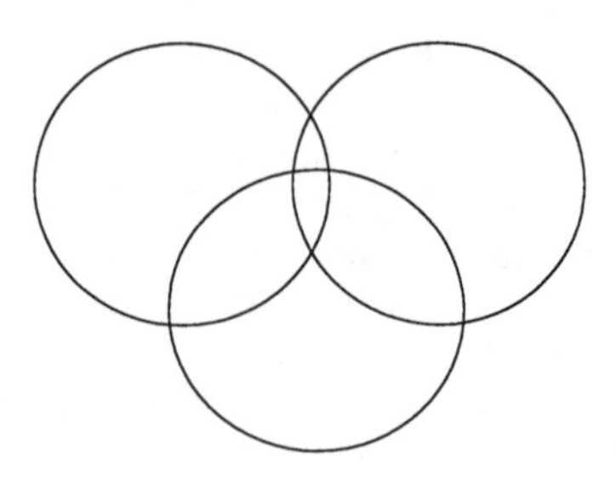

Figure 7.6 A normal deployment example.

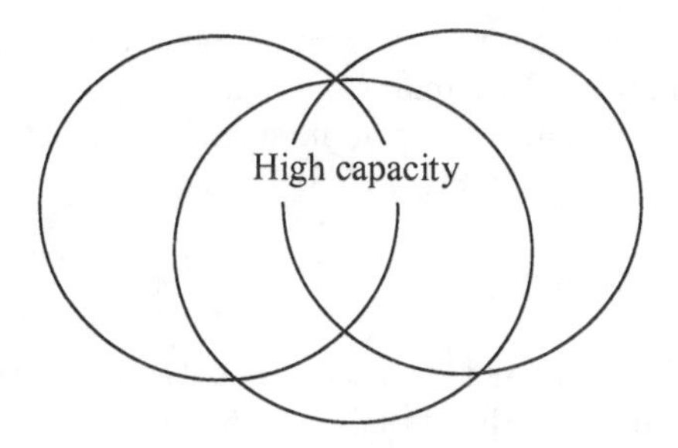

Figure 7.7 High capacity deployment example.

For most office environments, you can use a "normal density" pattern as pictured in Figure 7.6. This type of density pattern provides satisfactory results in normal density networks, where mobile stations roam at pedestrian speed.

Increasing the overlap of wireless cells as pictured in Figure 7.7 enables you to create a high capacity network, where numerous mobile wireless stations can engage in heavy traffic loads, or roam at velocities higher than pedestrian speed. If each of the wireless cells is assigned a distinct operating frequency, this type of network enables you to triple the average throughput capacity in a specific environment.

7.3.3.2 Vertical Overlap

To deploy a WLAN network in a multistory building the APs must not be placed on top of one another. One will leverage far more cost/throughput efficiency when one alternates placement of APs across the various floors, in combination with smart radio channel allocation.

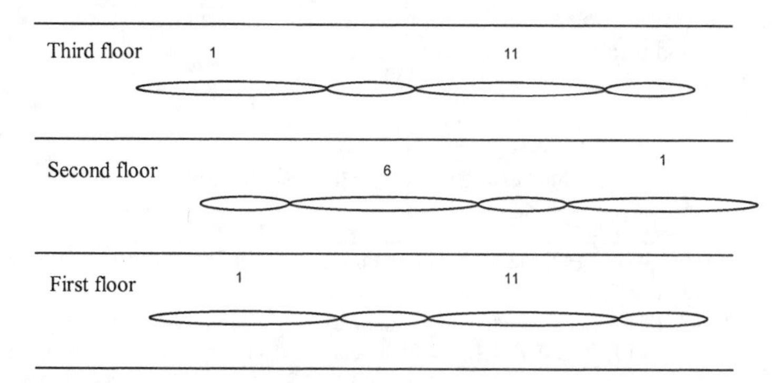

Figure 7.8 IEEE 802.11 in multifloor building.

In this way APs with the same frequency cannot hear each other, so they don't take each others bandwidth. In multistory buildings, one can also identify the vertical point-to-point distance, to determine the range of access points across multiple floors. This is shown in Figure 7.8.

7.3.4 Frequency Planning

Cell planning, which is focused on a certain data rate and based on the above mentioned radius for reliable operation, leads to coverage range values as given in Table 7.2. The 802.11 carrier sensing thresholds can be configured to find the optimum defer behavior with respect to medium reuse (related to the target cell size). Thus, there is no need to plan cell dimensions based on the maximum coverage range. In this way we could go for a cell radius allowing reliable operation at the higher data rate, or a larger cell that requires fallback in outer ring(s). The 802.11b adjacent channel rejection allows usage of channel with a spacing of 25 MHz with full cell overlap. For open environments (open air and open-plan buildings) neighbor cells only need a channel spacing of 15 MHz (see Figure 7.6). Therefore the number of available channel frequencies in the 2.4GHz band is sufficiently to fill up the two-dimensional space and there is no sharing required of channel frequencies between cells.

The channels and number of channels available for different regions of the world are given in Table 7.2. In Figure 7.9 the coverage around AP for different data rates are given together with overlap. The innermost ring is for 11 Mbps and the outermost for 1 Mbps.

Figure 7.10 gives an example of the channel frequency reuse pattern. This example is for Europe or Japan where 5 channels can be used. The aggregate net throughput that can be reached is the total of net throughput values for the cells considered. We have to be aware of only occasionally a lower throughput in the outer ring(s) with respect to margin for fading.

Enhanced provisions like load balancing and dynamic frequency selection provide automatic selection of optimum setting depending on the actual transmission conditions within a cluster of cells.

Table 7.2

Number of Channels for Different Regions

	Channels Available	*Number of Channels for Planning*
United States	2,412, 2,417, ... 2,462 MHz (11 channels)	4 channels with 15MHz spacing
Europe	2,412, 2,417, ...2,472 MHz (13 channels)	5 channels with 15MHz spacing
Japan	2,412, 2,417, ...2,484 MHz (14 channels)	5 channels with 15MHz spacing

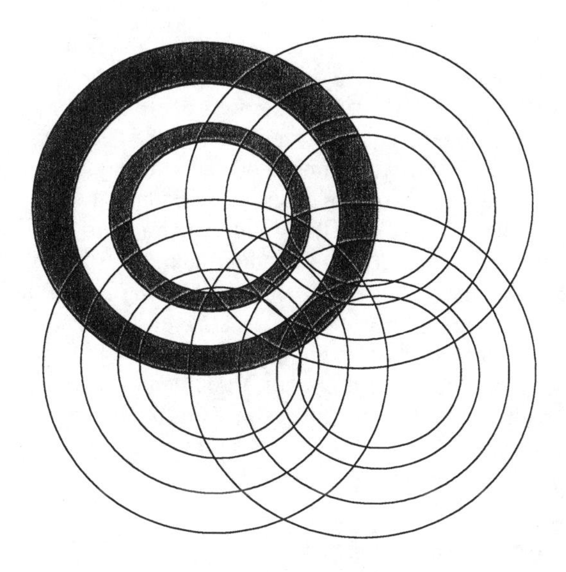

Figure 7.9 Coverage around the APs and overlap of "rings" for different data rates.

7.3.5 Cell Overlay Structure

For failure-tolerant systems where each spot has to be covered by at least two APs it can be worked by doubling the number of AP.

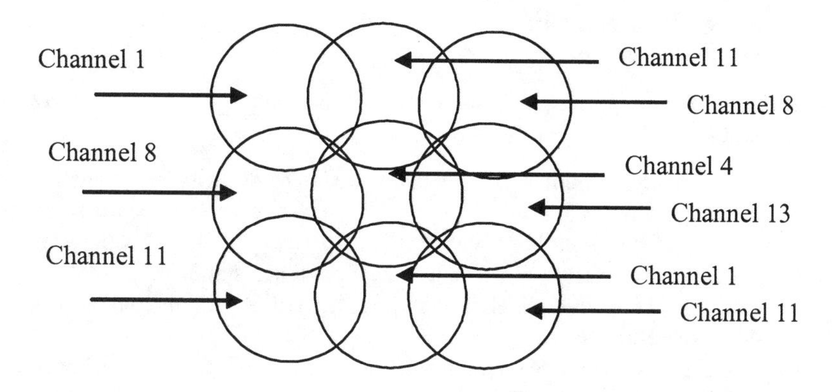

Figure 7.10 Channel frequency reuse pattern.

This can mean installing each new AP close to one of the first set, or installing each new one right in between (in the middle of diagonals). The first alternative means the same frequency for the new AP as the first one and more or less overlapping coverage (selecting another frequency would mean the need for more than 15MHz spacing). The second alternative means another frequency planning and exploiting the AP density configuration option with an individual threshold for a defer and sharing range and the carrier detect threshold, which relates the maximum range to be covered. In fact addition of APs at the mid of diagonals means that the AP grid is shrunk with 29% (1 : √2) and turned over 45°. Selecting the AP high density means limiting the defer range somehow while making the carrier detect threshold –85 dBm. The first alternative means no improvement of throughput capacity. The second alternative means a doubling of throughput capacity, but at high-load conditions and with a failure of an AP a station near the failing AP becomes an outer station of another AP and its access fairness is influenced somehow.

7.4 CORPORATE WLAN DEPLOYMENT

WLAN technology is applicable to all enterprises. Although the IEEE 802.1x extensible authentication protocol (EAP) [14, 15], also see Chapter 4, is considered the optimal deployment model, for most enterprise particular markets, such as the finance sector and corporate users are likely to prefer triple data encryption standard (3DES) [16] IPSec based security deployment models. Verticals applications within an organization may be using application-specific clients that can only support static wired equivalent protocol (WEP), and these clients will use the static WEP security deployment model. The security model has the largest impact upon the network design. There are three security models presented here [3]:

- IEEE 802.1x EAP security model;
- IPSec VPN security model;
- Static WEP security model.

The security design solutions presented in this section have the following characteristics:

- The security solution model depends upon the security requirements for the corporate LAN. The focus here is on the two most secure solutions namely EAP [14] and IP security (IPSec) virtual private network (VPN).
- Where EAP or IPSec VPNs are not possible, static WEP and access filtering are discussed although they are not a recommended security deployment model.
- The designs assume that one security model is used (EAP, IPSec, or Static WEP are not mixed within the one enterprise).

- WLAN APs should be on a dedicated subnet (not shared with wired LAN users).
- The wired LAN is not replaced by the WLAN. The WLAN is used to enhance the current network flexibility and accessibility by providing an extension to the existing network.
- Local area mobility within the corporate LAN.

Note that all security solutions are discussed in Chapter 4; here the deployment method is discussed.

7.4.1 IEEE 802.1x EAP Deployment

Such deployment addresses users that are operating in "nomadic" mode; these clients move from place to place, and need network connectivity when they are stationary [3]. For example, employees going from their desks to a meeting room normally do not access the network while traveling to the meeting room, but would need access once there. Figure 7.11 shows the EAP implementation of WLAN as dynamic layer-2 key generation with RADIUS. This delivers the features of the ideal WLAN with only the addition of RADIUS [17] authentication servers. The figure is given with VLANs that provide further security including access control. EAP is considered the optimal solution because of the following reasons:

- Provide per user authentication and accounting;
- Provide dynamic layer-2 key that can be used as WEP key or for IEEE 802.11i;
- No additional filtering or access control required;
- Multi protocol support and may carry protocols other than IP over the WLAN;
- Filtering requirements at the network access layer are the same as those for wired implementations.

While EAP is the recommended option it may not be suitable in all cases for the following reasons:

- EAP requires EAP-aware APs and WLAN clients.
- Security features offered by IPSec, such as 3DES encryption, one time password (OTP) support, or per-user policies.
- Where seamless roaming within a layer-2 domain is required, EAP clients may take longer to roam between APs; compared to those using static WEP, this may impact solutions such as VoIP over 802.11 (see Chapter 5).

7.4.2 IPSec Deployment

The IPSec solution requires users to connect to the network through a VPN client even though they are within the campus. A schematic of this is shown in Figure 7.11. Following are the characteristics of a WLAN using IPSec VPNs:

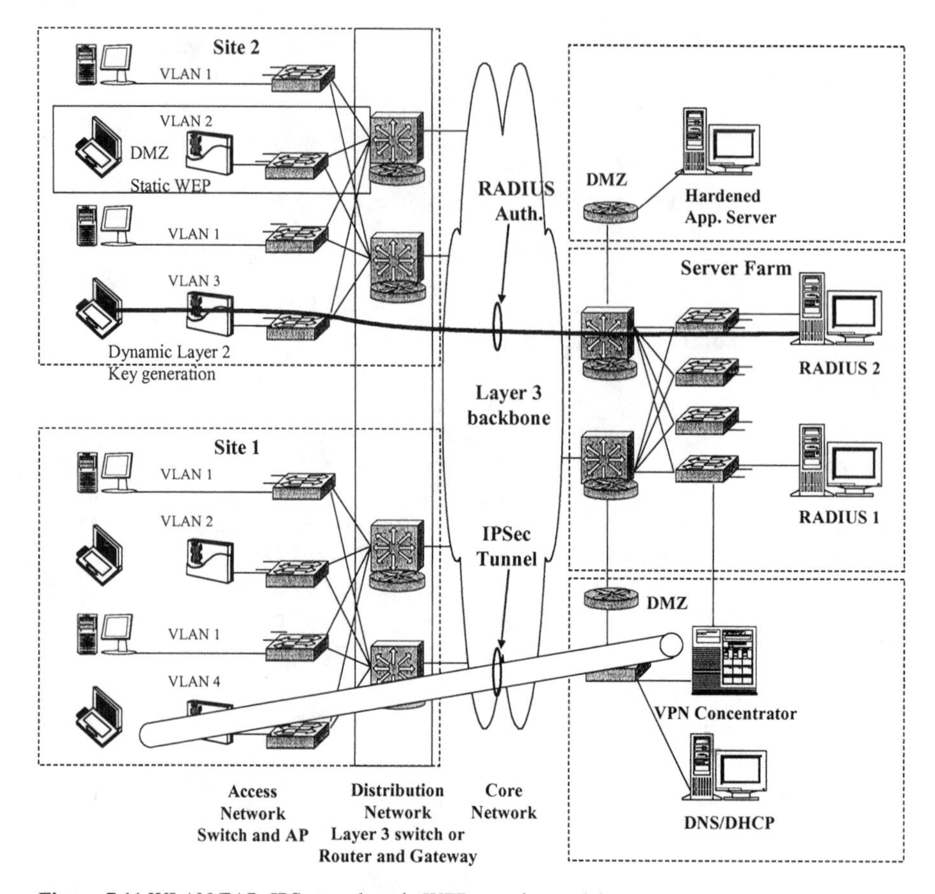

Figure 7.11 WLAN EAP, IPSec, and static WEP security model.

- WLAN with IPSec extension does not require the use of EAP, and allows any client adaptor to be used with a 3DES encryption.
- Allows the use of multifactor authentication systems OTP-systems.
- Requires the implementation of extensive filters on the network edge to limit network access to IPSec-related traffic destined to the VPN concentrator network.
- Requires user intervention, i.e., the users have to launch the VPN client before they attach to the network.
- Local traffic must still go through the VPN concentrator in the demilitarized zone (DMZ), causing traffic to cross the network multiple times, increasing traffic across the network and degrading performance.

7.4.3 Static WEP Deployment

WLAN static WEP addresses the specialized clients that are application-specific that support only static WEP. Within each enterprise, small application verticals exist that can benefit from WLAN applications. Applications requiring this type of solution may also require uninterrupted seamless coverage, as they are specialized mobility applications. Examples of potential WLAN applications that may use static WEP are:

- VoIP over 802.11;
- Messaging applications;
- Workflow and security applications.

Figure 7.11 shows the WLAN static WEP network. The DMZ notation indicates additional filtering required for securing the network, and the inclusion of the layer-2 backbone indicates a possible need to extend the layer-2 network to support campus-wide roaming. WLAN static WEP, shown in Figure 7.11, is a design that supports clients who are incapable of EAP or IPSec. The solution is considered less satisfactory than EAP or IPSec for the following reasons:

- Wireless privacy is provided by static WEP, which is vulnerable to attacks; the higher protocol layers should provide additional privacy.
- Introduces logistical problems with key management.
- Requires the implementation of extensive filters on the network edge to limit access to vertical-application related traffic destined for the clients and servers.
- May require the use of a firewall to secure the application protocols used.
- Requires that application server be hardened to prevent attacks from WLAN.

7.4.4 Selection Criteria Model

Table 7.3 summarizes the deployment solutions discussed earlier. Table 7.4 shows a detailed summary of the different security deployment options in the WLAN solution space, with regards to privacy and network access.

7.4.5 Corporate WLAN Deployment Issues

WLAN APs that are found today in the corporate or enterprise environment are individually responsible for traffic handling, radio frequency management, mobility, and to some extent authentication, e.g., static WEP [3]. These APs act in isolation making it difficult to perform critical functions such as secure seamless roaming, single sign-on, load balance among other APs, QoS support, and radio management across the entire network. In a small-scale deployment this is fine but in an enterprise environment where there are tens and hundreds of APs supporting hundreds of users with distributed control function assigned to individual AP is a major challenge for network managers to guarantee secure seamless mobility.

Table 7.3

Characteristics of WLAN Security Deployment Solutions

	EAP	IPSec	Static WEP
Protocols	Multi-protocol	IP unicast only	Multi-protocol
NIC cards	WPA- or 802.11i-compliant	802.11b-compliant	802.11b-compliant
Connection to network	Integrated with Win login. others enter usr id /pswd	The user must launch a VPN client and login	Transparent to user
Clients	Laptops and high-end PDAs. Range of OSs	Laptops and high-end PDAs. Range of OSs	Any 802.11 client
Authentica-tion	Username/password	OTP or username/password	Matching WEP key required
Privacy	Dynamic, WEP with time-limited keys and TKIP enhancements	3DES	Problematic key management
Impact on existing network architecture	Additional RADIUS Server required	Add. Infra. WLAN will be on a DMZ and require VPN concentrators, auth. servers, DHCP servers	Option of additional firewall software or hardware at access layer
Filtering	None required	Extensive filtering required - limiting network access until VPN authentication has occurred	Extensive filtering required - limiting wireless access to only certain predetermined apps
Layer-2 Roaming	Transparent: Automatically re-auths. without client intervention (may be slower than VPN or WEP)	Transparent: May be easier to extend layer-2 domain, due to reduced broadcast and multicast traffic	Transparent
Layer-3 Roaming	Requires IP address release/renew or Mobile IP solution	Requires IP address release/renew or Mobile IP solution	Requires IP address release/renew or Mobile IP solution
Management	Network is open to existing network management systems	Filtering must be adjusted to support management applications	May have application specific mgt. reqs. Filtering must be adjusted to support the mgt. applications
Multicast	Supports multicast	Currently cannot support IP Multicast	Supports multicast
Performance	WEP encryption performed in hardware	3DES performed in software. Throughput will be	WEP encryption performed in hardware

In order to guarantee function such as mobility across subnets, QoS, security, traffic policing, load balancing across wireless and wired networks, it is necessary for the APs to act more like a wireless access server (WAS) working at the networking layer instead of simple layer-2 devices. Such APs (or WAS) should have greater memory and processing power and should be capable of acting as a policy enforcement point (PEP) in order to guarantee the service level agreement (SLA) across the entire corporate network. Further, there is also a great need for a centralized network management system to deal with traffic management, authentication, encryption, and policy decisions (see Figure 7.12).

Although a WAS-based solution is one line of thought, there is another line of thought in the industry that is aiming at a split MAC solution leading to thin APs. The idea is to move several core functionalities to a switch from where the control and management can be done. This switch is known as the access controller (AC). Such switches will also serves as PEP. A study going on in the IETF control and provisioning of wireless access point (CAPWAP) group gives several possible

Table 7.4

Encryption and Network Access Options

	EAP (EAP-TLS)	IPSec	Static WEP
Key length (bits)	128	168	128
Encryption algorithm	RC4	3 DES	RC4
Packet integrity	CRC32/MIC	MD5-HMAC/SHA-HMAC	CRC32/MIC
Device authentication	None	Pre-shared secret or certificates	None
User authentication	Username/Password (PKI Certificates)	Username/Password or OTP	None
User Differentiation	No	Yes	No
Transparent user experience	Yes	No	Yes
ACL requirements	None	Substantial	N/A
Additional hardware	Authentication Server (Certificate Authority)	Authentication Server and VPN gateway	No
Per users keying	Yes	Yes	No
Protocol support	Any	IP Unicast	Any
Client support	PCs and high end PDAs.	PCs and high end PDAs	All clients supported
Open standard	No	Yes	Yes
Time based key rotation	Configurable	Configurable	No
Client hardware encryption	Yes	Available, software is most common method	Yes
Additional software	No	IPSec client	No
Per-flow QoS policy management	At access switch	After VPN gateway	At access switch

WLAN network architectures [29]. The goal of CAPWAP is to produce solutions for centralized management with intelligence at the switch instead of at the AP. AP is termed a wireless termination point (WTP). The schematic for a CAPWAP type network will be similar to Figure 7.12 except that the WAS should be WTP and thus without the router. The switches to the APs (WTPs) will be the ACs and can also be working in layer-3.

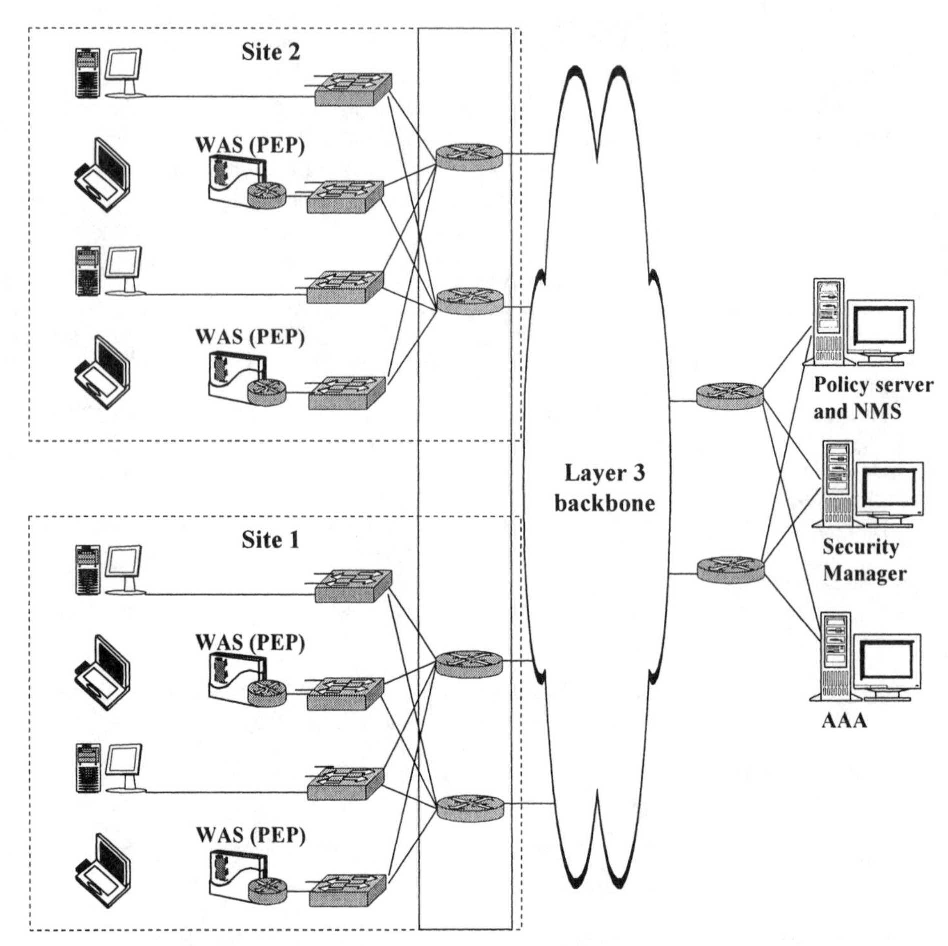

Figure 7.12 Corporate WLAN security using centralized security manager, AAA, and policy server.

7.5 PUBLIC WLAN DEPLOYMENT

There are two types of public WLAN (PWLAN) hotspots, secure and non-secure networks [3]. Non-secure PWLAN is based on the notation of just providing Internet access to the customers, and if a corporate user requires secure network access to his/her corporate network then the user needs to initiate an IPSec tunnel from his/her device to the corporate VPN gateway. An alternative method is to run a Web-based application layer secure socket layer (SSL) VPN. The latter is the most cost-effective solution and it suits an organization that mainly uses Web applications for its business. The other advantage of SSL VPN is that it does not require client software for the end user device nor does it require negotiating security association (SA) in advanced with the corresponding entity.

Deployment of cost-efficient and hundreds of secure PWLAN hotspots requires a service provider to take countermeasures against the security threat by WLAN and exposure of user data and network entities to potential hackers. Figure 7.13 shows a centralized security solution for a PWLAN hotspot using digital subscriber line (DSL) as backhaul transmission. A centralized architecture is chosen over a decentralized because remote network management such as a change in security policy or a software upgrade from the central service area is faster and more convenient than going out to change settings at every hotspot location. In the centralized security architecture the link between the hotspot access router (AR) and management router at the central service area is secured using IPSec ESP 3DES encryption [9]. This secure link is dedicated for network management and for operational support system. The secure link between the hotspot AR and edge access router (EAR) located at the central service area is secure using IPSec ESP null instead of AH. Since IPSec ESP null provides integrity protection and authentication services without covering the IP header in the message integrity check, IPSec ESP null is used in preference to IPSec AH. This link secures the traffic between the hotspot AR and the service provider's centralized service area. However, the link between the AP and end-user device is exposed to a potential security threat from hackers since WEP encryption is easy to crack. For a corporate user this is no issue because an end-to-end secure tunnel can be set between the client device and corporate VPN gateway. All other users may gain secure access between the client device and the central service area VPN gateway by creating an IPSec ESP 3DES encryption tunnel within a secure IPSec ESP null tunnel existing between AR and EAR (shown in Figure 7.13) for secure link between the client device and service provider's central service area. For this scenario the client is required to install preconfigured IPSec client software, which should be made available by the wireless service provider.

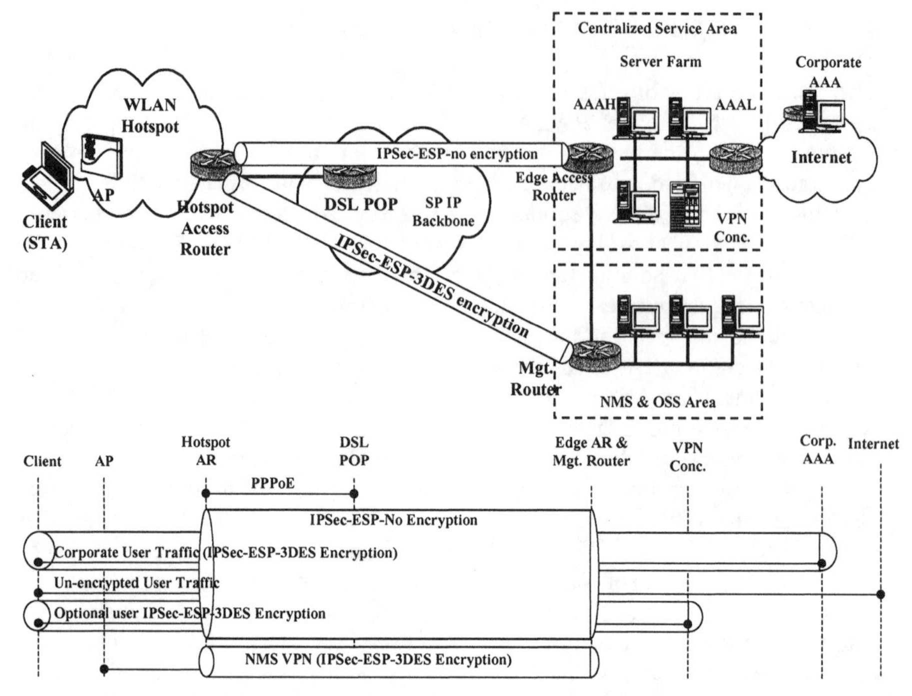

Figure 7.13 Centralized PWLAN security solution.

7.6 OPERATOR-OWNED PWLAN SOLUTIONS

For the deployment of PWLAN by mobile operators there are two fundamental interworking solutions, tight and loose interworking, and it depends on the level of integration required between the systems [3, 19, 24] (see Figure 7.14). The tight interworking solution is based on the idea of making use of the WLAN radio interface as a bearer for a cellular network, e.g., GPRS/EDGE/UMTS, [18] with all the network control entities in the core network integrated. A tight interworking solution would mandate the full 3GPP security architecture [19] and require the 3GPP protocol stacks and interfaces to be present in the WLAN system. For a loose interworking solution there is no need to make changes to the WLAN standard. This solution has the benefit of not needing a convergence layer and avoids link layer modifications; the authentication protocol is allowed to run at the link layer using EAP and AAA as transport mechanism. A fundamental requirement in 3GPP is that 3GPP-WLAN interworking shall not compromise the UMTS security architecture. Therefore, it is required that the authentication and key distribution be based on the EAP-enhanced GSM authentication (EAP-SIM) [20, 21] or EAP UMTS authentication and key agreement (EAP-AKA) [22] (see Chapter 5).

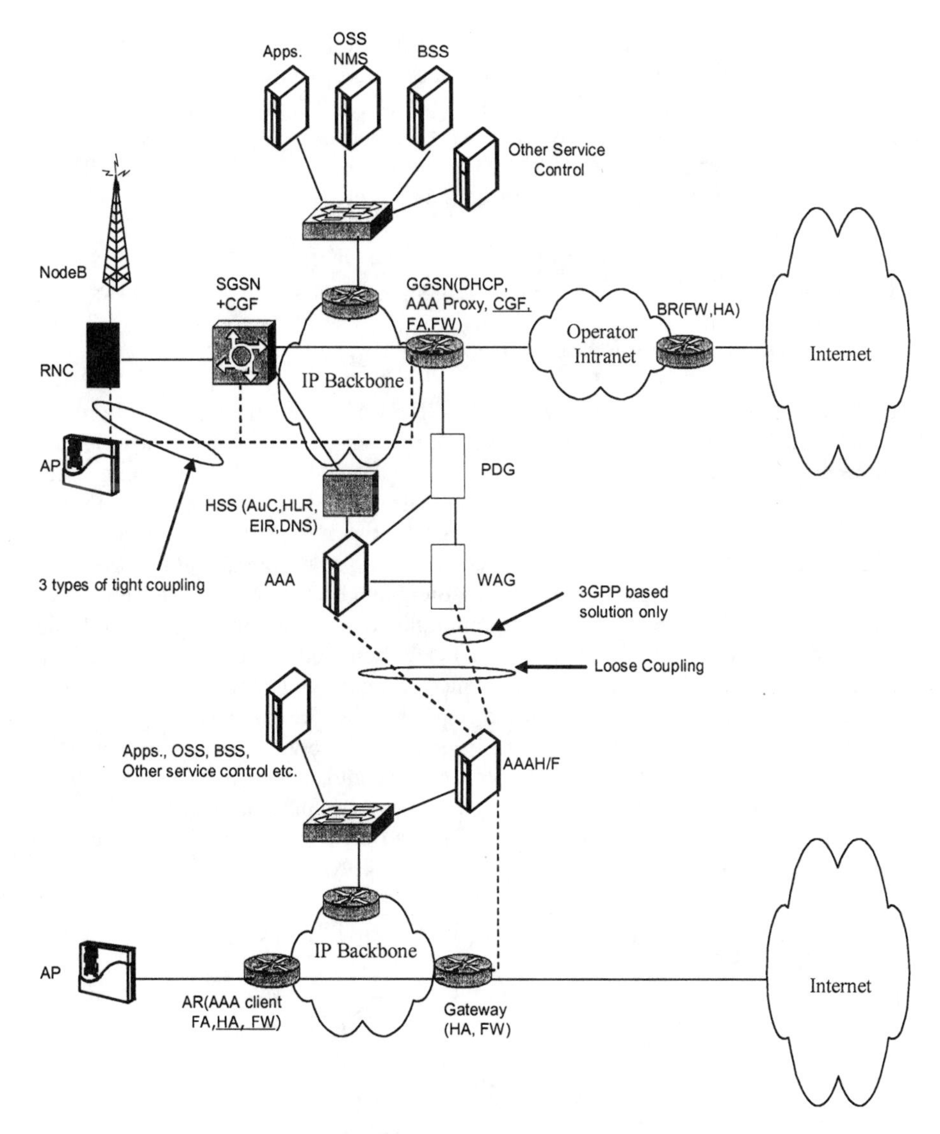

Figure 7.14 No, loose, and tight coupling between 3GPP and WLAN networks.

SIM-based authentication (EAP-SIM/802.1X) is based on open standards and allows a GSM operator to leverage its existing subscriber security database and network elements. The approach is fully aligned with the robust security networks (RSNs) being defined by the IEEE 802.11i. This includes the reuse of the GSM A8 algorithm for a secure per-user, per-session key exchange for implementing encryption over the WLAN air interface. This standard allows users to roam between different 802.1X WLAN networks, as the specifics of the SIM-based security mechanism are transparent to the visited WLAN network. Finally, SIM authentication based on EAP-SIM/802.1X does not require new WLAN NIC cards. Instead, the installed base of WLAN enabled 802.1X PCs and PDAs can be SIM-enabled by the simple addition of a SIM reader using the industry standard personal computer/smart card (PC/SC) interface or connection between GSM handset and mobile client through infrared, cable, or Bluetooth. This discussion is also valid for EAP-AKA.

Another possibility for the operator is to use short message service (SMS). Both SMS and SIM based PWLAN deployment by an operator is discussed below.

7.6.1 SMS-Based Public WLAN Deployment

Mobile operators may offer PWLAN access to postpaid clients on a monthly based subscription for limited or unlimited usage depending on the type of subscription. In order to offer granular WLAN access to prepaid, postpaid, and roaming users operators may use SMS bearer to communicate OTP to its customer (see Figure 7.15). As for the prepaid user it is important to check their credit balance before WLAN access is granted. Customers could also pay by credit card if they are connected to a foreign operator with no roaming agreement; in this case web server could be used. The web server should communicate in some form or other with the credit card company to debit the amount. Once the amount is debited successfully the user can access the PWLAN. A customer of mobile operators where a roaming agreement exists sends a SMS with the request of WLAN usage for a given time period. The OTP server receives the request and contacts the billing/prepaid charging system to check if the customer has an adequate deposit on his account. If the customer has adequate deposit, the OTP server computes a one time password and sends it via SMS to the customer, if not an error message will be returned. Afterwards the OTP server sends the password to an AAA server via LDAP. If the

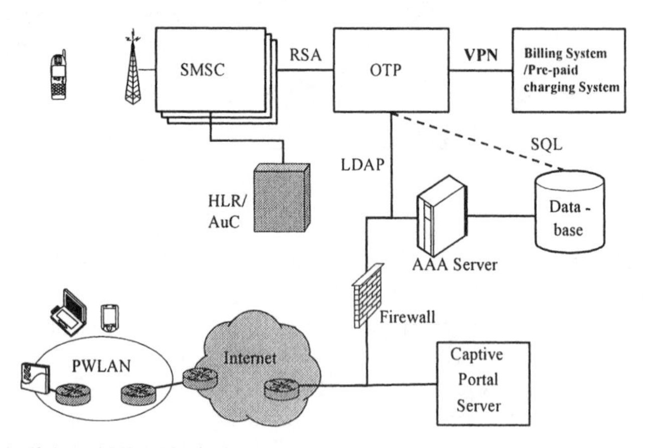

Figure 7.15 SMS-based PWLAN deployment.

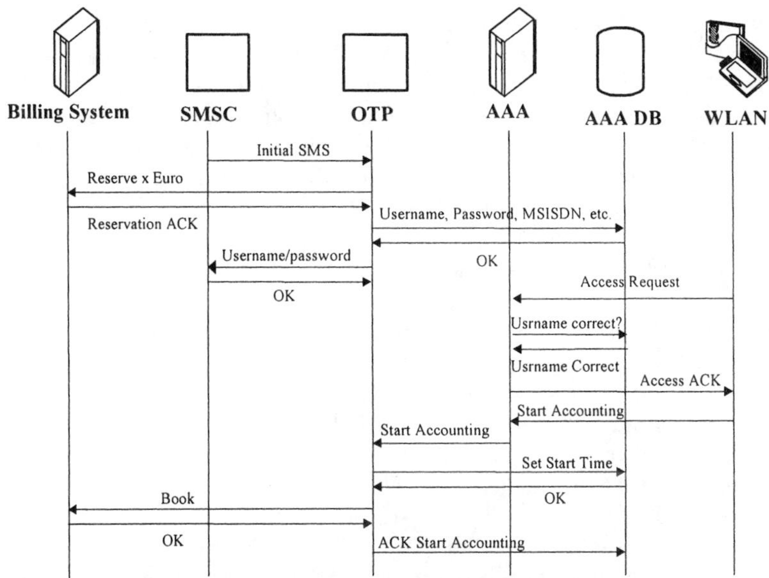

Figure 7.16 SMS-based prepaid procedure of a successful transaction.

customer uses the password, the AAA server tells the OTP server the use of the password and the OTP server tells the billing system the amount a user has to pay. Figure 5.11 shows the concept of using SMSC and OTP server to authenticate, authorize and bill users for WLAN usage. Figure 7.16 illustrates the procedure of a successful transaction of a pre-paid user for its WLAN usage.

7.6.2 SIM-Based Public WLAN Deployment

As service providers start to deploy PWLAN they are finding well known issues with techniques that do not incorporate IEEE 802.11 encryption. In particular, session hijacking cannot be prevented if encryption is not enabled, or supplementary security is applied (e.g., a client-initiated IPSec tunnel). This is becoming an issue with service provider deployments as they realize that they risk a tarnished image due to any bad press generated due to such well-known weaknesses. SIM-based security deployment is interesting for a GSM operator for the following reasons.

- It leverages an already deployed security association using a high-entropy shared key, K_i, stored in a tamper proof smart card.
- It leverages already deployed authentication and cipher key generation algorithms, A3 and A8.

Regarding new security threats compared with GSM, the rogue network is not considered a security threat for GSM but is clearly a threat with WLAN. In order to combat this, EAP-SIM enhances GSM A3 A8 security with network authentication. With regard to securing the WLAN 802.11 link, EAP-SIM reuses the A8 algorithm to support mutual key exchange for the negotiated cipher suite. According to the negotiated cipher suite, additional protection may be required to guard against initialization vector rollover. Hence, rekeying should be performed before rollover. In such cases the AAA session timeout attribute in the access accept is set to ensure rekeying before rollover.

7.6.3 Mobile and WLAN Roaming

Delivering a roaming service encompasses a number of key building blocks: signaling, billing, data clearing, financial clearing and settlement, contract management, testing, and fraud management. Because of the significant investment made by the mobile service providers in transferred account procedure (TAP), it is clearly advantageous to look to leverage this core competency when looking to build a WLAN roaming service. While much of the interest has been to use SIM-based authentication for WLAN users in order to enable roaming, this is not strictly necessary. It is highly unlikely that a single authentication standard will exist for all roaming users. Therefore a generic mapping from non international mobile subscriber ID (IMSI) to IMSI is required even for EAP-SIM authentication; this would allow an existing GPRS ticket for WLAN users. This then can allow an operator other than the home network to perform mediation of AAA-based accounting records into TAP tickets. These two techniques when combined offer the capability of using any authentication mechanism, including user name and password, the use of which can be mediated into a GPRS ticket. The choice of which authentication mechanisms to be supported and the mapping between non-IMSI and IMSI is then a matter for the home operator

AAA (e.g., RADIUS and Diameter) protocols are peer-to-peer [17, 23]. Both require shared secrets and both require secure transport. 3GPP and WLAN interworking [10–13, 19, 24] using AAA architecture is shown in Figure 7.14. Hence, there are recognized scalability issues with roaming using such functionality, e.g., compared with over 30,000 existing GSM bi-lateral roaming agreements. Consequently, scalable AAA requires broker functionality. Such a broker will have bi-lateral peering agreements with a number of WLAN providers and also with home networks offering roaming WLAN service to their subscribers. This roaming broker functionality can be performed by an established GSM operator, a GRX provider, a GSM clearinghouse, or a new entity.

One of the clear advantages of roaming broker functionality is that it allows the AAA interfaces to scale. One of the disadvantages is that it means that the home operator no longer has a direct (business) relationship with the WLAN provider. This is important because the WLAN provider may not be trusted by the home network provider. Hence, the broker must assume the risk of validating the correct operation of the inter-operator interfaces between the broker and the WLAN provider. This may include validating that the accounting records produced by the WLAN provider are authentic. Depending on both the inter-operator billing metric (time/volume/or flat fee) and the level of trust between the broker and the WLAN provider, this assumption of risk by the roaming broker may mean that broker equipment is placed in the user plane for roaming subscribers supported using this broker. This equipment can then be used to detect fraud by the WLAN provider.

7.7 SECURE NETWORK MANAGEMENT

Simple network management protocol (SNMP) [25–28] provides management capabilities for TCP/IP-based networks and because of its simplicity and the achievement of interoperability of the SNMP module from different vendors it became a de facto standard for the management of network based equipment. SNMPv3 [28] includes three important services: authentication, privacy, and access control as shown in Figure 7.17. To deliver these services in a flexible and efficient manner, SNMPv3 introduces the concept of a principal, which is the entity on whose behalf services are provided or processing takes place. A principal can be an individual acting in a particular role; a set of individuals each acting in a particular role; an application or set of applications; or a combination thereof. In essence, a principal operates from a management station and issues SNMP commands to agent systems. The identity of the principal and the target agent together determine the security features that will be invoked, including authentication, privacy, and access control. The use of principals allows security policies to be tailored to the specific principal, agent, and information exchange, and gives human security managers considerable flexibility in assigning network authorization to users.

SNMPv3 is defined in a modular fashion, as shown in Figure 7.18. Each SNMP entity includes a single SNMP engine. An SNMP engine implements functions for sending and receiving messages, authenticating and encrypting/decrypting messages, and controlling access to managed objects. These functions are provided as services to one or more applications that are configured with the SNMP engine to form an SNMP entity. This modular architecture provides several advantages. First, the role of an SNMP entity (see Table 7.5) is determined by the modules that are implemented in that entity. For example, a certain set of modules is required for an SNMP agent, whereas a different (though overlapping) set of modules is required for an SNMP manager. Second, the modular structure of the specification lends itself to defining different versions of each module. This, in turn, makes it possible to (1) define alternative or enhanced capabilities for certain aspects of SNMP without needing to go to a new version of the entire standard (for example, SNMPv4), and (2) clearly specify coexistence and transition strategies.

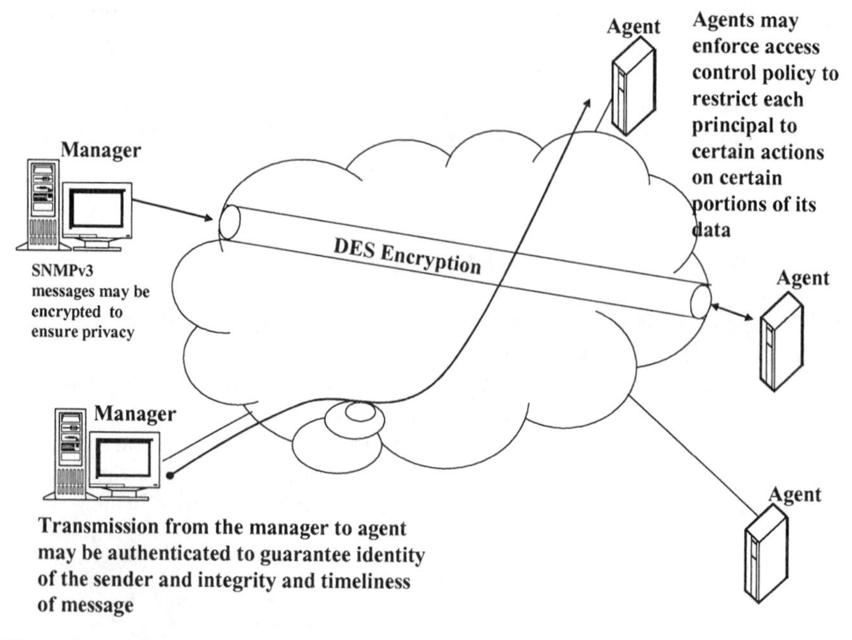

Figure 7.17 SNMPv3 security features.

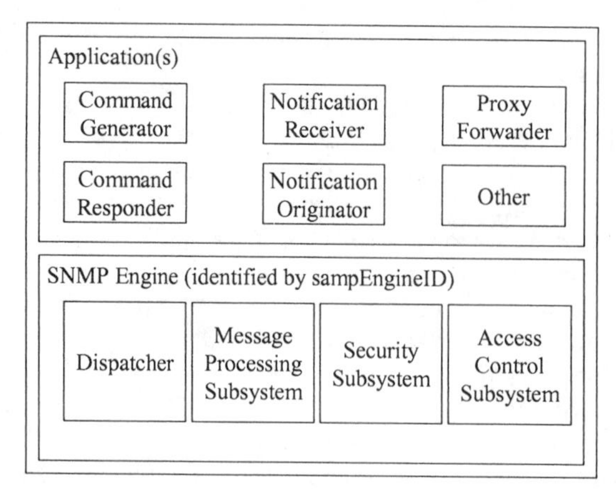

Figure 7.18 SNMP entity (RFC 2271).

Table 7.5

Components of SNMP Entity (RFC 2273)

Component	Description
Dispatcher	Allows for concurrent support of multiple versions of SNMP messages in the SNMP engine. It is responsible for (1) accepting protocol data units (PDUs) from applications for transmission over the network and delivering incoming PDUs to applications; (2) passing outgoing PDUs to the message processing subsystem to prepare as messages, and passing incoming messages to the message processing subsystem to extract the incoming PDUs; and (3) sending and receiving SNMP messages over the network.
Message processing subsystem	Responsible for preparing messages for sending and for extracting data from received messages.
Command responder	Receives SNMP Get, GetNext, GetBulk, or Set request PDUs destined for the local system as indicated by the fact that the contextEngineID in the received request is equal to that of the local engine through which the request was received. The command responder application performs the appropriate protocol operation, using access control, and generates a response message to be sent to the originator of the request.
Security subsystem	Provides security services such as the authentication and privacy of messages. This subsystem potentially contains multiple security models.
Access control subsystem	Provides a set of authorization services that an application can use for checking access rights. Access control can be invoked for retrieval or modification request operations and for notification generation operations.
Command generator	Initiates SNMP Get, GetNext, GetBulk, or Set request PDUs and processes the response to a request that it has generated.
Notification originator	Monitors a system for particular events or conditions, and generates trap or inform messages based on these events or conditions. A notification originator must have a mechanism for determining where to send messages, and which SNMP version and security parameters to use when sending messages.
Notification receiver	Listens for notification messages, and generates response messages when a message containing an inform PDU is received.
Proxy forwarder	Forwards SNMP messages. Implementation of a proxy forwarder application is optional.

7.7.1 Secret Key Authentication

The authentication mechanism in SNMPv3 assures that a received message was, in fact, transmitted by the principal whose identifier appears as the source in the message header. In addition, this mechanism assures that the message was not altered in transit and that it was not artificially delayed or replayed. To achieve authentication, each pair of principal and remote SNMP engines that wishes to communicate must share a secret authentication key. The sending entity provides authentication by including a message authentication code with the SNMPv3 message it is sending. This code is a function of the contents of the message, the identity of the principal and engine, the time of transmission, and a secret key that should be known only to the sender and the receiver. The secret key must initially be set up outside of SNMPv3 as a configuration function. That is, the configuration manager or network manager is responsible for distributing initial secret keys to be loaded into the databases of the various SNMP managers and agents. This can be done manually or by using some form of secure data transfer outside of SNMPv3. When the receiving entity gets the message, it uses the same secret key to calculate the message authentication code again. If the receiver's version of the code matches the value appended to the incoming message, then the receiver knows that the message can only have originated from the authorized manager, and that the message was not altered in transit. The shared secret key between sending and receiving parties must be preconfigured.

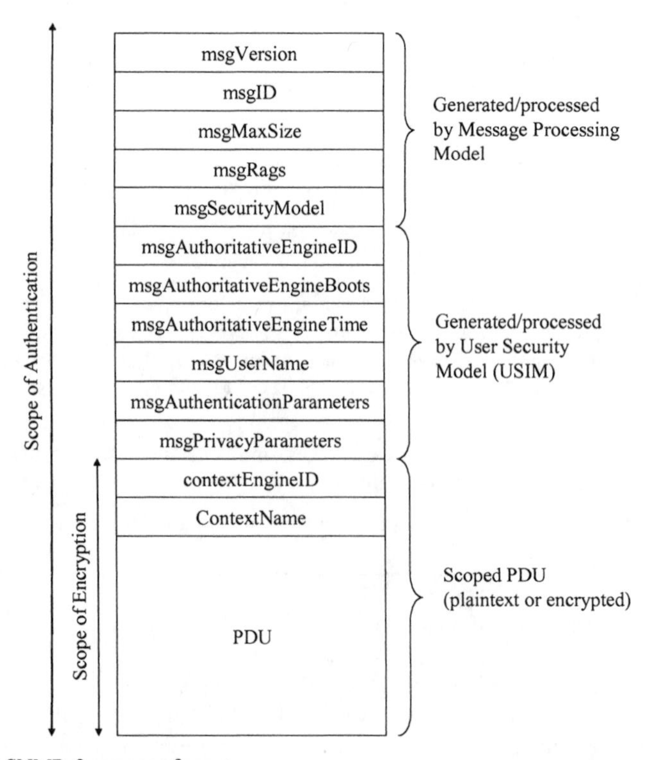

Figure 7.19 SNMPv3 message format.

User-based security model (USM) authentication is used for timeliness verification. USM is responsible for assuring that messages arrive within a reasonable time window to protect against message delay and replay attacks. Two functions support this service: synchronization and time-window checking. Each authoritative engine maintains two values, snmpEngineBoots and snmpEngineTime, that keep track of the number of boots since initialization and the number of seconds since the last boot. These values are placed in outgoing messages in the fields msgAuthoritativeEngineBoots and msgAuthoritative EngineTime. A non authoritative engine maintains synchronization with an authoritative engine by maintaining local copies of snmpEngineBoots and snmpEngineTime for each remote authoritative engine with which it communicates. These values are updated on receipt of an authentic message from the remote authoritative engine. Between these message updates, the non authoritative engine increments the value of snmpEngineTime for the remote authoritative engine to maintain loose synchronization. These values are inserted in outgoing messages intended for that authoritative engine. When an authoritative engine receives a message, it compares the incoming boot and time values with its own boot and time values. If the boot values match and if the incoming time value is within 150 seconds of the actual time value, then the message is declared to be within the time window and, therefore, to be a timely message.

7.7.2 Privacy Using Conventional Encryption

The SNMPv3 USM privacy facility enables managers and agents to encrypt messages to prevent eavesdropping by third parties. Again, manager entity and agent entity must share a secret key. When privacy is invoked between a principal and a remote engine, all traffic between them is encrypted using the DES. The sending entity encrypts the entire message using the DES algorithm and its secret key and sends the message to the receiving entity, which decrypts it using the DES algorithm and the same secret key. Again, the two parties must be configured with the shared key. The cipher-block-chaining (CBC) mode of DES is used by USM. This mode requires that an initial value (IV) be used to start the encryption process. The msgPrivacyParameters field in the message header contains a value from which the IV can be derived by both sender and receiver.

7.8 3GPP – WLAN DEPLOYMENT ARCHITECTURE AND STANDARD

The architecture is an integration of WLAN and 3GPP access networks belonging to different stakeholders. This architecture supports the following solutions [3]:

1. SIM-based authentication for mobile postpaid and prepaid and roaming users;
2. For prepaid users real-time charging and billing;

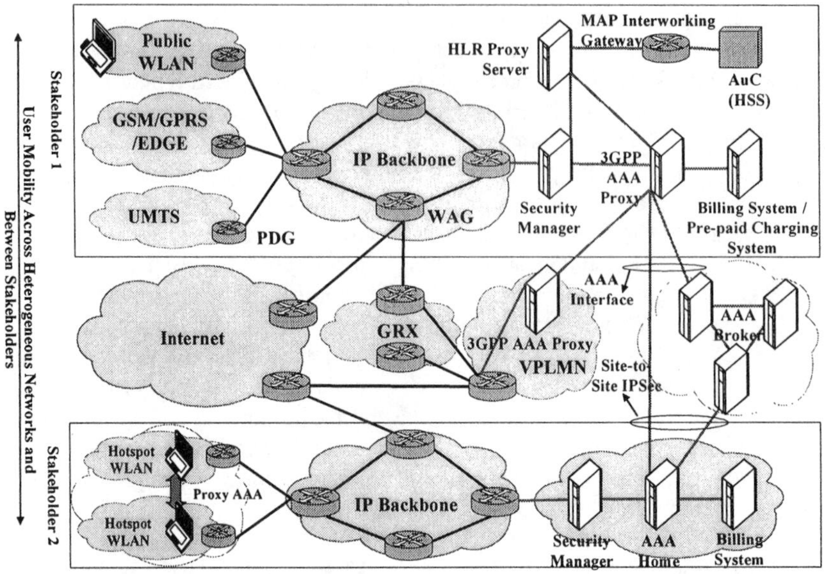

Figure 7.20 3G-WLAN deployment architecture.

3. SMS using OTP for postpaid and prepaid users;
4. Subscription-based billing using username/password credentials;
5. Supports Internet roaming users.

Secure mobility across different stakeholders is achieved using IPSec with Mobile IP [6–8].

The architecture as shown in Figure 7.20 is also compliant with the six levels of interworking that have been defined by 3GPP spanning from the simple common billing (scenario 1) to the seamless service continuity when moving from the 3GPP access network to the WLAN access network and vice versa as shown in Table 7.6.

The intention of scenario 3 is to provide access to all 3GPP packet switched (PS)-based services; the ones that are available now over GPRS, and the ones that will be in future provided for PS access, namely IMS-based services. The architecture considered here is based on end-to-end VPN tunneling from WLAN User Equipment (UE) to the packet data gateway (PDG), which can be located in the home or visited network depending on where the service is provisioned. The WLAN access gateway (WAG) which enforces the routing of the user traffic from the WLAN access network to the PDG (in the case of roaming through the inter-operator interface/network), is located in the home network in case of home-service access, and in the visited network in case of visited-service access as shown in Figure 7.21.

Although scenario 3 is not supposed to deal with mobility issues, it is understood that the chosen tunneling solution needs to be future-proof, in the sense

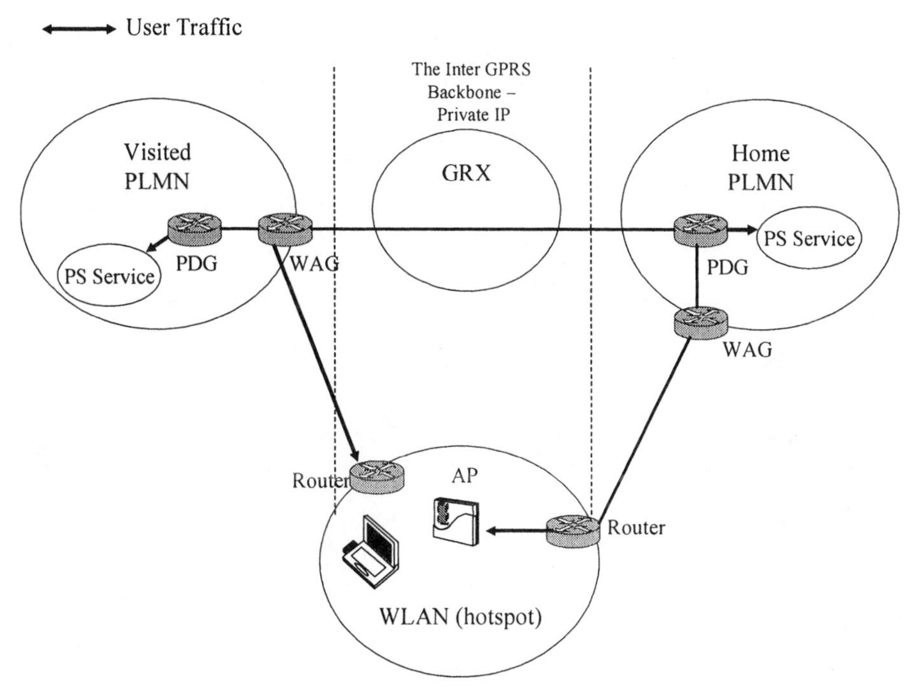

Figure 7.21 End-to-end tunneling architecture.

that it must be possible to migrate to mobility scenarios 4 and 5 without changing the standardized architecture of scenario 3. Possibly by using off-the-shelf MIP solutions. The support of seamless mobility may also depend on the terminal capacity of being able to connect to WLAN and GPRS/UMTS simultaneously.

7.9 CONCLUSIONS

The purpose of this chapter was to present WLAN deployment methods and the issues involved. The first part of the chapter showed wireless deployment methods and system design that affect the deployment. The next part of the chapter showed deployment of WLAN in enterprises, public areas, and by mobile operators. For all deployment environments security was taken as the most important requirement. The reason for choosing security was due to the known issues of IEEE 802.11. Now solving security issues in deployment is the main issue. Following security we will need to consider mobility and QoS.

The issue of mobility was not given as much attention as it deserves. We hope that the discussions in this chapter and those in Chapter 6 will give enough information to our readers to understand the available mobility solutions for WLAN deployment and their issues.

Table 7.6
Levels of Interworking Between WLAN and 3GPP Network

Scenario Services and operational capabilities	1 Common billing and customer care	2 3GPP system-based access control and charging	3 Access to 3GPP system PS-based services	4 Service continuity	5 Seamless services	6 Access to 3GPP system CS-based services
Common billing	X	X	X	X	X	X
Common customer care	X	X	X	X	X	X
3GPP system-based access control		X	X	X	X	X
3GPP system-based access charging		X	X	X	X	X
Access to 3GPP system PS-based services from WLAN			X	X	X	X
Service continuity				X	X	X
Seamless service continuity					X	X
Access to 3GPP system with seamless mobility						X

Another topic to which we have not given adequate attention is QoS. It is extremely important that the QoS in WLAN and mobile networks are mapped to each other correctly. Also mapping of QoS to the IP layer should be done in the correct fashion. Here the correct fashion basically means that each network and system has the same or at least an understanding of QoS and that they should have similar policy enforcement for QoS.

Similar to mobility and QoS, accounting and billing play very important roles. After all without money from customers there would not be a business. Some discussion has been done on metering, accounting, charging, and billing in the chapter. Existing solutions using a clearinghouse or AAA infrastructure should be used for this purpose. Again this topic by itself merits readers' attention; we are unable to fully cover this in this chapter [29–32].

In the future we foresee that there will be a need for WLAN interworking with IEEE 802.16 or WiMax. Some thoughts on such interworking of various technologies will be discussed in Chapter 8.

REFERENCES

[1] Prasad, N. R., and A. R. Prasad, eds., *WLAN Systems and Wireless IP for Next Generation Communication*, Norwood, MA: Artech House, 2002.

[2] Prasad, A. R., *Wireless LANs: Protocols, Security and Deployment*, Ph.D. Thesis, Delft University of Technology, Delft University Press, 2003.

[3] Prasad, N. R., *Adaptive Security for Heterogeneous Networks*, Ph.D. Thesis, University of Rome "Tor Vergata," Rome, Italy, 2004.

[4] Murhammer, M., et al., *IP Network Design Guide*, Redbooks, IBM, 1999.

[5] Prasad, A. R., et al., "Wireless LANs Deployment in Practice," *International Journal on Wireless Personal Communications*, Kluwer Academic Publishers, 2001.

[6] Perkins C., ed., "IP Mobility Support for IPv4," IETF RFC 3344, August 2002.

[7] Solomon, J. D., *Mobile IP The Internet Unplugged*, Upper Saddle River, New Jersey: Prentice Hall, 1998.

[8] Kent S., and R. Atkinson, "IP Authentication Header," RFC 2402, November 1998.

[9] Kent, S., and R. Atkinson, "IP Encapsulating Security Payload (ESP)," RFC 2406, November 1998.

[10] Loughney, J., and G. Camarillo, "SIP-AAA Requirements" (draft-loughney-sip-aaa-req-00.txt), April 2002.

[11] Arkko, J., et al., "Security Mechanism Agreement for the Session Initiation Protocol (SIP)," RFC 3329, January 2003.

[12] Camarillo, G., W. Marshall, and J. Rosenberg (eds.), "Integration of Resource Management and Session Initiation Protocol (SIP)," RFC 3312, October 2002.

[13] Rosenberg, J., et al., "SIP: Session Initiation Protocol," RFC 3261, June 2002.

[14] Blunk, L., J. Vollbrecht, and Bernard Aboba, "Extensible Authentication Protocol (EAP)," October 2002. IETF pppext working group draft draft-ietf-pppext-rfc2284bis-07.txt.

[15] IEEE Std 802.1X-2001 "IEEE Standard for Local and metropolitan area networks—Port-Based Network Access Control," 14 June 2001.

[16] Schneier, B., *Applied Cryptography*, John Wiley & Sons, Inc., 1996.

[17] Rigney, C., et al., "Remote Authentication Dial In User Service (RADIUS)," RFC 2138, January 1997.

[18] Pejanovic, M., and N. R. Prasad, "Optimized Deployment of 3G/4G Mobile Systems," World Wireless Congress 2003, San Francisco, May 27-30, 2003.

[19] 3G Security, Wireless Local Area Network (WLAN) Interworking Security, 3GPP TS 33.234 V1.0.0 (2003-12), (Release 6).

[20] Buckley, A., et al., "EAP SIM GMM Authentication August 2002," IETF personal draft draft-buckley-pppext-eap-simgmm-00.txt.

[21] Haverinen, H., and J. Salowey, "EAP SIM Authentication," draft-haverinen-pppext-eap-sim-07.txt, November 2002.

[22] Arkko, J., and H. Haverinen, "EAP AKA Authentication," draft-arkko-pppext-eap-aka-11.txt, October 2003.

[23] DIAMETER: http://www.diamter.org

[24] "3GPP system to Wireless Local Area Network (WLAN) Interworking (Release 6)," 3GPP TS 23.234 v2.3, November 2003.

[25] Wijnen, B., D. Harrington, and R. Presuhn, "An Architecture for Describing SNMP Management Frameworks," RFC 2571, April 1999.

[26] Case, J., et al., "Message Processing and Dispatching for the Simple Network Management Protocol (SNMP)," RFC 3412, December 2002.

[27] Levi, D., P. Meyer, and B. Stewart, "Simple Network Management Protocol (SNMP) Applications," RFC 3413, December 2002.

[28] Blumenthal, U., and B. Wijnen, "User-based Security Model (USM) for version 3 of the Simple Network Management Protocol (SNMPv3)," RFC 3414, December 2002.

[29] Yang, L., "Architecture Taxonomy for Control and Provisioning of Wireless Access Points (CAPWAP)," IETF draft, draft-ietf-capwap-arch-04, July 28 2004.

[30] IEEE 802.11 Report, "Cellular/WiFi convergence to drive hot-spot growth," 20 August 2004.

[31] Iyer, P., et al, "Public WLAN Hot-spot Deployment and Interworking," *Intel Technical Journal*, Vol. 07, Iss. 03, pp. 10 - 19, August 19, 2003.

[32] Koutsopoulu, M., et al., "Charging, Accounting and Billing Management Schemes in Mobile Telecommunication Networks and the Internet," *IEEE Communications Survey*, Vol. 6, No. 1, pp. 50 - 58, First Quarter 2004.

Chapter 8

Future Generation Communications

The future generation of communications or the next generation of communications is something that stays with us like our shadow. Fortunately for whatever we develop there is always a next step to it. After all, that keeps our world going. Currently we are again at that stage of work on future generation communications where these words have not yet achieved a consensus. In Chapter 1 we have already alluded to what the future will be; in this chapter let us look at the crystal ball and try to materialize at least in words what we see in it.

We first define fourth generation (4G) and beyond 3G (B3G). Then we look at the requirements for future generation communications from the perspective of the users, the operators, and the service providers. Next the technologies that should and are being developed to materialize the future generation are discussed. A dip is also taken into the ongoing standardization or prestandardization efforts. At first the introduction section tries to raise questions on the "future."

8.1 INTRODUCTION

Telecommunications is in its infancy; we have said it [1] and heard it but what does it entail?

Let us step back and look at what telecommunications provides; at first with wired telephones people could communicate without letters or visits to each other. This was the start of voice communications; next came analog mobile phones, which made this voice mobile, and then the digital era increased the capacity. Third generation (3G) communications came next and stumbled before even taking the first step. The reason for this stumble was that 3G was aiming at services other than voice communications. What does it mean? The infant is growing? What shall we do? Which path should we take?

The other side was the boom of the Internet, which opened a new world and new ways of communications (e.g., e-mailing, peer to peer) and made a plethora of data available to us. WLANs came in and made the Internet wireless and mobile. WLANs grew at an amazing pace with a short product lifetime and required continued standardization effort towards improving the capabilities and functionalities of the technology. WLANs have been successful exactly where mobile communications, to a reasonable extent, failed. Now voice over WLAN is the vision and that too with mobility; in this arena mobile communications has already set a high standard. Voice over IP (VoIP), although gaining momentum now has been stumbling for a long time and as yet it has no mobility. WLAN deployments have a small footprint, and mobility is still being worked on. Seamless mobility for WLAN will be ever complex. These points from mobile systems and the WLAN side are represented in Figure 8.1. So what to do? How to proceed? Not only technology-wise but market acceptance-wise.

There is other standardization work going on. 3G standards are giving solutions for interworking with WLAN. IEEE 802.16 and IEEE 802.20 are looking at mobility with high bandwidth and then there is IEEE 802.21, which is, like 3G, also looking at interworking and thus handover between various technologies. There is another word looming in the air and that is known as 4G; with it is an aura, an aura that creates confusion, the aura is also known as the B3G. Now what will happen in the future? What are these different Gs and interworking? How do we step forward? How should we approach the market? What technologies should we develop? How can telecommunications become such an integrated part of human life that breathing will become its synonym?

In this chapter we try to tackle most of the questions mentioned above. We look into what 4G, B3G, and to add to the confusion, next or future generation communications will be.

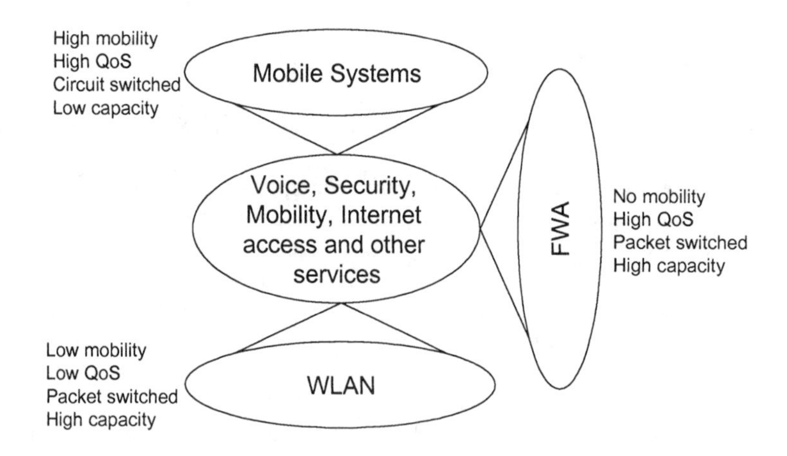

Figure 8.1 Technologies moving towards a common goal.

8.2 THE NEED FOR FUTURE GENERATION COMMUNICATIONS

Let us first have a look at the need for future generation communications [2]. Note that we have evaluated none of the information given in this section; text represents our opinions or information taken from the references.

8.2.1 What Will Sell?

One can never say what the user will want in the future. Whatever you say will be way off the reality. So what can one do? Learn from past. This is the logical tactic mostly used but using this tactic we can go way off the mark when it comes to such futuristic predictions [3]; 3G proves this logic. Nevertheless voice is the best bet for a future. What else? Well, person-to-person communication in any possible way is the service that really sells, seamless service provisioning being a part of it. Next is machine-to-machine communications; this will pick up strongly in the future. In general a product is bought (or sold) if any of the following is fulfilled:

- The product solves a problem being faced by the customer (e.g., the Intuit software by Scott Cook that gave us an easy way to fill out tax forms [4]).
- There is a hole in the market for a given product (e.g., the WCDMA test equipment developed by Quintillion Technologies from Japan) [5].
- The product is cheaper than others in the market and is within a quality limit (e.g., fixed and mobile operators' voice service cost in the United States).

Besides the above, there are several reasons a product sells, such as brand name. Here, however we will try to find the need of the user based on the three principles given above.

8.2.2 Is It Common Sense?

Let us look at two examples that defy the conventional wisdom of what will or should sell.

The first is example is Telenor's electronic shepherd project. The shepherds with sheep grazing in forests needed some way of locating lost sheep or at least determine whether the sheep were alive and then find them. In response to this request Telenor developed a solution where each sheep has a radio connected to it; these sheep create an ad hoc network with the "mother sheep" as the central controller. The information about each sheep can thus be accessed by the shepherd. Thus Telenor has expanded its market out of the already saturated market of mobile services in Norway.

Another example is i-mode. It is said very often that i-mode or other mobile Internet services in Japan made it because of the lack of space in Japanese houses for computers. The fact is that Japan is good at miniaturizing electronic goods and

Japanese people like new electronic goods including computers, so the lack of space is simply a non-issue. Let us look at it differently. Until almost the end of the 1990s Japan was lagging behind in Internet connectivity so the Japanese government started promoting broadband access in the country; a major share of work was Japan Gigabit Network (JGN). Now i-mode had come before that and naturally captured the market. Today in Japan one can get xDSL as high as 48Mbps for barely $10; this might be the cheapest in the world. Anyway, back to the olden days; another reason for i-mode was the improved voice quality; i-mode terminals use full-rate instead of half-rate speech codec as in standard personal digital cellular (PDC) terminals. Higher speech quality leads to longer conversations and thus the beginning of i-mode showed growing revenue for voice. Other i-mode services such as searches for restaurants also led to extra voice revenues (i.e., one finds a good restaurant through i-mode and calls 10 friends). Only later did a new service provision increase the data revenue, some of the services being gaming, photo or short video clips, messaging, location services, and ring tone download. Japan had found a new source of revenue and thus the price war over voice services that occurred in the United States did not happen; operators were competing on service types. The increased usage of networks led to the lack of spectrum and thus the need for 3G systems came to Japan (even 3G spectrum might not be sufficient). So one would say the operators had achieved their goal: maximize the use of networks or increase the traffic on their network. Now let us look at the other side; in 3G networks flat-rate services have started in Japan. This now changes the whole concept. The goal of operator as always will be to keep as many users as possible, create customer loyalty, and attract new customers, and decrease the churn rate. However, the operator would prefer to decrease the traffic on the network simply because an increase in traffic now does not increase revenue. At the same time operators will have to keep their services attractive for the users. So now where should the operators go? They have to find a way to keep the customers and increase revenue with attractive services not requiring as much use of the network.

This section has given views from two totally different markets for mobile communications. It should be noted that the business models in Europe and Japan are different. In Europe the customers usually stick to the brand name of the vendor as the standard defines all the details. On the other hand in Japan the customers stick to the operator as the standard gives space for operators to develop new services or solutions. So we see the operator being dependent on the vendor in Europe and other way round in Japan.

8.2.3 How to Know What Will Sell

There is a Hindi saying, "Koop Mandook," meaning a frog from a well. Similarly mobile communications was in a well, the well of voice service. Mobile communications came out of the well in the form of 3G, and WLANs are moving towards QoS-based services. 3G stumbled; WLAN—not as visibly due to lower costs—is also stumbling but stumbling is good for the lessons it gives. How does one learn from the lessons? People have learned to become "user centric"; that is good but its meaning is not understood. The way out is to ask the user but even what should be asked is unclear. Asking the users about service need and products to appear in 10 years time is surely difficult to answer. The trick is to see how the user uses a product, maybe even lives; this will give us a better idea about the possible needs of the user, and thus the product. The amazing fact is that this line of thought brings us to a junction where art and technology seem to meet. E. H. Gombrich in his famous book [6] discusses how people see things and perceive them. He says that people should learn how to see, that artists see more, and we appreciate a piece of art if it is close to the nature that is visible to us and we can thus associate with it. This way of seeing the customer is something the mobile business should pay attention to. Another side is the work by C. M. Christensen [3] that says one should try out new products and then learn from the sales and marketing experience. Lessons learned should tell the team what to do to make the product successful, keeping in mind that the initial targeted market might not ultimately use it. This method of Christensen gives a way for the industry to see, as Gombrich said, the mobile market.

The discussion in the previous paragraph also brings us to the junction where we have to say that focusing on one thing can lead to single-mindedness. The point is also to have a broad vision and simple thought. So what does this mean? Let us have a sneak preview. In most developing countries the society is getting old and thus services like telemedicine and, due to lack of teachers, tele-education would make sense. On the other hand for fast-developing countries like India where more than 50% of the population is under 20 years of age a different approach will be required. Of course here we have not talked about culture and the effect of globalization at all.

Another line of thought is to learn how children use things and behave. This has been proposed by several people. After all, something that will come in the future will eventually be used by the children.

Let us see a few examples based on observations: Barely at the age of one, my daughter used to pick up small toys that would fit in her hand and try to mimic a phone conversation. This shows the need of connectivity, and the solution is ubiquitous communications with context awareness. Another one happened just a day or two back during a German lesson, when I found out that people hardly remembered what a turntable is; today kids only know DVDs. So it is also the technology the future generation is interfacing with. Objects such as mobile phones and DVDs (already DVD is being replaced by hard disk drives) are normal for them, at least in the developed world.

We should also observe things in our own daily life; I for example do not like the phone ringing during dinner. It should be possible for the phone to know that I am having dinner and allow only emergency calls or calls from people of high priority in my priority list to come in.

8.2.4 Different Perspectives

Let us now look at future generation needs from the point of view of the user, the operator, and the vendor. Many of these needs have appeared in an earlier text in one form or another. Although the term user is used in this text, it should be noted that the subscriber or someone else could be the user. The point is that whoever the subscriber is, the user is the one using the service at the end and thus holds the right to decide the fate.

Looking from the user perspective it is but natural to have a reasonably cheap and easy to use solution. The solution from the user perspective should allow access to services at all times.

The vendor on the other hand will look for solutions that will be easy to implement and such that one design is reusable for different products. Similar to the user the vendor will want the solution to be cheap so as to maximize profit. The solution should be such that it minimizes implementation errors. The vendor will have to implement the mobile device such that it is easy to use.

From the operator point of view once again the main thing that comes out is the reduction of the cost: the cost of network elements, the cost of deployment, and finally the cost of operations and management of the network. These are just a few standard points. The operator will provide the services that the user will need and will be easy to use. In future generations the operator will have to communicate with business in daily usage by users (microwave oven, the doorbell, the home entertainment companies, consumer electronics, car industry, etc.; see Figure 8.2). Note that we do not say DVD player, VCR, and music system because they will not survive the coming 10 years.

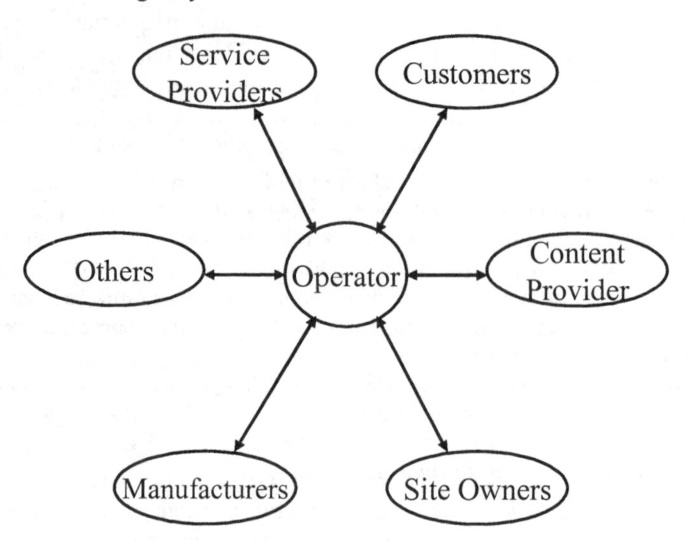

Figure 8.2 Operator's role in the future.

8.3 DEFINING THE FUTURE

In this section we have taken the simple path of defining, based on what is being said.

Since 3G did not launch as it was envisaged, using the term 4G had become a taboo. It seems this "taboo" gave birth to the term B3G, which should have been an underground synonym for 4G but now has developed its own meaning. There are different views on the definition of 4G; some say that any technology that provides data rates above 100 Mbs is 4G. In the following our definitions, which are generally accepted in Europe, of B3G and 4G are given.

Already in Chapter 1 a definition was given of B3G and 4G. The definition was generated based on the ITU-R vision, which calls for interworking/integration of available technologies and development of a new air interface that would work at 100 Mbps. The new interface is 4G and the interworking/integration of technologies is B3G, which also includes 4G (see Figure 8.3).

Today there are several operators or stakeholders for different types of technologies and networks; thus there is a need for handover between them. A common layer for all these technologies and stakeholders will be IP but IPv4 as we know is fragmented due to the lack of addresses. Even though IPv6 is available the v4 is here to stay. All this together forms B3G and with the variety come several technical issues. Issues related to security, QoS, and mobility are discussed in Chapters 4–6; in this chapter they will be summarized and thoughts towards solutions will be presented.

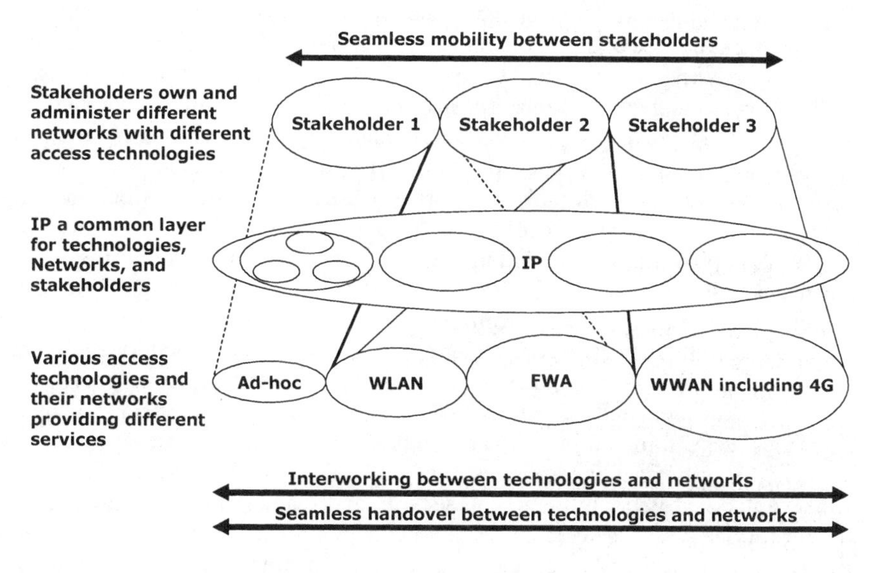

Figure 8.3 Defining B3G.

Achieving seamless mobility between different stakeholders and technologies and defining B3G and 4G is just a part of future generation communications. There is more to it. The goal of all this development is to achieve ubiquitous communications [7] or UbiCom.

To be precise, 4G basically gives a higher data rate with mobility while B3G gives continuous connectivity. The generation beyond that will give the user real ubiquity. The future generation will provide ubiquitous communication through ubiquitous networking and thus bring ubiquitous services to the user.

8.4 TECHNOLOGIES

The previous section gave a definition of B3G and what will come beyond that. Let us now have a deeper look at the definitions.

8.4.1 B3G

B3G or interworking of different technologies is a thought born around 1998 when the first IEEE 802.11 products had just started shipping and i-mode was just picking up (or planning to).

Figure 8.4 shows the envisaged development in stakeholder of various networks and technological development for the short-, mid-, and long-term future. The figure also points out several technological issues that should be worked on. Arrows between two cells of the figure show the possibility of handover between the two technologies while the shade of the arrow (gray scale) shows the expected extent of handover. Research work should be done on seamless handover bringing in the study of several issues like security and QoS, which should be done at each protocol layer and network element. This topic itself will require further study on development methods and technologies including hardware, software, and firmware and technologies like application specific integrated circuits (ASICs). Another important research topic is software defined radio (SDR), which includes reconfigurability at every protocol layer.

WLANs provide roaming within LANs and work is going on towards further enhancement in this field. While WWANs provide roaming too, the challenge now is to provide seamless roaming from one system to another, from one location to another, and from one network provider to another. In terms of security again both WLANs and WWANs have their own approach. The challenge is to provide the level of security required by the user while roaming from one system to another. Users must get end-to-end security independent of any system, service provider, or location. Security also incorporates user authentication that can be related to another important issue: billing. Both security and roaming must be based on the kind of service a user is accessing. The required QoS must be maintained when a user roams from one system to another. Besides maintaining the QoS it should be possible to know the kind of service that can be provided by a particular system, service provider, and location. Work on integration of the WLANs and the WPANs must also be done. The biggest technical challenge here will be the coexistence of the two devices as both of them work in the same frequency band.

FWA is a technology that should be watched as it develops; depending on its market penetration and development of standards it should also be integrated together with other technologies.

Figure 8.4 does not show the development of terminals and terminal technology, nor does it show development of satellite communications. Terminals will be integrated with various technologies: computer and mobile integration with cameras, for example.

8.4.2 Beyond

There are several technological developments that will take us towards the true era of ubiquitous communications. In this section such developments for the future are discussed. At first a definition of UbiCom is given.

8.4.2.1 UbiCom

Ubiquitous computing (ubicomp) is a term coined for a situation, where small computational devices are embedded into our everyday environment in a way that allows them to be operated seamlessly and transparently [7–10]. This means that many small objects/devices have the power of computing — computing

	Stakeholder (for handover; of one or more access networks)	IP	Broadcast (DVB-T,DAB etc.)	WWAN (3G, 2.5G etc.)	WLAN (IEEE 802.11)	FWA (802.16 etc.)	WPAN and Ad-hoc (IEEE 802 .15 etc.)
Short Term (2-3 yrs.)	Same (not broadcast)	v4	Similar to TV or Radio	3G and 2.5G, handover: maybe	b,g,a,n,s, MAC enh.	WiMax	
Mid Term (3-5 yrs.)	Same (maybe few different, surely not broadcast)	v4&6	As above	3G → 3.5G and 2.5G, handover: possible	g,a,n,s,NG QoS etc.	WiMax, MBWA	
Long Term (5-10 yrs.)	Same and Different	v4&6	SDR	2.5G, 3G, 3.5G and 4G handover: must	g,a,n,s,NG, QoS etc.	WiMax, MBWA, NG	UbiCom

NG(+) Next Generation and beyond
↔ Handover
Gray For handover (arrows in 3 levels) it signifies the expected extent of handover. Darker the arrow the more common the handover between the concerned technologies.
For technologies (e.g. 2.5G) it signifies lesser or decreased used of technology.

Figure 8.4 Envisaged technology development in short-, mid-, and long-term.

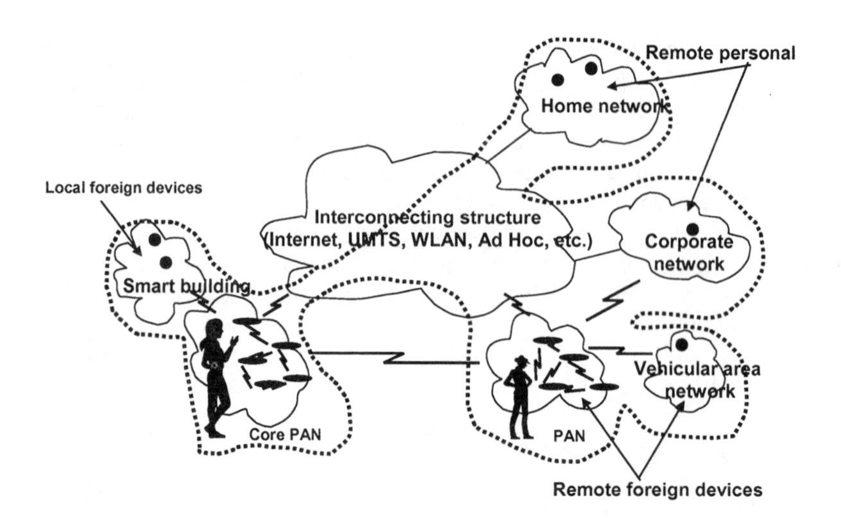

Figure 8.5 Personal network [11].

everywhere.These devices are suggested to be active and aware of their surroundings so that they can react and emit information when needed — the devices can communicate with others. Ubicomp is not observable, though it is perceivable to human beings (i.e., it is unnecessary to make humans aware of its existence). However, interaction/communication between devices and humans should not be excluded.

Ubicom means that a human being or a device can do communication everywhere, anytime, whether it is a person, or a mobile phone controlled by human or autonomous devices, through either wired or wireless link, satellites, or even ad hoc networks, or other means [7]. This is the description from an individual's perspective. Communication/interface between human users and systems (devices), if necessary, should be as natural as possible — like human talking or conversation, supporting audio/visual and other natural means of information exchange. (Mobile) devices can talk to other devices everywhere when necessary; communications and computing should also be context-aware (e.g., knowing the location, time). Ubicom requires interworking solutions that enable users to use, and roam between heterogeneous (multiple types of) wireless networks.

Ubiquitous networking is one of the means to support ubicomp: Communication networks exist everywhere.

Ubiquitous service could be used to describe the new services developed based on ubicomp. It could also be used to describe the services that are accessible everywhere at anytime.

8.4.2.2 Personal Networks

Another area of research for the next generation communications will be in the field of PNs [11]. A PN provides a virtual space to the users that spans over a variety of infrastructure technologies and ad hoc networks. In other words PNs provide a personal distributed environment where people interact with various companions, embedded or invisible computers not only in their vicinity but potentially anywhere. Figure 8.5 portrays the concept of PNs. Several technical challenges arise with PNs, besides interworking between different technologies, some of which are security, self-organization, service discovery, and resource discovery [11].

8.4.2.3 Ad Hoc

Ad hoc wireless networks do not need any infrastructure. In these systems mobile stations may act as a relay station in a multi hop transmission environment from distant mobiles to base stations. Then they do need infrastructure if they've got base stations. Mobile stations will have the ability to support base station functionality. The network organization will be based on interference measurements by all mobiles and base stations for automatic and dynamic network organization according to the actual interference and channel assignment situation for channel allocation of new connections and link optimization. These systems will play a complement role to extend coverage for low-power systems and for unlicensed applications. A central challenge in the design of ad hoc networks is the development of dynamic routing protocols that can efficiently find routes between two communication nodes. A mobile ad hoc networking (MANET) working group has been formed within the Internet Engineering Task Force (IETF) to develop a routing framework for IP-based protocols in ad hoc networks. Another challenge is the design of proper MAC protocols for multi hop ad hoc networks. There are several other research activities going on in the field of ad hoc networks.

8.4.2.4 HAPs

High altitude platforms (HAPs) have been proposed for a variety of applications ranging from communications to monitoring and sensing [12]. From a communications perspective, the relatively low altitude of these platforms (15–30 km) enables ultra high capacity communication to small ground-, air-, and sea-based terminals. Links to satellites are desirable for connectivity between metropolitan areas and islands of terrestrial cellular networks, as well as providing global area networks, where infrastructure is otherwise thin or lacking.

HAPs offer the potential for ultra high capacity extremely high frequency (EHF) HAP-ground links (to 100s of Mbps dependent on terminals), due to the low altitude of the platform. This enables high-capacity communication with extremely small, potentially mobile terminals (e.g., consistent with handsets). Optical crosslinks and satellite uplinks are largely above atmospheric and rain attenuation that would otherwise substantially degrade link performance and availability. Additionally, HAPs are well suited for providing full, high-capacity, multimedia information services over small densely populated areas. Due to relatively low delay and delay variations, they can be more readily integrated with existing networks than satellite links, enabling reuse and optimization of use of existing infrastructure and technology.

8.5 A LESSON TO LEARN

One of the common behaviors in any research or development is the thought of components. Here component can be a study on QoS or security or simply a different protocol layer. A developer, for example, of a application layer thinks that all lower layers will work perfectly fine. The fact is that there is a severe need for understanding and working from a system perspective.

Looking component vice is of course beneficial but only at the very starting stages of research. This stage can be called the organic stage when thoughts are still developing and should be given space to develop without hindrance. Once this stage is past, which should not take long, boundaries should be set and reality should be brought in to play. Here the reality is that each component has to work in a system and the boundary consists of the technological limitations and the system level view.

Many of the issues that are to be tackled for future generation communications can be solved by good communication between human beings and hard work—it does not require "rocket science."

8.6 OTHER TECHNOLOGIES

There are several other developments ongoing besides the development of mobile communications [13]. These developments will have an effect on next generation communications. In this section a few of these technologies are described briefly.

One such technology is nanotechnology, a technology of building things small and atom by atom. This will affect the medical side of computing. There is also development in fuel cells (already available in the market), and fusion cells, which will make energy very efficient. Another development is the paper battery, which is very flexible and even foldable.

On the other hand there is also development in the field of display technology —not only foldable displays but work is ongoing towards holographic ones. There are also solutions existing for user interfaces where a keyboard, for example, is projected and the user can simply type on the projected image.

Meanwhile the development of holographic memory should allow a lot more memory space in smaller space than available today. Along the same lines there is development in the field of 3D imaging.

Another field of development is quantum computing, communication, and cryptography. This technology hopes to deliver tremendous computing power. Quantum cryptography is already available.

When talking about small, we should also discuss sensor technology including smart dust. The use of smart dust might be in different arenas; one of them is, for example, in cars, including the tires. Talking about smart also brings us to smart fluids, which can change shape depending on the electric charge applied to them. Changing shape while being strong certainly opens many doors.

Then there is the field of virtual reality, which is still active; also research work going on in artificial intelligence (not mixing the two). This also brings us to robotics. Robots are already in industrial usage and are developing faster to bring solutions for the home market.

There is also tremendous growth in the field of genetics. Solutions will emerge for various diseases while research on genetics is ongoing; there are also developments in the biotechnology field focusing on internal organs and body parts.

A lot of work is going on in the field of complex systems in terms of analyzing and modeling. This research can help us to make better predictions.

8.7 FUTURE DEVELOPMENT

In this section some technological developments required for future generation communications are discussed [1, 14–16]. There are topics we have not covered, including multiple input multiple output (MIMO)-based technology and software and adaptive antenna technologies.

8.7.1 MAC

The MAC protocols do not properly address the important issue of energy consumption and ignore the behavior of the wireless channel mainly by assuming it is completely "hidden" by the physical layer interface.

The main power consumption sources from the access protocol perspective include the radio transceiver and the CPU. Maximum power is consumed in transmit mode, followed by the receive mode and standby mode. One of the options for an energy-efficient MAC design is the minimization of the number of transitions. Furthermore, due to the high power consumption for data transmission, there is a desire to minimize unsuccessful actions of the transceiver such as collisions and errors that result in retransmissions. From this point of view, MAC protocols that use reservation and polling are more attractive even if in some of them the receiver has to be turned on for longer periods. Furthermore, valuable channel resources can be allocated inefficiently during the time the channel is impaired. If transmissions are avoided during periods of bad error conditions by scheduling no traffic during these periods, energy saving can be achieved.

Key elements of a proper MAC layer design, such as the exploitation of the peculiarities and knowledge of the wireless channel, are emerging, both for improving the channel utilization and reducing the energy consumption. The exploitation of the capture effect in some receivers is an example of the fact that fading can be properly managed to improve performance. Fading can selectively disable a subset of the contending users thus allowing a group of users to efficiently access a single base station. Furthermore, the correlation in channel errors could be exploited for reducing the energy consumption by a proper design of error recovery mechanisms at the link layer [e.g., automatic repeat request, (ARQ)].

Channel-sensitive scheduling policies in MAC protocols can dramatically improve the channel utilization efficiency of MAC protocols. Power control commands represent an implicit link quality feedback from the receiver that may be used by the MAC layer protocols to drive the scheduling process.

If a proper design of MAC protocols requires the study of the interaction with physical layer processes, it is also true that a proper design of lower layers should take into account MAC layer mechanisms. Transmit diversity has been introduced at the physical layer to improve the overall capacity of the system by improving the downlink per-user performance. Results show that the combined performance of scheduling and transmit diversity does not lead to the same conclusions on the capacity improvements as a physical layer — only study of transmit diversity. The study of these interactions among different layers of the protocol stack requires a proper modeling of the error process. For instance, a characterization of the radio channel based only on the BER is incomplete to predict the performance of the MAC protocol. Higher-order statistical information is needed. The study of the effect of wireless channel errors on the performance of protocols represents an interesting and still open research area.

8.7.2 IP

Mobile IP has been proposed for network-level mobility support. In cellular systems, like GSM and UMTS, mobility within the system is required to be handled by the system itself.

From an IP network point of view mobility is handled by the link layer and does not affect the IP layer as long as the user stays in the same network type. However, when a user connects to the network via different network types or through a different operator network, roaming mechanisms are needed. The Mobile IP version 4 represents a simple and scalable global mobility solution but is not appropriate in support of fast and seamless handover. Mobile IP version 6 offers a more robust, secure, and scalable set of features with a significantly increased address space.

What can be envisaged about the future of IPv6? Because the technology is so new, the protocol evolution could take various directions. While the marketing strategy of many IPv6 proponents was focused on convincing wireline ISPs to adopt the protocol, the focus has now moved towards wireless technology. IPv6 may be most widely deployed in mobile phones, PDAs, and other wireless terminals in the near future. 3GPP, which is working on next generation wireless-network technology, has specified IPv6 as the standard addressing scheme for mobile IP multimedia. So far it has received support only in some geographical areas, like Asia. Equipment and OS vendors have just begun supporting IPv6. There is a relatively little support for IPv6, particularly in North America, from the ISPs and network administrators. There are contrasting opinions on the reason for such little support. Industry observers believe that it is partly due to the cost and effort required to migrate to IPv6, as well as the protocol shortcomings and a perception that standard advantages are not so necessary. Others claim that the widespread adoption will take place in years, because IPv6 represents a major change in Internet technology.

The open issues in Mobile IPv6 certainly are listed as follows:

- Triangle routing and inefficient direct routing;
- Local management of micro-mobility events;
- Seamless intra-domain handoff;
- Efficient Mobile IP-aware reservation mechanisms;
- RSVP reservations on Mobile IP triangle route situations;
- The use of one single subscription for all service types;
- Firewall support.

Resource reservation is necessary for providing guaranteed end-to-end performance for multimedia applications. The issues of QoS and resource reservation have been studied in great detail, even if the first and key step in resource reservation is routing. A communication can be undertaken if a path between sender and receiver that meets the specific requirements the application has set can be found. Resources can, hence, not be reserved unless the routing protocol can find a suitable path. Current routing algorithms used in IP networks are transparent to any particular QoS that different flows could require. Routing decisions are made without referring to the QoS of the flow. This means that flows are often routed over paths that are unable to support their requirements, while alternate paths with sufficient resources exist. Finding the shortest path to the destination is not enough anymore. New routing algorithms have to be developed with the aim of finding a path in the network that satisfies the given requirements.

Although important research areas can be found in the framework of the provision of QoS in IP networks, critical elements already exist for implementing QoS-enabled IP networks. What is needed is a practical architecture able to combine all the established technologies and technologies under development in a more and more effective heterogeneous environment.

8.7.3 TCP

The data link and transport layer protocols work on a number of successive data blocks, and, hence, the simple specification of the average error rate is no longer sufficient to assess protocol performance. The common layered design approach relies on the assumption that, whatever scheme or protocol is going to be implemented at the data and the transport layer, lowering the average error rate at the physical layer will always result in better performance for the upper layer protocols. Therefore, the main task of the physical layer designer is to lower the average error rate; the data link layer designer focuses on the reliable transfer of frames; routing and mobility management are the most important issues for the network layer designers, while transport layer designers focus on the end-to-end performance. This approach is no longer valid in a wireless scenario with highly time-varying channels. It is widely recognized that link characteristics, such as channel characteristics as well as physical and data link layer mechanisms, have to be carefully considered in the assessment of end-to-end performance. The first step of the assessment is a proper characterization of the wireless channel at the packet level. The average error rate is an incomplete performance metric. Several studies have shown the strong dependence of error statistics, especially of *second-order* statistics, on data link and transport layer protocol performance. Two consequences of neglecting the channel auto-correlation are the following:

1. The inability to predict protocol performance, coming up with the wrong conclusions in some cases;
2. The waste of the opportunity to optimize data and transport layer protocol design, according to the knowledge of the channel.

Good end-to-end wireless network performance is not possible without a truly optimized, integrated, and adaptive network design, where each level in the protocol stack should adapt to wireless link variations in an appropriate manner, taking into account the adaptive strategies at the other layers.

In what follows some of the design protocol proposals that try to exploit the knowledge of the error statistics are presented.

8.7.4 RRM

The first very important RRM algorithm for supporting multimedia communication is CAC. The CAC is crucial for protecting the QoS of ongoing connections and optimal usage of radio resources. For a proper operation of CAC, it is crucial that the network can estimate as accurately as possible the consumption of the radio resources of a service and the current state of the wireless system in terms of traffic load, interference conditions, and capacity costs. These basic requirements for CAC are valid for all wireless systems. The wireless network planners will have to design the appropriate thresholds for the CAC decision logic in order to have satisfactory radio resource utilization and at the same time the desired grade of service and coverage.

Other important functions of the RRM are traffic scheduling, transport channel allocation and switching, handover control, and link adaptation. The trade-off that characterizes the RRM is between satisfying the QoS requirement and optimal usage of the radio resources.

GPRS has limited multimedia support due to the relatively low transmission rate of the radio link and thus the requirements for RRM are minimal. On the other hand, RRM mechanisms are very important in UMTS and beyond. Further research is needed to find the optimal radio resource allocation procedures in B3G. For example, finding the optimal mapping between multimedia application and transport channels, based on the achieved end-to-end QoS and resource utilization, is an issue that requires further investigation. Radio resource allocation in the MAC layer for WLANs and MAC design complemented with QoS-based routing in ad hoc wireless networks are also promising topics for future research.

Finally, the role of common radio resource management is foreseen as crucial in heterogeneous wireless networks. Here issues such as traffic addressing, handover control, allocation of wireless link over the most optimal radio access network based on capacity consumption, and the resulting cost for the end user represent interesting fields for further research.

8.7.5 Source Coding

One of the main challenges related to the delivery of high-quality video and audio in future communication systems, or what makes all the other technical issues more challenging, is the heterogeneity of the communication network. For example, the Internet includes clients and servers with widely different capacities as well as diverse and dynamic network connections between them. This diversity causes the amount of resources available between video servers and clients to vary, both from network to network, and dynamically at a single network. If insufficient resources are available anywhere along the video pipeline, quality rapidly degrades to unacceptable levels. One approach to avoid this outcome is to have the applications use reservations to ensure that enough resources are available (RSVP). Furthermore, due to the large variety of existing network technologies, it is most likely that hybrid networks are used to support multimedia services. Different networks have different characteristics. Optimizing the performance over such heterogeneous networks requires more flexible and scalable video coding that still has high compression efficiency and low complexity. One promising solution for meeting all these conflicting requirements is to combine or integrate several video coding techniques in a layered structure. The video information is decomposed in layers in different dimensions: bit rate dimension, delay tolerance dimension, and error resilience dimension. The core video information is the most visually significant data requiring the minimum bandwidth, the least error resilient, and the lowest delay. The core layer can only carry key frames with an aggressive compression and it has no adaptability to any fluctuation. Adding layers/levels along one or more dimensions increases adaptability. Moreover, a new design approach is needed that jointly considers data compression techniques and communication protocols.

8.7.6 Channel Coding

Adaptivity is one of the main technologies for future wireless communication multimedia systems. Information theoretical results have shown that when CSI is made available to the transmitter, adaptive techniques can greatly improve performance in systems with stringent delay constraints, without the need to interleave or the exploitation of diversity techniques. Adaptive modulation can provide a five-fold increase in the spectral efficiency and 3–6 dB of coding gain are provided by adaptive coded modulation with Trellis codes with respect to uncoded transmission. Exploiting error control mechanisms with feedback, such as ARQ, can reduce the delay needed to achieve a given performance in terms of BER. Turbo codes (or other forms of concatenated coding) with iterative decoding appear to be a promising solution for hybrid ARQ schemes. However, the behavior of iterative decoders in the presence of decoding errors should be better characterized in order to exploit it for error detection.

The gain that can be achieved by channel adaptive techniques strictly depends on the knowledge of the current channel fading value. Therefore, new algorithms for fast and reliable long-range channel prediction are under investigation. At the same time, solutions for extending the applicability of adaptive techniques to systems where prediction techniques are not effective (i.e., systems with users characterized by high mobility) are being proposed.

The knowledge of CSI at the transmitter requires signaling information between the transmitter and receiver, which can be included at the MAC and higher layers. The signaling process should be undertaken in a capacity-efficient way. A promising research area is represented by distributed coordination mechanisms where some decision logic provides inherent coordination between transmitter and receiver.

The availability of inexpensive software radio platforms for a multiplicity of wireless applications will be one of the enablers of future multimedia communication systems.

Much work remains in developing good adaptive strategies. For multiuser systems adaptive modulation can be combined with other adaptive resource allocation policies like dynamic channel and base station assignment. Adaptive joint source and channel coding strategies that combine adaptive compression with adaptive modulation may also lead to good performance in time-varying channels. Furthermore, the cross-layer design is a methodology that requires further investigation, in order to lead to a scalable, robust, and simple enough implementation.

Although the adaptive approach can be seen as a lower complexity alternative to the diversity technique, it can also be applied to diversity techniques. Transmit diversity techniques that utilize CSI at the transmitter have shown considerable performance improvement over nonadaptive techniques. The adaptive antenna technology, which is one of the enabling technologies of future communication systems, has not been discussed in this chapter.

8.7.7 Physical Layer

Multicarrier techniques will play important roles in 4G systems; however, to make the multicarrier a core physical layer technique in 4G systems, there are many future research areas we should investigate further.

For instance, recently there has been intensive research on variants based on the MC-CDMA scheme all over the world; on the other hand, few works have been dedicated to pure OFDM-based schemes aiming at 4G systems, of course, with an emphasis on the signal format to cope with Doppler shift due to mobile motions.

In terms of access protocols, no one knows whether CDMA is really suited for the specification in 4G systems. OFDM-TDMA, OFDM-CSMA/CA, as well as MC-CDMA systems are all candidates. The performance comparison of these systems in multiple and isolated cell environments will be required.

Adaptive array antennas can enhance the transmission performance for OFDM-based systems. There are many different ways to configure on array antenna and OFDM demodulator. This has been really a recent hot topic on 4G systems; however, further investigation taking into consideration the trade-off between the receiver complexity and the attainable performance will be required.

For MIMO-OFDM, this has also been a recent hot topic in wireless communications in conjunction with adaptive array and diversity antennas. However, we have never seen the capacity analysis of a MIMO-OFDM system that can jointly suppress cochannel interference from other cells.

Finally, for the LINC method, there are a lot of ways for gain/phase mismatch cancellation. Examination of the trade-off between hardware complexity and attainable performance will be important.

Besides the MC methods discussed in this section for high data rates there is also work needed and ongoing in the field of ultra-wideband (UWB).

8.7.8 QoS

For wireless communications there are two major goals: to maximize the usage of the wireless resource and minimize the usage of energy. These two goals should be fulfilled while providing the best possible quality to the user. The issue of energy is of course directly linked to the radio resource when it comes to the transmit power but it is also related to the battery lifetime of the device.

In this section first the method for quality measurement and adjustment/control is discussed. Then the issues related to providing good quality while fulfilling the two goals is presented.

8.7.8.1 Quality Measurement and Adjustment

So as to provide QoS all protocol layers must understand the quality. Although quality can be understood and provided by different layers, it must be measured and adjusted during the communication. In Figure 8.6 protocol stack and quality measurement and adjustment/control parameters for different layers are given.

For layer-1 and layer-2 (PHY and MAC layer) the quality is usually measured in terms of the signal-to-interference ratio (SIR), the packet error rate (PER) or the frame error rate (FER), and the bit error rate (BER). The adjustment of quality in layer-1 can be done by adjusting the transmit (Tx) power, changing the data rate/channel coding, and changing the channel or frequency, while for layer-2 the packet size can be changed or the automatic repeat request (ARQ) mechanism can be used.

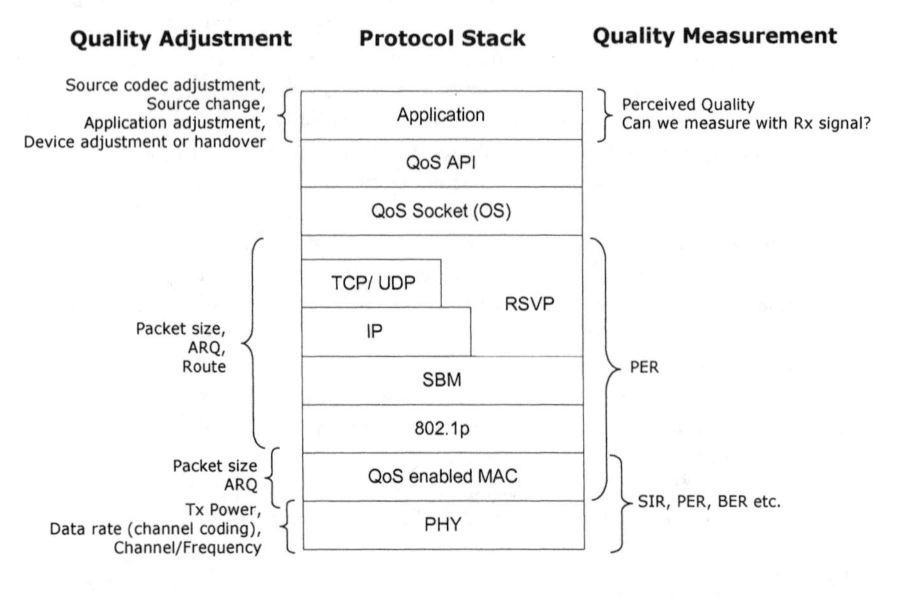

Figure 8.6 QoS protocol stack, quality measurement parameters, and quality adjustment methods.

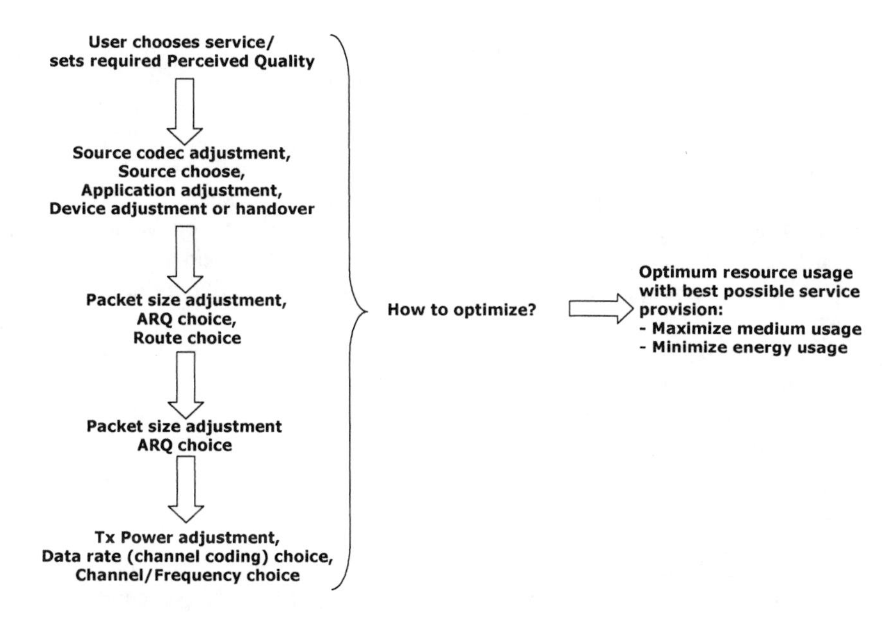

Figure 8.7 Quality adjustment issues.

While for layer-2.5 [subnet bandwidth manager (SBM) and layer-3 (IP), transmission control protocol (TCP)/user datagram protocol (UDP)] resource reservation protcol (RSVP)] the quality can be measured in terms of the PER and adjusted by varying the packet size, the ARQ, or simply by changing the route or path of the communication.

The application layer quality measurement can be done by measuring the perceived quality (i.e., the objective perceived quality); of course the issue is how to measure the perceived quality just from the received (Rx) signal. The adjustment of quality at the application layer can be done by changing the source codec rate or other variables in the source codec (e.g., for video coding I frame interval); it is also possible to simply change the source, adjust the application (e.g., changing the size of the application window or by adjusting the device to give a better perceived quality).

8.7.8.2 Quality Adjustment Issues

It is understandable that the quality of the system and the resource adjustment should be done based on the application layer quality measurement. The application layer quality can be measured by measuring the perceived quality from the received signal; this is not yet done and is a very difficult but important problem to be solved.

Once the perceived quality is known, the quality in lower layers can be adjusted (i.e., a study should then be done on adjusting the packet size, the transmit power, and the data rate). Optimization of all the parameters to achieve a good QoS while maximizing resource usage and minimizing energy and power usage is a very big but important issue.

Current wireless systems mostly adjust the quality independently in different layers. The first step could be to study methods for the optimum quality adjustment in each layer. Then one can pursue the optimum quality adjustment in two layers (layer-1 and 2), going to three (layer-1, 2, and 3) and then finally to the application layer.

In Figure 8.7 the preliminary adjustment per layer based on the user quality requirement is given; preliminary because each layer should find the optimum quality adjustment value so as to provide the requested quality. These parameters will be adjusted as the communication continues.

8.7.9 Security

The current trend in the wireless industry is towards heterogeneous networks; the most visible example is the work going on towards WLANs and 2.5G/3G interworking. The future will see further development of such interworking that brings heterogeneity in networks.

In the case of heterogeneous networks the first thing to understand is the trust level required for inter-domain (or inter-stakeholder) and the vertical handover, issues of authentication in case of the handover, the key management, the accounting and billing, the location privacy, the infrastructure security, the secure attachment and detachment, the lawful interception, the scalability, and the management. Besides requirements from security issues, other requirements like maintaining the security level during mobility, impact on the network, and impact of the security solution on the resources of the network and the device must also be considered. The whole dimension of security issues changes when the issue of more than one stakeholder is considered (i.e., 3G can be operated by one stakeholder and the WLAN by another stakeholder). Some of these issues are discussed in further detail below.

Handover can happen between different technologies and even between unknown operators. This will require the creation of trust between the two operators. The creation of trust can be done beforehand or on-the-fly during the handover.

Authentication mechanisms may vary depending on the heterogeneous technologies and/or on the network policies. When a device is handed over, authentication will take place once; again this may be very bad for some types of services. Similarly different key management mechanisms will be deployed by different networks; how to handle it and how to do key management is a major issue especially for a device moving from place to place.

When a user moves from one network to another the SLA must be communicated between the two networks. This negotiation phase can give the possibility to an intruder to cause changes or even highjack a connection. The SLA will be related to the cost of the service, also thus bringing in the issues related to the accounting during handover and the billing. If handover happens during an active session of a service especially between different administrative domains, some questions arise, like how to keep the integrity and the consistency of the accounting record and how to assure non-repudiation.

A firewall is often used by the networks and it becomes especially an issue for IPSec, Mobile IP, and network address translator (NAT). NAT is considered to be of use for IPv4 and in the future IPv6 will be used but it will be long before IPv4 is completely replaced by IPv6.

Lawful interception in an operator network is required by most countries and is based on the laws of the country. Location privacy on the other hand is a double-edged sword; the users do not want their location to be known while the governments are setting requirements on the location information for emergency cases.

Although not necessarily related directly to the security, it will be very important that security mechanisms at different layers talk to each other and maybe share security needs. During handover it will be necessary that lower layers inform, for example, layer 3 when handover is taking place.

Header compression is important for efficient use of the wireless medium. A malicious header compressor could cause the header decompressor to reconstitute packets that do not match the original packets but still have valid IP, UDP, and RTP headers and possibly also valid UDP checksums. Such corruption may be detected with the end-to-end authentication and the integrity mechanisms, which will not be affected by the compression. Denial-of-service attacks are possible if an intruder can introduce packets onto the link. The encrypted or the authenticated packets cannot be compressed. Compressing such packets will not provide end-to-end security but will lead to network-to-end or network-to-network security. Also, most protocols available for the security and the mobility for example IPSec and any Mobile IP, will require a lot of message exchange, which uses valuable capacity and battery life. Especially if the user is very mobile the total message exchange will be too much. This again needs to be studied while keeping the security requirements in mind.

Besides a security attack, it is possible that an authorized user can misuse the resources; this brings in the need for the secure access control, monitoring of the use of the resources, and the management of resources. There is also a need for security of the infrastructure, which means security of the network elements like router, the server, and the customer information. An attack on these elements of the network can lead to a crash of the whole network and the business.

Attaching securely to a network is a very important issue. When one first tries to attach to a network a lot of information is sent and received. Normally the username and the password are communicated but besides issues of such well known items one of the major security issues for the IP network is the dynamic host configuration protocol (DHCP), which is not secure. Another point is the secure detachment from the network. It is possible that an intruder can simply observe (sniff) the detachment process or can simply continue using the connection after a user has left. This can lead to such problems as a misuse of the network resources or an increase in the cost for the legitimate user.

Negotiation of the service, be it in terms of the quality requirement or the payment, should be done securely. Intruding in this phase can lead to overcharging or a decrease in quality or even denial of service. The negotiation of service will also lead to a SLA and this should be maintained throughout the communication period.

A network needs to be managed and security holes exist in management solutions. Insecure management methods can lead to control of a network by an intruder.

The scalability of the security solutions is very important. High mobility will require that the security solutions are not only limited to a place or number of people. The scalability also means that the security solution should be flexible enough to modify and to fulfill the security needs as the network changes.

Ad hoc networks are expected to play an important role in the future. There are several security issues that the ad hoc networks will face; they also face issues related to the variety of devices that can communicate with each other; some of them might have a powerful CPU or a lot of memory, while others might exist that have severely restricted (peanut) CPU and battery power.

8.7.10 Mobility

As also discussed in Chapter 6 and alluded to in Section 8.3, mobility will be of the utmost importance in the future. Here mobility is with seamlessness. The user should not perceive any change in service quality when performing handover from one network to another or from one technology to another.

What we thus see is that mobility is intertwined with QoS and security and these three cannot be separated. This brings in the need for a study of the complete system instead of one component (mobility, QoS, security, or routing).

8.8 IEEE 802 ACTIVITIES TOWARDS THE FUTURE

Obviously it is clear from the activities of IEEE 802 that a lot of work is ongoing towards what is discussed in this chapter as 4G, B3G, and future generation communications.

IEEE 802.11 is looking towards very high data rates as IEEE 802.11n. While a separate group .11s is developing mesh networking solutions. The IEEE 802.11 standard is also developing solutions for fast handover and voice services. IEEE

802.11e on the other hand provides QoS. There are other IEEE 802.11 activities on security and measurements and solutions for RRM.

IEEE 802.16 already provides mesh networking and is now looking into mobility. Such a high data rate and mobility are something close to 4G, which of course IEEE 802.11 is also going to do. On the other hand IEEE 802.20 is working on mobile broadband wireless access (MBWA). MBWA is a clean solution, free of previous developments as is the case for IEEE 802.16. The solution plans to provide around 1 Mbps for high-speed devices. This solution is not close to what 4G is supposed to be but still is in the right direction. IEEE 802.21 on the other hand is looking into heterogeneous networks, which in itself is B3G.

There is also standardization work ongoing in IEEE 802.15 that is looking into very high data rate and very low data rate solutions for WPANs. These solutions bring us closer to ubiquitous communications.

8.9 STANDARDIZATION AND REGULATIONS

As future generation communications work is still in its infancy the standardization work is not really happening. Still one can see different things.

IEEE 802.21 is working towards B3G and is the only true standardization work in this field. 3GPP and 3GPP2 are also looking into 3G and WLAN integration and in a way can be seen as standardization work for B3G. It remains a question if 3G standardization activities will become the base for 4G standardization. Although 3GPP activities in the future towards All-IP and seamless handover between 3G and WLAN does make one think so.

Now different countries or regions have started work on 4G like India, China, Korea, Japan, and Europe. In fact China, Japan, and Korea (CJK) have formed a cooperative to develop next generation solutions. In Japan a Mobile IT Forum (mITF) has formed to work towards 4G. In Europe WWRF is working on 4G and so are the European 6th Frame Work Projects. WWRF is now becoming a worldwide forum. FUTURE in China is doing similar work. A Next Generation Mobile Communications (NGMC) group formed to discuss 4G and B3G especially in CJK [16].

FCC has provided the 255MHz unlicensed band in 5 GHz and there is a proposal for the 3.65GHz band as an unlicensed band. There are also thoughts to make licensed band available at 70, 80, and 90GHz.

ITU-R is now talking about further development of IMT-2000 to 30Mbps possibly by 2005. There is also thought of 100 Mbps for high mobility and 1 Gbps for low mobility systems. Further the work on interworking between different air interfaces remains active.

8.10 CONCLUSIONS

IEEE 802.11-based WLAN technology only defines solutions for layer 1 and layer 2 while it should work with higher-layer protocols like the IP layer. Keeping that in mind, this book has given an overview of security, QoS, and mobility in WLAN and IP networking. The market and business of WLAN has also been discussed. The most important point is to deploy the technology, and today the issue is also to deploy WLAN with mobile networks.

This final chapter discussed the next generation communications technology. Definitions of 4G, B3G, and future generation communications where given in the chapter. The technology for the future should be acceptable to the users and sellable by vendors and operators. It is also important to understand the various technologies that are being developed in other fields as they also affect the usage of telecommunications-related products; these too were briefly discussed in this chapter. The chapter also discussed various technical challenges and problems to be solved for next generation communications.

REFERENCES

[1] Prasad, A. R., *WLANs: Protocols, Security and Deployment*, Ph.D. Thesis, Delft University Press (DUP), Delft, The Netherlands, December 2003.

[2] Lauridsen, O. M., and A.R. Prasad, "User needs for services in UMTS," *International Journal on Wireless Personal Communications*, Kluwer Academic Publishers, August 2002, vol. 22, nr. 2, pp. 187-197.

[3] Christensen, C. M., *The Innovator's Dilemma*, New York, Harper Collins Publishers, January 2003.

[4] Innovators: http://www.fortune.com/fortune/fsb/specials/innovators/cook.html.

[5] Quintillion Technologies: http://www.quintillion.co.jp/.

[6] Gombrich, E. H., *The Story of Art*, Boston, MA: Phaidon Press, April 1995.

[7] Prasad, A. R., P. Schoo, and H. Wang, "An Evolutionary Approach towards Ubiquitous Communications: A Security Perspective," *SAINT 2004*, 26-30 Jan., 2004, Tokyo, Japan.

[8] Weiser, M., "The Computer of the Twenty-First Century," *Scientific American*, vol. 265, nr. 3, Sept. 1991, pp. 94 - 104.

[9] Weiser, M., "Hot Topics-Ubiquitous Computing," *Computer*, Volume: 26 Issue: 10, Oct. 1993, pp. 71 – 72.

[10] WWRF, *The Book of Visions 2001, Visions of the Wireless World*, v1.1b, December 2000.

[11] Niemegeers, I. G., and S. M. Heemstra de Groot. "Research Issues in Ad-Hoc Distributed Personal Networks," *International Journal on Wireless Personal Communications*, Kluwer Academic Publishers, September 2003.

[12] Farserotu, J. et al., "Scalable, Hybrid Optical-RF Wireless Communication System for Broadband and Multimedia Service to Fixed and Mobile Users," *International Journal on Wireless Personal Communications*, Kluwer Academic Publishers, Jan. 2003, vol 24, nr. 2, pp. 327 - 339.

[13] The Economist, *Technology Quarterly*, March 11, 2004, June 10, 2004, and September 4, 2003.

[14] Prasad, R., and M. Ruggieri, *Technology Trends in Wireless Communications*, Norwood, MA: Artech House, April 2003.

[15] Hara, S., and R. Prasad, *Multicarrier Techniques for 4G Mobile Communications*, Norwood, MA: Artech House, June 2003.

[16] *International Conference on Beyond 3G Mobile Communications-2004*, 26-27 May 2004, Tokyo, Japan.

List of Abbreviations

μs	Microsecond
2G	Second generation
3DES	Triple DES
3G	Third generation
3GPP	Third generation partnership protocol
4G	Forth generation
5GSG	5 GHz study group

A

AAA	Authentication, authorization, and accounting
AAAB	Broker AAA
AAAF	Foreign AAA
AAAH	Home AAA
AAAH/F	Authentication, authorization, and accounting home/foreign
AAD	Additional authentication data
AAL	ATM adaptation layer
AC	Access category
AC	Admission control
AC	Access category
ACK	Acknowledgment
ACL	Access control list
ACM	Admission control manager
ACO	Authentication ciphering offset
ADDTS	Add traffic stream
AES	Advanced encryption standard
AF	Assured forwarding
AGC	Automatic gain control
AH	Authentication header
AIFS	Arbitration IFS
AK	Authorization key
AKA	Authentication and key agreement
AKMP	Authentication and key management protocol
AMPS	Advanced mobile phone system

AP	Access point
APF	Access point functionality
API	Application program interface
APME	AP management entity
AR	Access router
ARF	Automatic rate fallback
ARP	Address resolution protocol
ARQ	Automatic repeat request
AS	Authentication server
ASIC	Application specific integrated circuit
ATM	Asynchronous transfer mode
AuC	Authentication center
AVP	Attribute value pair

B

B3G	Beyond 3G
BA	Behavior aggregate
BB	Bandwidth broker
BD_ADDR	Bluetooth address
BER	Bit error rate
BKR	Broadcast key rotation
BLER	Block error rate
BR	Border router
BS	Base station
BSA	Basic service area
BSS	Basic service set
BSS	Business support system
BW	Bandwidth
BWA	Broadband wireless access
BWIF	Broadband wireless Internet forum

C

CA	Certification authority or Central authority
CAP	Controlled access phase
CAPWAP	Control and provisioning of wireless access point
CARD	Candidate access router discovery
CBC	Cipher-block chaining
CBC-MAC	Cipher block chaining-message authentication code
CCA	Clear channel assessment
CCF	Charging collection function
CCK	Complementary code keying
CCMP	Counter mode with cipher-block chaining message authentication code protocol
CCoA	Collocated care-of-address
CDH	Computational Diffie-Hellman
CDMA	Code division multiple access
CDT	Carrier detect threshold
CEPT	Conference of European Post and Telecommunication administrations
CFB	Cipher feedback

CFP	Contention-free period
CGF	Charging gateway function
CIP	Cellular IP
CIR or C/I	Carrier-to-interference ratio
CJK	China, Japan, Korea
CKSN	Ciphering key sequence number
CL	Controlled load
cm	Centimeter
CMF	Channel matched filter
CN	Correspondent node
CoA	Care-of-address
COPS	Common open policy service
CP	Contention period
CPU	Central processing unit
CQ	Communication quality
CRC	Cyclic redundancy check
CSMA/CA	Carrier sense multiple access with collision avoidance
CSMA/CD	Carrier sense multiple access with collision detection
CSRC	Contributing source
CT	Cordless telephone
CTP	Context transfer protocol
CTR	Counter
CTS	Clear-to-send
CW	Contention window

D

DA	Destination address
DARPA	Defense advanced research projects agency
dB	Decibel
DB	Database
DBPSK	Differential binary phase shift keying
DCA	Dynamic channel assignment
DCF	Distributed coordination function
DCS	Dynamic channel selection
DECT	Digital enhanced cordless telephone
DES	Data encryption system
DFS	Dynamic Frequency Selection
DFWMAC	Distributed Foundation Wireless MAC
D-H	Diffie-Hellman
DHCP	Dynamic host control protocol
DiffServ	Differentiated Service
DIFS	Distributed coordination function IFS
DLC	Data link control
DLL	Dynamic link library
DMA	Direct memory access
DMHA	Dormant mode host alert
DMZ	Demilitarized zone
DNS	Dynamic name server
DoS	Denial of service

DQPSK	Differential quadrature phase shift keying
DRARP	Dynamic reverse address resolution protocol
DS	Distribution system
DS	Differentiated service
DS	(802.11) Distribution system
DSAP	Destination service access point
DSCP	Differentiated services code point
DSCP	Diffserv code point
DSL	Digital subscriber line
DSM	Distribution service medium
DSP	Digital signal processor
DSRC	Dedicated short range communication
DSSS	Direct sequence spread spectrum
DT	Defer threshold
DTBS	Distributed time-bounded services
DTIM	Delivery TIM
DVD	Digital video disc

E

EAP	Extensible authentication protocol
EAPoL	EAP over LAN
EAR	Edge access router
ECB	Electronic code book
EDCA	Enhanced distributed channel access
EDCF	Enhanced DCF
EDGE	Enhanced data rates for global evolution
EF	Expedited forwarding
EHF	Extremely high frequency
EIR	Equipment identity register
ERP	Extended rate PHY
ESP	Encapsulating security payload
ESS	Extended service set
ESSID	Extended service set ID
ETSI	European Telecommunications Standards Institute

F

FA	Foreign agent
FAST	Flexible authentication secure tunneling
FCC	Federal Communications Commission
FDDI	Fiber distributed data interface
FEC	Forward error control
FF	Fixed-filter
FFT	Fast Fourier transform
FHSS	Frequency hopping spread spectrum
FIFO	First in first out
FR	France
FSM	Finite state machine
FTP	File transfer protocol
FW	Firmware

FWA	Fixed Wireless Access

G

GGSN	Gateway GPRS support node
GHz	Gigahertz
GMK	Group master key
GPRS	General packet radio service
GRX	GPRS roaming exchange
GS	Guaranteed service
GSM	Global system for mobile communication
GTK	Group transient key
GTKSA	GTK security association
GTP	GPRS tunneling protocol
GW	Gateway

H

HA	Home agent
HAP	High altitude platform
HC	Hybrid coordinator
HCCA	HCF controlled channel access
HCF	Hybrid coordination function
HSCSD	High speed circuit switched data
HDR	High data rate
HIPERLAN	High performance radio local area network
HiSWAN	High speed wireless access network
HLR	Home location register
HMIP	Hierarchical mobile IP
HO	Handover
HR	High rate
HSS	Home subscriber server
HTTP	Hypertext transport protocol

I

IANA	Internet assigned numbers authority
IAPP	Inter-access point protocol
IARP	Intrazone routing protocol
IBSS	Independent basic service set
IC	Integrated circuit
ICV	Integrity check value
ID	Identity
IDEA	International data encryption algorithm
IDS	Intrusion detection system
IE	Information element
IEEE	Institute of electrical and electronic engineers
IETF	Internet engineering task force
IFFT	Inverse fast Fourier transform
IFS	Inter-frame spacing
IKE	Internet key exchange
IMSI	International mobile subscriber identity

IntServ	Integrated service
IP	Internet protocol
IPSec	IP security
IR	Infrared
ISAKMP	Internet security association and key management protocol
ISM	Industry scientific and medical
ISP	Internet service provider
ITS	Intelligent transportation system
ITU	International telecommunications union
IV	Initialization vector

J

JGN	Japan gigabit network
JP	Japan

K

K_c	Ciphering key
Kbps	Kilobits per second
KCK	Key confirmation key
KEA	Key exchange algorithm
KEK	Key encryption key
K_i	Individual subscriber authentication key

L

LAI	Location area identity
LAN	Local area network
LDAP	Lightweight directory access protocol
LFSR	Linear feedback shift register
LLC	Logical link control
LMDS	Local multipoint distribution system
LMSC	LAN MAN Standardization Committee
LRU	Least recently used

M

MAC	Medium access control
MAC	Message authentication code
MAHO	Mobile assisted HO
MAN	Metroplitan area network
MANET	Mobile ad hoc networking
max	maximum
Mbps	Megabits per second
MBS	Mobile broadband system
MBWA	Mobile broadband wireless access
Mchips	Mega chips
MCHO	Mobile controlled HO
MCU	Multipoint control unit
MD5	Message digest 5
MDR	Medium data rate
MF	Multifield

MGCP	Media gateway control protocol
MGCP	Media Gateway Control Protocol
MHz	Mega Hertz
MIC	Message integrity check
MIMO	Multiple input multiple output
MIP	Mobile IP
mITF	Mobile IT forum
MitM	Man in the middle
MK	Master key
MKK	Japanese frequency assignment authority
MLME	MAC sub layer management entity
MMAC	Multimedia mobile access communication system
MMDS	Multichannel multipoint distribution service
MMS	Mobile multimedia service
MMUSIC	Multiparty multimedia session control
MN	Mobile node
MNO	Mobile network operator
mobike	Mobile IKE
MOS	Mean opinion score
MoU	Memorandum of understanding
MPDU	MAC protocol data unit
MS	Mobile station
MSC	Message sequence chart
MSDU	MAC service data unit
MSISDN	Mobile station integrated services digital network number
MTBF	Mean time between failure
MTTR	Mean time to repair
mW	milli Watt

N

NA	North America
NAI	Network address identifier
NAPT	Network address port translation
NAS	Network address server
NAT	Network address translator
NAV	Network allocation vector
NCHO	Network controlled HO
NesCom	New standards committee
NGMC	Next generation mobile communications
NIC	Network interface card
NIST	National Institute of Standards and Technology
NMS	Network management system
NNTP	Network news transport protocol
ns	nano second

O

OFDM	Orthogonal frequency division multiplexing
OMS	Operation and maintenance subsystem

OSA	Open systems authentication
OSI	Open system integration
OSP	Open settlement protocol
OSPF	Open shortest path first
OSS	Operation support system
OTP	One time password

P

PAC	Protected access credential
PAC	Public access controller
PAE	Port access entity
PAP	Peak-to-average power
PAR	Project authorization request
PC	Point coordinator
PCF	Point coordination function
PDC	Personal digital cellular
PDG	Packet data gateway
PDP	Policy decision point
PDU	Protocol data unit
PEAP	Protected EAP
PEP	Policy end-point
PHB	Per hop behavior
PHOP	Previous hop
PHS	Personal handyphone system
PHY	Physical layer
PIFS	Point coordination function IFS
PIN	Personal identification number
PKI	Public key infrastructure
PLCP	Physical layer control protocol
PMK	Pairwise master key
PN	Personal network
POE	Power over Ethernet
POP	Point of presence
POP3	Post office protocol
PPDU	PHY packed data unit
PPP	Point to point protocol
PPPoE	Point to point protocol over Ethernet
PRF	Pseudorandom function
PRN	Pseudorandom number
PRNG	PRN generator
PS	Power save
PS	Packet switched
PSB	PATH state block
PSK	Pre-shared key
PSTN	Packet switched telephone network
PTK	Pairwise transient key
PTKSA	PTK security association
PWLAN	Public WLAN

Q

QAM	Quadrature amplitude modulation
QAP	QoS enhanced AP
QBSS	QoS supporting BSS
QoS	Quality of service
QPSK	Quadrature phase shift keying
QSTA	QoS enabled STA

R

RADIUS	Remote authentication dial-in user service
RAND	Random number
RAS	Registration admission status
RESV	Reserve
RevCom	Review committee
RFC	Request for comment
RIP	Routing information protocol
RNC	Radio network controller
RR	Radio regulatory
RR	Receiver report
RREP	Route Reply message
RRM	Radio resource management
RSA	Rivest, Shamir, Adleman
RSB	RESV state block
RSN	Robust security network
RSNA	RSN association
RSpec	Reservation specification
RSS	Received signal strength
RSSI	RSS indicator
RSVP	Resource reservation protocol
RTCP	Real-time control protocol
RTP	Real-time protocol
RTS	Request-to-send
RTSP	Real-time signaling protocol
RX	Receive

S

SA	Security association
SA	Source address
SAD	Security association database
SAID	Security association identities
SAP	Service access point
SAP	Session announcement protocol
SAR	Specific absorption ratio
SBM	Subnet bandwidth manager
SCTP	Stream control transmission layer protocol
SDB	Services database
SDP	Session description protocol
SDR	Software defined radio
SE	Shared-explicit

Seamoby	Seamless mobility
SEC	Sponsor executive committee
SFD	Sunc field delimiter
SG	Study group
SGSN	Serving GPRS support node
SHA	Secure hash algorithm
SIFS	Short IFS
SIM	Subscriber identity module
SIP	Session initiation protocol
SKEME	Secure key exchange mechanism
SLA	Service level agreement
SLS	Service level specification
SMS	Short message service
SMTP	Send mail transport protocol
SNEP	Security network encryption protocol
SNMP	Simple network management protocol
SOI	Son of IKE
SoS	Security of service
SP	Service policy
SP	Service period
SPD	Security policy database
SPI	Security policy identifier
SQL	Structured query language
SR	Sender report
SRES	Signed response
SRP	Secure remote password
SS	Subscriber station
SSID	Service set ID
SSL	Security socket layer
SSRC	Synchronization source
STA	Station
STS	Station-to-station
SYNC	Synchronization

T

TA	Transmitter address
TACS	Total access communications system
TAG	Technical advisory group
TAP	Transferred account procedure
TC	Traffic category
TCP	Transmission control protocol
TCS	Traffic conditioning agreement
TDMA	Time division multiple access
TEK	Traffic encryption key
TGS	Ticket granting service
TID	Traffic ID
TIM	Traffic indication map
TK	Temporal key
TKIP	Temporal key integrity protocol

TLS Transport layer security
TMSI Temporary mobile subscriber identity
TPC Transmit power control
TSC TKIP sequence counter
TSF Timing synchronization function
TSID Traffic stream ID
TSN Transitional security network
TSPEC (TSpec) Traffic specification
TTLS Tunneled TLS
TX Transmit
TXOP Transmission opportunity

U
UAM Universal access method
ubicom Ubiquitous communication
ubicomp Ubiquitous computing
UDP Universal datagram protocol
UE User equipment
UHF ultra high frequency
UMTS Universal mobile telecommunications system
UNII Unlicensed national information infrastructure
UP User priority
URL Uniform resource locator
US United States
USIM UMTS SIM
USIM UMTS subscriber identity module
USM User-based subscriber module
USM User-based security model
UWB Ultra-wideband

V
VCR Video cassette recorder
VLR Visitor location register
VoIP Voice over IP
VoWLAN Voice over WLAN
VPLM Visited public land mobile network
VPN Virtual private network
VPNC VPN consortium

W
WAG WLAN access gateway
WAP Wireless access protocol
WAPI WLAN authentication and privacy infrastructure
WAS Wireless access server
WAVE Wireless access for the vehicular environment
WCDMA Wideband code division multiple access
WECA Wireless Ethernet compatibility alliance
WEP Wired equivalent privacy
WG Work group

WIBS	Wireless integrated billing and security
WIEN	Wireless internetwok and external network
Wi-Fi	Wireless-fidelity
WINS	Wireless integrated network sensor
WISP	Wireless Internet service provider
WISPr	Wireless Internet service provider roaming
WLAN	Wireless local area networks
WMAN	Wireless metropolitan area network
WMM	Wi-Fi multimedia
WNM	Wireless network management
WNG	Wireless next generation
WPA	Wi-Fi protected area
WPAN	Wireless personal area networks
WTLS	Wireless TLS
WTP	Wireless termination point
WWAN	Wireless wide area network

About the Authors

Anand Raghawa Prasad (The Netherlands), senior researcher, DoCoMo Euro-Labs, Munich, Germany was born in Ranchi, India. He received his Ph.D. degree on "WLANs: Protocols, Security, and Deployment," and MSc (Ir.) degree in the field of "Self Similarity in ATM Network Traffic" from Delft University of Technology, The Netherlands in 2003 and 1996, respectively. From 1996 to November 1998 he worked as a research engineer and later project leader in the Uniden Corporation, Tokyo, Japan. From 1998 to 2000, he worked as a systems architect for IEEE 802.11-based WLANs in Lucent Technologies in The Netherlands. Subsequently, he was technical director at Genista Corporation, Tokyo, Japan. He is an active participant of the IEEE 802.11 standardization committee. In addition to his publications in journals, international conferences, and chapters in books, he has several patent applications in the field of wireless communications and has coedited a book titled *WLAN Systems and Wireless IP for Next Generation Communication* published by Artech House. He has been a member of organizing committee of various International Conferences including VTC and WPMC. He was a guest editor of a special issue on *Security for Next Generation Communication* of the *Kluwer International Journal on Personal Wireless Communications*. His research interests include software radio, next generation wireless systems, networks and access, security, mobility, and QoS for wireless systems and IP networking. He is a senior member of IEEE.

Neeli Rashmi Prasad was born on September 14, 1970, in Ranchi, Bihar, India. She received her Ph.D. from the University of Rome "Tor Vergata," Rome, Italy, in the field of "Adaptive Security for Wireless Heterogeneous Networks" in 2004 and M.Sc. (Ir.) degree in electrical and electronics engineering from Delft University of Technology, The Netherlands, on "Indoor Wireless Communications using Slotted ISMA Protocols" in 1997. She joined Vodafone, Maastricht, The Netherlands as a radio engineer in 1997. From November 1998 until May 2001, she worked as systems architect for wireless LANs in the Wireless Communications and Networking Division of Lucent Technologies, Nieuwegein, The Netherlands. From June 2001 to July 2003, she was with T-Mobile Netherlands, The Hague, The Netherlands as senior architect for core network. Subsequently, from July 2003 to April 2004, she was senior research manager at PCOM:I^3, Aalborg, Denmark. Since April 2004, she has been associate professor at the Center for TeleInfrastruktur (CTIF), Aalborg University, Aalborg, Denmark. In addition to her publications in journals, international conferences and chapters in books, she has coedited a book titled *WLAN Systems and Wireless IP for Next Generation Communications* published by Artech House. In December 1997 she won best paper award for her work on Inhibit Sense Multiple Access (ISMA) protocol. Her current research interest lies in wireless security, mobility, mesh networks, low data rate networks, and heterogeneous networks.

Index

The Artech House Universal Personal Communications Series

Ramjee Prasad, Series Editor

802.11 WLANs and IP Networking: Security, Qos, and Mobility, Anand R. Prasad and Neeli R. Prasad

CDMA for Wireless Personal Communications, Ramjee Prasad

IP/ATM Mobile Satellite Networks, John Farserotu and Ramjee Prasad

Multicarrier Techniques for 4G Mobile Communications, Shinsuke Hara and Ramjee Prasad

OFDM for Wireless Communications Systems, Ramjee Prasad

OFDM for Wireless Multimedia Communications, Richard van Nee and Ramjee Prasad

Practical Radio Resource Management in Wireless Systems Sofoklis A. Kyriazakos and George T. Karetsos

Radio over Fiber Technologies for Mobile Communications Networks, Hamed Al-Raweshidy and Shozo Komaki, editors

Simulation and Software Radio for Mobile Communications, Hiroshi Harada and Ramjee Prasad

Space-Time Codes and MIMO Systems, Mohinder Jankiraman

TDD-CDMA for Wireless Communications, Riaz Esmailzadeh and Masao Nakagawa

Technology Trends in Wireless Communications, Ramjee Prasad and Marina Ruggieri

Third Generation Mobile Communication Systems, Ramjee Prasad, Werner Mohr, and Walter Konhäuser, editors

Towards a Global 3G System: Advanced Mobile Communications in Europe, Volume 1, Ramjee Prasad, editor

Towards a Global 3G System: Advanced Mobile Communications in Europe, Volume 2, Ramjee Prasad, editor

Universal Wireless Personal Communications, Ramjee Prasad

WCDMA: Towards IP Mobility and Mobile Internet, Tero Ojanperä and Ramjee Prasad, editors

Wideband CDMA for Third Generation Mobile Communications, Tero Ojanperä and Ramjee Prasad, editors

Wireless IP and Building the Mobile Internet, Sudhir Dixit and Ramjee Prasad, editors

WLAN Systems and Wireless IP for Next Generation Communications, Neeli Prasad and Anand Prasad, editors

WLANs and WPANs towards 4G Wireless, Ramjee Prasad and Luis Muñoz

For further information on these and other Artech House titles, including previously considered out-of-print books now available through our In-Print-Forever® (IPF®) program, contact:

Artech House
685 Canton Street
Norwood, MA 02062
Phone: 781-769-9750
Fax: 781-769-6334
e-mail: artech@artechhouse.com

Artech House
46 Gillingham Street
London SW1V 1AH UK
Phone: +44 (0)20 7596-8750
Fax: +44 (0)20 7630-0166
e-mail: artech-uk@artechhouse.cc

Find us on the World Wide Web at:
www.artechhouse.com